Perfluorinated Ionomer Membranes

Perfluorinated Ionomer Membranes

Adi Eisenberg, EDITOR
McGill University

Howard L. Yeager, EDITOR
University of Calgary

Developed in advance of the

Topical Workshop on

Perfluorinated Ionomer Membranes,

Lake Buena Vista, Florida,

February 23–26, 1982.

ACS SYMPOSIUM SERIES180

AMERICAN CHEMICAL SOCIETY

WASHINGTON, D. C.   1982

Library of Congress CIP Data
Perfluorinated ionomer membranes.

(ACS symposium series, ISSN 0097-6156; 180)
Includes bibliographies and index.

1. Membranes (Technology)—Congresses. 2. Ionomers—Congresses. 3. Organofluorine compounds—
Congresses.
I. Eisenberg, A. (Adi). II. Yeager, Howard L.,
1943-   III. American Chemical Society. Division
of Polymer Chemistry. IV. Series.

TP159.M4P47   660.2'8423   81-20570
ISBN 0–8412–0698–8   AACR2
                        ACSMC8 180 1–500 1982

Copyright © 1982

American Chemical Society

PRINTED IN THE UNITED STATES OF AMERICA

ACS Symposium Series

M. Joan Comstock, *Series Editor*

FOREWORD

The ACS SYMPOSIUM SERIES was founded in 1974 to provide a medium for publishing symposia quickly in book form. The format of the Series parallels that of the continuing ADVANCES IN CHEMISTRY SERIES except that in order to save time the papers are not typeset but are reproduced as they are submitted by the authors in camera-ready form. Papers are reviewed under the supervision of the Editors with the assistance of the Series Advisory Board and are selected to maintain the integrity of the symposia; however, verbatim reproductions of previously published papers are not accepted. Both reviews and reports of research are acceptable since symposia may embrace both types of presentation.

CONTENTS

PREFACE

Perfluorinated ionomer membranes represent a major advance in membrane technology. Although a variety of applications have been found for these chemically inert polymers since their appearance about fifteen years ago, the main driving force for their development has been their potential application as membrane separators in the commercial electrolysis of brine to produce chlorine-caustic. This major chemical technology is in the process of being revolutionized by the use of these materials, a remarkable accomplishment for such a small group of polymers.

In spite of their importance, relatively little information exists in the literature concerning the fundamental physicochemical properties of these membranes. This is due both to their proprietary nature and to the rapid growth of the technology. Also, what information is available is dispersed in a wide range of sources, ranging from the literature specializing in macromolecules to that devoted to the chlor-alkali industry. This work attempts to provide, in one volume, an overview of both the fundamental properties and the technological aspects of the field.

Because of the rapid growth of this area of research, this book, unlike most of its predecessors in the series, has been assembled without the prior occurrence of a symposium. However, in response to the dramatic advances in this field in several disciplines, the Polymer Division of the American Chemical Society has scheduled a workshop on perfluorinated ionomer membranes to be held in February of 1982. It is the aim of the workshop to provide workers in the electrochemical and polymer fields with a summary of current knowledge and directions for future development. Many of the contributors to this volume plan to participate in the workshop.

Finally, it is a pleasure to acknowledge the special efforts of the Books Department of the American Chemical Society in assuring the publication of this volume in time for the workshop.

A. Eisenberg
McGill University
Department of Chemistry
Montreal, Quebec, Canada

H. L. Yeager
University of Calgary
Department of Chemistry
Calgary, Alberta, Canada

September 8, 1981.

Introduction

H. L. YEAGER
Department of Chemistry,
University of Calgary, Calgary, Alberta T2N 1N4, Canada

A. EISENBERG
Department of Chemistry, McGill University, Montreal, Quebec, Canada

The past 15 years have witnessed an explosive growth in the literature on ionomers, i.e. ion-containing copolymers in which up to 15 mol % of the repeat units contain ionic groups. Two monographs have appeared on the subject[1,2], as well as proceedings of several symposia[3], in addition to a large number of individual publications and patents. Furthermore, Chemical Abstracts has started to issue, as part of their "CA Selects", the series entitled "Ion-Containing Polymers" ca. 2 years ago, and a biennial Gordon Conference on this topic started in 1979.

In 1984, the Macromolecular Secretariat of the American Chemical Society will devote 4 days at the Philadelphia ACS meeting to a comprehensive symposium on coulombic interactions in polymers of which a major part will be devoted to the ionomers.

This major research effort devoted to these macromolecules can be understood if one realizes that the incorporation of ions into organic polymers can modify the properties of the materials profoundly. Increases in the glass transition by five hundred degrees[4], increases in the modulus by over three orders of magnitude[5] and increases in the viscosity by over four orders magnitude[6] have been observed, among many other effects. Of direct relevance to this symposium, the incorporation of ions, even in small amounts (0-10 mol %), can yield polymeric membranes in which the diffusion coefficient for water is orders of magnitude greater than in the non-ionic parent polymer while the membrane, at the same time, becomes permselective.

Most of the research effort on the ionomers has been devoted to only a small number of materials, notably the ethylenes[7], the styrenes[8], the rubbers[9], and those based on poly(tetrafluoroethylene), the last of which is the subject of the present volume. As a result of these extensive investigations, it has become clear that the reason for the dramatic effects which are obsverved on ion incorporation is, not unexpectedly, the aggregation of ionic groups in media of low dielectric constant. Small angle X-ray and neutron scattering, backed up by a wide range of other techniques, have demonstrated clearly the existence of ionic

0097-6156/82/0180-0001$05.00/0

scattering centers in the ionomers which are not present in the
parent polymer of the non-ionic acid copolymers. Specifically
in the styrene carboxylates, dielectric[10] and Raman[11] studies
have suggested the existence of two types of aggregates, i.e.
small multiplets consisting of several ion pairs held together
by strong coulombic interactions, and much larger clusters resulting from weaker interactions. While the multiplets are
stable over the entire temperature range investigated, the clusters
breack down progressively into multiplets with increasing temperature.

The majority of research which is discussed in this volume
deals with the Nafion   brand perfluorosulfonate polymers,
manufactured by E. I. du Pont de Nemours and Co.. These materials
were developed during the middle 1960's, and have been available
in various forms for study during the past few years. The synthesis and general properties of Nafion membranes are summarized
below[12-14].

Tetrafluorethylene reacts with SO_3 to form a cyclic sultone.
After rearrangement, the sultone can then be reacted with hexafluoropropylene epoxide to produce sulfonyl fluoride adducts,
where

$$CF_2{=}CF_2 + SO_3 \longrightarrow \underset{\substack{| \quad \quad |\\ O - SO_2}}{CF_2 - CF_2} \longrightarrow FSO_2CF_2C{=}O$$

$$FSO_2CF_2C{=}O + (m+1)\underset{\substack{|\\ CF_3}}{CF_2 - CF} \longrightarrow FSO_2CF_2CF_2\underset{\substack{|\\ CF_3}}{(OCFCF_2)_m}\underset{\substack{|\\ CF_3}}{OCFC{=}O}$$

$m \geqslant 1$. When these adducts are heated with sodium carbonate, a
sulfonyl fluoride vinyl ether is formed. This vinyl ether is
then copolymerized with tetrafluoroethylene to form XR Resin:

$$FSO_2CF_2CF_2\underset{\substack{|\\ CF_3}}{(OCFCF_2)_m}OCF{=}CF_2 + CF_2{=}CF_2 \longrightarrow$$

$$-(CF_2CF_2)_n - \underset{\substack{|\\ CF_2\\ |}}{CFO}(CF_2-\underset{\substack{|\\ CF_3}}{CFO})_mCF_2CF_2SO_2F \qquad \text{XR Resin}$$

This high molecular weight polymer is melt fabricable and
can be processed into various forms, such as sheets or tubes, by
standard methods. Upon base hydrolysis, this resin is converted
into Nafion perfluorosulfonate polymer:

XR Resin + NaOH $\longrightarrow$

$$- (CF_2CF_2)_n\text{-}\underset{\underset{CF_2}{|}}{C}FO(CF_2\text{-}\underset{\underset{CF_3}{|}}{C}FO)_m CF_2CF_2SO_3^- \, Na^+$$

Nafion

The sodium counterions in this ion exchange polymer can be readily exchanged for other metal ions or hydrogen ion by soaking the polymer in an appropriate aqueous electrolyte solution. For commercial materials, m is probably equal to one and n varies from about 5 to 11. This generates an equivalent weight range of about 1000 to 1500 grams of dry hydrogen ion form polymer per mole of exchange sites. Membranes are produced in nominal thicknesses from about 0.1 mm to 0.3 mm.

Various modifications can be made to a basic Nafion homogeneous polymer film to produce materials with special characteristics. Open weave Teflon fabric can be laminated into the polymer film for increased strength. Also, composite membranes have been made in which layers of two different equivalent weights of polymer film are laminated together. This is useful because higher equivalent weight polymers show increased anion rejection under electrolysis conditions, which is desirable in certain applications such as in chlor-alkali membrane cells. Higher equivalent weight films also exhibit larger electrical resistances though. Composite membranes are thus used to optimize permselectivity and electrical properties.

Surface treatment has also been employed to generate membranes with improved hydroxide ion rejection capability for chlor-alkali applications. In this procedure, one surface of a sulfonyl fluoride XR resin film is treated with an amine such as ethylenediamine. After hydrolysis, a thin barrier layer of weakly acidic sulfonamide exchange sites is formed. When this treated surface faces the cathode solution, improved hydroxide rejection is realized in a membrane chlor-alkali cell.

An interesting and important practical aspect of Nafion perfluorosulfonate membranes is their ability to sorb relatively large amounts of water and other solvents. The polymers typically sorb 10-50% of their dry weight in water, depending upon polymer equivalent weight, counterion form and temperature of equilibration. Counterions with large hydration energies increase water uptake, as do low equivalent weights. It is interesting that up to 50% more water is taken up if the materials are boiled in water, compared to room temperature equilibration, and that this increased water is permanently retained after cooling[15]. Increases in water sorption are accompanied by decreases in electrical resistance and tensile strength, as might be expected[12]. Nafion membranes sorb even larger amounts of other solvents as well, particularly alcohols and other protic solvents[12-16].

Chemically, the perfluorosulfonate Nafion polymers are quite unreactive. They are stable in strong bases, strong oxidizing and reducing acids, cholorine, oxygen, hydrogen, and hydrogen peroxide at temperatures at least up to 125°C. Thermal stability is excellent to 200°C or higher, depending on counterion form.

Other perfluorinated cation exchange membrane materials have also been produced for chlor-alkali cell and other applications. These are the Flemion   membrane products (Asahi Glass Co. Ltd.), Neosepta-F   membranes (Tokuyama Soda Co. Ltd.), and the perfluorinated membranes produced by the Asahi Chemical Industry Company. These membranes are discussed in detail in chapters 14 to 17.

Because of the technological importance of the perfluorinated ionomers, as well as the novel structural features encountered in these materials, a wide range of physical and physico-chemical tools have been brought to bear on the problems related to the structure of these polymers.

Structurally, the materials are quite complex. In addition to a small amount of crystallinity (discussed in chapters 10 and 11) two distinct non-crystalline regions are present, the hydro-phobic fluorocarbon phase and the hydrophilic ionic areas. Hydration further complicates the picture because of the small sizes of the regions. The various techniques described in the first two sections of the volume probe different aspects of these structural features, and, because of their complexity as well as the different regimes to which the various tools address them-selves, no single agreed-upon view of the structure of these polymers has emerged as yet. Among the classical techniques, small angle X-ray scattering (chapters 10 and 11) focuses on differences in electron density and is thus particularly useful in detecting heterongenieties due to heavy metal atoms in the ionic clusters. Small angle neutron scattering (chapter 12), by contrast, is sensitive to the presence of protons, and has thus been used extensively in the study of hydrated materials. Quasi-elastic neutron scattering, in addition, provides some information on the mobility of the water molecules on the characteristic time scale of the experiment.

The spectroscopic techniques, on the other hand, probe individual species which make up the various regions. Infrared (chapter 8) and nuclear magnetic resonance (chapter 7) address themselves to water and the interactions of water with the various species with which it is in contact. Mossbauer spectroscopy (chapter 9), in addition, provides valuable information on the proximity of the cations and their environment. Mechanical (chapter 6) and transport (chapter 4) properties provide more indirect insight into the structural aspects, which is sup-plemented by thermodynamic studies (chapters 2 and 5) of the interaction between the polymer and water or other liquids. All these techniques are discussed in the present volume, and from these studies several structural models have emerged (chapter 13).

The major impetus for the development of the perfluorinated

sulfonate and carboxilate membranes which are the subject of this monograph has been their application in the chlor-alkali industry. The high strength and chemical stability of these polymers has been coupled with excellent permselectivity characteristics for several materials.   This achievement has enabled a new process to be developed for the manufacture of chlorine and sodium hydroxide, two of the largest tonnage chemicals produced in the world.   The membrane cell method is now beginning to be adopted, due to advantageous economic and envrionmental considerations in comparison with the mercury cell and asbestos diaphragm processes.   The importance of this advance in synthetic membrane technology is comparable to the development of asymmetric reverse osmosis membranes for water desalination in the early 1960's.   In both cases, a membrane was designed with a series of specific properties to solve an imortant problem.   Additional accomplishments of this kind are needed in membrane science if the inherent advantages of membrane separation systems are to be fully utilized in the chemical industry.

Thus the study of these perfluorinated ionomers is important not only because of the fundamental significance of their ion clustered morphologies, but also becuase this research will help to provide a scientific foundation for future developments in membrane science.   The work which is discussed in this monograph represents the efforts of many workers to establish this foundation.

References

1.   Holiday, L., Ed. "Ionic Polymers", Halstead Press, Wiley, New York 1975.
2.   Eisenberg, A. and King, M. "Ion Containing Polymers", Academic Press, New York 1977.
3.   a) ACS Polymer Preprints, Am. Chem. Soc. Div. Polymer Chem. $\underline{9}$(1) 505-546, 583-622, 1968.
     b) Bikales, N. M., Ed. "Water Soluble Polymers", Plenum Press, New York 1973.
     c) Eisenberg, A., Ed. J. Polymer Sci. $\underline{C}$, Polymer Symposium $\underline{45}$ (1974).
     d) Rembaum, A. and Selegny, E., Eds. "Polyelectrolytes and their Applications", VII Reidel, Dordrecht 1975.
     e) Eisenberg, A., Ed. "Ions in Polymers", Adv. in Chem. Series $\underline{187}$, Am. Chem. Soc. 1980.
4.   Reference 2, p. 57.
5.   Reference 2, p. 149.
6.   Eisenberg, A.; Ovans, K.; and Yoon, H. N., Reference 3e, Chapter 17.
7.   Reference 1, Chapter 2.
8.   Reference 2, p. 141-162.
9.   Reference 1, p. 173-207.
10. Hodge, I. M. and Eisenberg, A., Macromolecules $\underline{11}$, 283(1978).

11. Neppel, A.; Butler, I. S. and Eisenberg, A., Macromolecules 12,948(1979). J. Poly. Sci., Polym. Phys. 17, 2145(1979), J. Mocromolecules Sci. B19,61(1981).
12. Grot, W. G. F.; Munn, G. E.; Walmsley, P. N. "Perfluorinated Ion Exchange Membranes", presented at the 141st National Meeting of the ElectrochemicalSSociety, Houston, Texas, May 7-11, 1972.
13. Vaughan, D. J. Du Pont Innovation 1973, 4(3),10-13.
14. Grot, W. G. F. Chem. Ing. Tech. 1978, 50, 299-301.
15. Steck, A.; Yeager, H. L. Anal. Chem. 1980, 52, 1215-18.
16. Yeo, R. S. Polymer 1980, 21, 432-35.

RECEIVED October 27, 1981.

THERMODYNAMIC, MECHANICAL, AND TRANSPORT PROPERTIES

Thermodynamic Studies of the Water—Perfluorosulfonated Polymer Interactions

Experimental Results

M. ESCOUBES
Laboratoire de Chimie Appliquée et de Génie Chimique, Université Lyon I–43 Bd. du 11 Novembre 1918, 69621 Villeurbanne, France

M. PINERI
Equipe Physico-Chimie Moléculaire, Section de Physique du Solide, Département de Recherche Fondamentale, Centre d'Etudes Nucléaires de Grenoble, 85 X–38041 Grenoble Cedex, France

The analysis of the sorption isotherms is the most common way to study the interactions of water with polymers. Mathematical models can be fitted to the experimental results and give information about these water–polymer interactions which can be directly obtained from enthalpimetric analysis. It is possible to get the heat of sorption of the water molecules during different sorption isotherms corresponding to different humidity levels. It is also possible to check the phase transformations of the absorbed water by differential scanning calorimetry.

The water–polymer interaction depend on the polymer free volume, crystallinity, porosity, chemical structure, etc... It is known that strong interactions between water and polymer can produce important modifications of the solid polymer like swelling or crystallisation. If these interactions are not homogeneous inside the polymer matrix it may result in some "clustering" of the water molecules with formation of holes inside the polymer.

Thermodynamic measurements must define both the water–polymer interaction and the structural change of the polymer. This information can be given from the direct measurement of the heat of sorption of the water molecules.

- An incremental increase of the relative water pressure is realized and during each increment both the amount of water adsorbed in the specimen and the total energy involved in this absorption are recorded. The average energy per water molecule corresponding to these water molecules absorbed during the increment can therefore be deduced. This energy value depends strongly on the nature of hydrogen bonding and also on the number of hydrogen bonds involved in this interaction. Changes of this value during the water absorption may reflect the existence of different sites of absorption with different energies of interaction. For instance in collagen we have shown the existence of different regimes of absorption (-16 Kcal.mole^{-1} between 1 and 10 % of water, -13 Kcal. mole^{-1} between 24 and 48 %). From these energy values and also from the number of water molecules corresponding to each regime it has been possible to propose a model of water absorption corresponding

to well defined sites (1). This kind of measurement is the only
one which is able to define the amount of bound water. This term
of bound water has been overused during these last years. It has
first to be noted even with hydrophilic polymers, hydrogen bonding with hydroxyl, amide or carbonyl groups are relatively low in
energy and sometimes lower than the energy of bonding of water molecules in liquid water (2). Another point is the indirect way
usually used to define both energy and the amount of bonded water
(3). Most of the enthalpy values are deduced from the Clausius
Clapeyron equation using the absorption isotherms even when the
reversibility conditions are not observed. The amount of bound
water is usually obtained from the difference between the total
water content and the amount of water giving a phase transition
at low temperatures. In fact in many cases such phase transition
is not possible because of steric limitations inside the water
clusters. The mobility of the bound water has also been shown by
NMR or dielectric measurements to be pretty close to the observed mobility in liquid water.

     – Low values of the heat of water absorption can be obtained.
These values smaller than the liquefaction energy may result from
the superposition of an endothermal mechanism : breaking of some
bonds (4), further cristallization of the polymers (5), expansion
of the macromolecules with change of volume. For this last case,
which does not correspond to an expansion mechanism described by
the Flory model one has to take into account the enthalpy and
free energy of expansion (6,7) or the internal pressure due to
the polymer (8).

Heat of absorption measurements

     This measurement is realized by coupling a microbalance and a
differential microcalorimeter (9). Before the sorption experiment,
the two identical samples are vacuum dried (10^{-4} torr) in situ.
Different increasing humidity levels are then obtained by changing
the temperature of a water cell (between – 80°C and 20°C). During
each water pressure increment, weight and energy changes are recorded and the molar energy of interaction is obtained. The MTB
10.8 Setaram microbalance has a sensibility better than 1 µg. The
limit of detection for the microcalorimeter is 80 µW. For a hydration energy around 10 Kcal.mole^{-1}, a 10 % precision is obtained for
water sorption larger than 1 mg per hour. The measured differential molar energy of absorption (q_f) is :

$$q_d = e_a - e_g + n_a \frac{de_a}{dn_a}$$

in which

e_a : internal molar energy of adsorbed H_2O
e_g : internal molar energy of adsorbed gas
n_a : number of H_2O molecules adsorbed
dn_a : change in the number of adsorbed water-
         molecules during the pressure increment.
de_a : change in the internal molar energy of the
         adsorbed H_2O molecules during the same increment

A detailed analysis of the thermodynamics involved in this measurement is given (13). It has to be noted that this internal energy change which is measured is not very different from the enthalpy change.

Experimental results of the heat of absorption measurements

Nafion 120 in the acid and neutralized form has been studied (11). Isotherms have been obtained at 10°C and at lower temperatures.

From the 10°C isotherms the main results are :

1. A well defined amount of water, depending on the cation, is kept after vacuum drying (10^{-5} torr) at room temperature (fig. 1). This residual water (1 to 2 H_2O/SO_3^-) is desorbed after heating above the glass transition of the Nafion matrix. During rehydration of the high temperature vacuum dried sample the initial isotherm sorption curve is reobtained only above a well defined relative water pressure (fig. 2).

2. This residual water has the same interaction energy as the first water molecules absorbed in the room temperature dried sample (fig. 3). Only two hydration regimes are observed during the room temperature sorption isotherm (fig. 3 and 4).

- The first regime corresponds to the first 8 per cent of absorbed water ($\sim 5/6$ moles/SO_3^-). The molar interaction energy is constant and characteristic of the cation

$$
\begin{array}{ll}
- 13 \text{ Kcal.mole}^{-1} \text{ for } Fe^{++} \\
- 13.5 \quad " \qquad\qquad H^+ \\
- 9.5 \quad " \qquad\qquad Na^+ \\
- 8.5 \quad " \qquad\qquad Cu^{++}
\end{array}
$$

The hydration energy of the cations in solution corresponds to the same order and this is consistent with the existence of ionic clusters in the dehydrated state.

- During the second regime the energy is decreasing and is always lower than the value corresponding to the liquefaction. This implies an endothermal contribution corresponding to the expansion of the cluster inside the organic phase.

An important decrease of the water content is obtained with the low temperature isotherms (fig. 5). At saturation the relative water contents at different temperatures are given in the following table:

TABLE I

Temperature	Water content	interaction energy Kcal/mole
0°C	14,75 %	- 12,5
20°C	12	- 12,7
-10°C	10	- 8
-13°C	8.5	

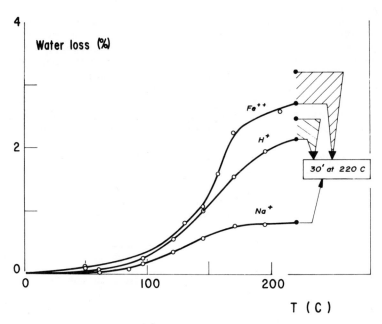

Figure 1. Water loss during high temperature heating for different Nafion forms; heating rate, 3° C/min; vacuum, 10⁻⁴ torr.

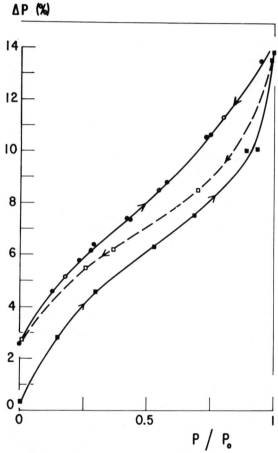

Figure 2. Room temperature sorption–desorption isotherms of acid Nafion. Key: ●, *room temperature dried absorption;* ■, *220° C dried primary absorption;* □, *220° C dried desorption;* ○, *220° C dried secondary absorption.*

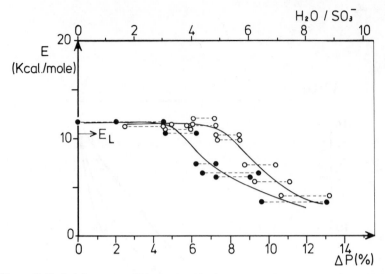

Figure 3. Enthalpic energy of absorption for the water molecules during an isotherm absorption at room temperature. Key: ○ - - - ○, *room temperature dried sample;* ● - - - ●, *220° C dried sample.*

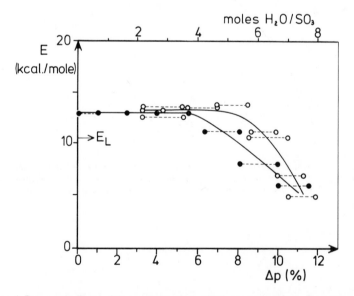

Figure 4. Same as in Figure 3 for an iron salt (∼65% neutralization). Key: ○ - - -○, *room temperature dried sample;* ● - - - ●, *220° C dried sample.*

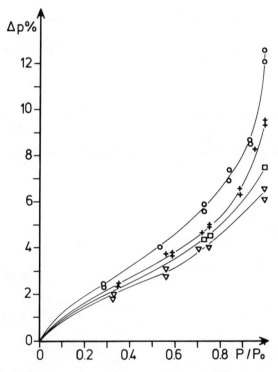

Figure 5. Different temperature isotherms of the acid Nafion sample. Key: ○, *20° C;* +, *0° C;* □, − *10° C;* − *12.6° C.*

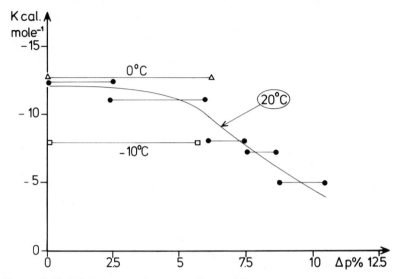

Figure 6. Enthalpic energy absorption during different temperature isotherm absorption.

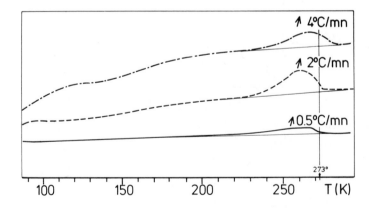

Figure 7. Influence of the heating speed on the acid Nafion polymer containing 12% by weight of water.

The interaction energy wich is given in this table has to be compared with liquefaction energy of water = -12.7 Kcal.mole^{-1} at $0°C$ and with the sublimation energy at $-10°C$: -12.5 Kcal.mole^{-1}. An important change in the polymer structure must therefore occur when the temperature is lowered (fig. 6).

The first conclusion which is apparent from these results is the absence of strong water–Nafion polymer interactions. The first water molecules which are absorbed correspond to the solvatation of the ions. For water contents larger than about 8 % ($\sim$ 5 water molecules) one observes a decrease in the energy which can be explained by an elastic deformation of the polymer involving the motion of the hydrophobic chains out of the hydrated zone. The amount of water absorbed at saturation and the corresponding hydration energy strongly depend on the temperature.

Differential scanning calorimetry (DSC)

DSC has been found to be the simplest method to define the relative amounts of freezing and non freezing water (14). A wet sample is hermetically kept in a sample pan. During cooling or heating runs the empty reference cell and the specimen-containing cell are kept at the same temperature. The relative difference of power necessary to do so is plotted versus temperature. From the peak surface corresponding to freezing or melting it is therefore possible to define the relative amount of freezing and non freezing water.

A CPC 600 calorimeter (15) has been used with heating or cooling speeds between 0.5 and 4°/mn. Acid Nafions 120 containing different water percentages have been studied.

The influence of the heating speed is shown in fig. 7 for a Nafion specimen containing 12 % by weight of water. The samples have been first slowly cooled at 1°C/mn down to liquid nitrogen temperature. In table (II) are given the integrated values of the endothermic peaks. For the 2 /mn sample two different runs have been done. The average value is 432 mcal within 5 % depending on the base line and the corresponding water – with 80 cal/g as the heat of melting – is 5.4 mg. In this case with such an hypothesis we would have $\sim$ 30 % of freezing water.

The influence of the cooling speed is shown in fig. 8 for the same sample run at a constant heating speed of 2°C/mn. No change is observed between samples quenched in liquid nitrogen or rapidly cooled with He gas. In this case some exothermic contribution is apparent.

In fig. 9 are shown the curves obtained during a cooling process at 1°/mn for two different water content samples (12 an 15 % by weight). An exothermic peak is apparent in both samples, the position and the form of this peak depend on the sample. Both peaks are located at temperature well below the temperature corresponding to the normal water freezing.

No endothermic or exothermic peak is apparent for samples containing less than 8 % of water.

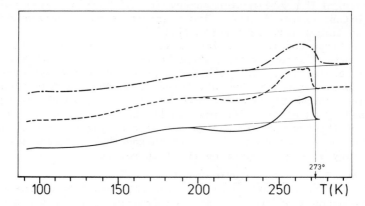

Figure 8. Influence of the cooling speed on the same specimen as in Figure 7. Cooling speeds: -·-·-·-, 1°/mn; - - -, He gas; ————, liquid nitrogen quenching.

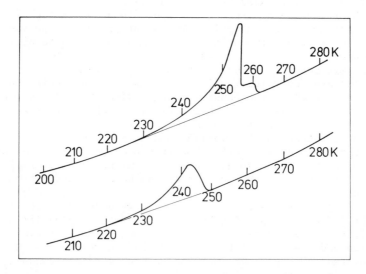

Figure 9. Influence of the water content on the thermograms: upper curve, 15% water; lower curve, 12% water.

TABLE II

Heating rate	Peak position K	H mcal
0.5°/mn	260/270°	450
2°/mn	263	416 433
4°/mn	268	430

From these results the first interpretation which comes to mind is freezing and melting of water in small clusters which only appear for water contents larger than 8 %.

The analysis of solid-liquid phase transformation in small pores can be done by thermoporometry. For a liquid contained in a porous material the solid-liquid interface curvature depends closely on the size of the pore and the solidification temperature is therefore dependent on this size (16). From the solidification thermogram it is therefore possible to obtain both the size of the pores from the solidification temperature position and also the total volume of water involved in this transformation from the measurement of the energy corresponding to this phase transformation.

By using a microcalorimeter the thermograms are obtained. The cooling and heating speeds are between 6 and 8°/hour. In fig. (10) and (11) are shown the thermograms corresponding to heating or cooling for two different Nafion water systems. In both samples an endothermic peak appears during heating which extends over a larger temperature range than in the previous experiment. Such a behaviour is similar to what is observed in porous materials like γ-alumina with spherical water containing pores (16).

A numerical relationship has been obtained between the freezing temperature depressions of a capillary condensate saturating a porous material and the radii.

$$Rp_{(nm)} = \frac{-64.67}{\Delta T} + 0.57$$

$$0 > \Delta T > -40$$

which gives $Rp = 21.8 \overset{\circ}{A}$ for $\Delta T = -40°$

$Rp = 38 \overset{\circ}{A}$ for $\Delta T = -20°$

If such interpretation is valid these $20 \overset{\circ}{A}$ and $40 \overset{\circ}{A}$ would correspond to the order of magnitude for the radius of the water clusters inside the 9 and 14 % water Nafion systems.

Another possibility which would explain such depression in the freezing and melting temperature would be the presence of ions in the water. It is possible to calculate this depression

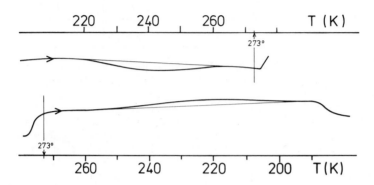

Figure 10. Low speed thermograms of a 9% H_2O content sample obtained during cooling (lower curve) and heating (upper curve).

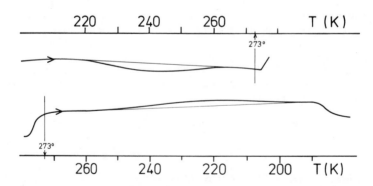

Figure 11. Low speed thermograms of a 14% water content sample obtained during heating (upper curve) and cooling (lower curve).

$$\theta = \frac{RT_o^2}{\Delta H} \cdot c$$

in which c is the ion concentration and ΔH the liquefaction energy
of water : 80 cal/g. The results are not consistent with such hypothesis both because the change in θ versus the water concentration
would be smaller and also because no such temperature difference
would be observed for the peak position during heating or cooling.

It is now important to report some NMR results obtained with
an acid 120 Nafion sample containing an excess of water. A Nafion-
water mixture is quenched from room temperature down to liquid nitrogen temperature and then rapidly put into the NMR spectrometer
at a well defined temperature below 0°C. The amplitude of the line
corresponding to the mobile water protons at this temperature is
then recorded versus time as shown in fig (12). The observed decrease in amplitude of the line corresponds to a change in the
number of the mobile water protons. During the annealing time some
desorption occurs and initially mobile water molecules are frozen
either outside the sample or in small holes inside this sample.

Conclusions

Two very well defined regimes of water absorption have therefore been evidenced. The first regime correspond to the first water
molecules which fill the hydration shell ; between four and six
water molecules are necessary to do so for the acid sample. A similar behaviour has been observed from NMR (11), Mössbauer (19).
These first water molecules which are absorbed onto these ionic
sites have an hydration energy which correspond to the value obtained for the corresponding cations in solution. The observed
decrease in the absolute value of the interaction molar energy for
further water sorption may involve a deformation of the hydrophobic
matrix. It has also to be pointed out that a rapid exchange occurs
between all the water molecules giving rise to a single line in
NMR.

Another important conclusion is obtained from the coupled
DSC/NMR experiments. The water content of the Nafion membranes
strongly depends on the temperature. Therefore the analysis of a
possible water phase separation cannot be done with experiments
involving temperature changes like DSC. This is pretty different
from what is obtained with γ-alumina which represent a relative
fixed and non temperature dependent hydrophobic matrix. The endothermic and exothermic peaks observed during heating and cooling
runs of the water–Nafion systems may be interpreted in two ways

- either the peak itself corresponds to the sorption-desorp-
tion thermal manifestation
- or the peak corresponds to the melting or solidification of
water in small pores which are formed during the thermal cycle.
Such a behaviour has already been observed in polyethylene (17).

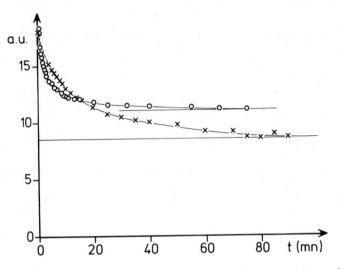

Figure 12. Change of the NMR line amplitude vs. time during annealing at different temperatures. Key: ○, − 30° C; ×, − 50° C.

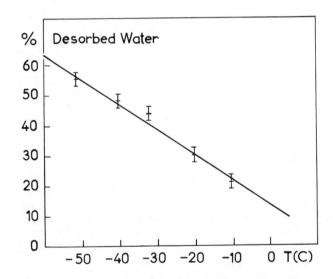

Figure 13. Amount of desorbed water vs. the annealing temperature.

A more detailed analysis of these results and of the possible interpretations will be given in a further publication (18).

Another interesting result is given in fig. 13 in which is plotted the amount of desorbed water versus the annealing temperature for the 15 % H_2O - acid Nafion system. From this figure it is shown that around 60 % of the total water content can be desorbed. We therefore have around 9 % desorbable water and around 7 % fixed water. These values are in close agreement with the two absorption regimes and also with the fact that no DSC peak has been observed for water content velow $\sim$ 8 % water.

Bibliography

1. M.H. PINERI, M. ESCOUBES, G. ROCHE, Biopolymers, 17, 12, 2799, (1978)
2. C.A.J. HOEVE - A.C.S. Symposium Series, 127, 7, 135 (1980)
3. J.A. RUPLEY, P.H. YANG, G. TOLLIN - A.C.S. Symposium Series, 127, 6, 11 (1980)
4. J. GUILLET, G. SEYTRE, A. COUILLARD, M. ESCOUBES - Die Angewandte Makromol. Chemie, 68, 1017, 149-162 (1978)
5. M. ESCOUBES, P. MOSER, P. BERTICAT - Die Angevandte Makromol. Chemie, 67, 991, 45-60 (1978)
6. H.J.C. BERENDSEN - Water in disperse systems, 15, 293, Franks Editor, Plenum Press (1975)
7. M. BRENER, E.M. BURAS, Jr and A. FOOKSON - A.C.S. Symposium Series, 127, 18, 311 (1980)
8. E. SOUTHERN, A.C. THOMAS - A.C.S. Symposium Series 127, 22, 375 (1980)
9. M. ESCOUBES, J.F. QUINSON, J. GIELLY, M. MURAT - Bull. Soc. Cim. Fr., 5, 1689 (1972)
10. Cl. LETOQUART, Fr. ROUQUEROL, J. ROUQUEROL - J. Chim. Phys. 3, 559 (1973)
11. R. DUPLESSIX, M. ESCOUBES, B. RODMACQ, F. VOLINO, E. ROCHE, A. EISENBERG, M. PINERI - A.C.S. Symposium Series, 127, 28, 470-486 (1980)
12. G. BELFORT, N. SINAI - A.C.S. Symposium Series, 127, 19 (1980)
13. M. ESCOUBES, M. PINERI, A. EISENBERG, S. GAUTHIER, to be published
14. S. DEODHAR, P. LUNER, A.C.S. Symposium Series 127, 28, 273-286 (1980)
15. E. BONJOUR, M. COUACH, J. PIERRE, Cahiers de la Thermique, n° 1, B 134-150 (1971)
16. M. BRUN, A. LALLEMAND, J.F. QUINSON, C. EYRAUD, Thermochimica Acta, 21, 59-88 (1977)
17. H.E. BAIR, G.E. JOHNSON, Analytical Calorimetry. Plenum Press Vol. IV, 219-225 (1977)
18. M. PINERI, C. BEN SAID, F. VOLINO, M. ESCOUBES, J.F. QUINSON, M. BRUN, to be published
19. B. RODMACQ, M. PINERI, J.M.D. COEY, A. MEAGHER, J. Polym. Sci. to be published.

RECEIVED October 13, 1981.

Cation Exchange Selectivity of a Perfluorosulfonate Polymer

HOWARD L. YEAGER

Department of Chemistry, The University of Calgary, Calgary, Alberta T2N 1N4, Canada

The ratio of the permeabilities of two cations in a cation exchange membrane is equal to the product of the ion exchange equilibrium constant and their mobility ratio (1). Therefore it is important to characterize the equilibrium ion exchange selectivity of ion exchange polymers in order to understand their dynamic properties when used in membrane form. Nafion (E.I. du Pont de Nemours and Co.) perfluorinated sulfonate membranes have found wide use in a variety of applications, many of which involve exchange of cations across membranes that separate solutions of different ionic composition. The inherent cationic selectivity of the polymer is an important consideration for such applications. Results of ion exchange selectivity studies of Nafion polymers are reviewed in this chapter, and are compared to those of other sulfonate ion exchange polymers.

The general properties of sulfonate ion exchange materials have been well-characterized (2). Divinylbenzene cross-linked polystyrene sulfonate resins, perhaps the most commonly used of organic ion exchangers, exhibit selectivity sequences which are successfully treated by Eisenman's theory of ion exchange (1, 3). In this approach the electrostatic field strength of the anionic exchange site is seen to be the principal factor in determining cation selectivity. Sulfonate exchange sites have a relatively low charge site density, and in such cases the order of cation affinities is determined by the relative magnitudes of their respective free energies of hydration. Cations with smaller hydration energies gain relatively more energy from electrostatic interaction with the exchange site, and are preferred as exchange counterions. Thus the order of selectivity for alkali metal ions is $Cs^+ > Rb^+ > K^+ > Na^+ > Li^+$ for sulfonate resins. The amount of sorbed water in the resin phase affects the relative magnitudes of selectivity coefficients but not the sequence. Large amounts of sorbed water are expected to dilute the effect and reduce the relative selectivity differences. Also, selectivity coefficients are affected by increased resin cross-linking (2). This is partly due to nonuniformity of exchange site spacing,

0097-6156/82/0180-0025$05.00/0

which produces overlap of sulfonate fields. This generates
higher effective field densities for some sulfonate sites.
Ultimately affinity reversals can occur. Thus the overall selectivity properties of the resin are determined by a combination of
the interrelated effects of ion-exchange capacity, cross-linking,
and water sorption.

   Although the Nafion perfluorinated sulfonate polymer is expected to demonstrate selectivity patterns which are similar to
those of cross-linked polystyrene sulfonates, there are several
notable differences between the two types of polymers. Nafion
polymers are presumably not cross-linked; thus the solvent swelling of this polymer is far more dependent on counterion and on
polymer pretreatment than the swelling of the more rigid sulfonate resins. Secondly, the phenomenon of exchange site clustering
(4) is expected to be an important factor in the selectivity
properties of Nafion. Although the exact nature of these ion
clusters is not fully resolved, convincing evidence now exists
to indicate that exchange sites, counterions, and sorbed water
exist as a separate microphase in the polymer. This phenomenon is
of course not possible for cross-linked polymers. Finally, the
ion exchange capacities for commonly used forms of Nafion (1100-
1500 equivalent weights) are about four times smaller than those
of commercial sulfonate resins.

Cation Exchange Selectivity Coefficients for Nafion Polymers

   General Considerations. In order to study cation exchange
equilibria it is necessary to determine the rate at which exchange equilibrium is attained. Exchange rates are relatively
rapid for 1200 EW Nafion. A H^+-form membrane, when immersed in
aqueous NaCl solution, attains 90% conversion to the Na^+-form in
less than two minutes (5). This time interval increases to 40
min for conversion to the Cs^+-form. This increase in equilibration time is attributable to the anomalously low diffusion coefficient of Cs^+ in the polymer phase (6). Even in this case
equilibration times of a few hours are sufficient to ensure
complete reaction. Another factor to consider is whether all
sulfonate sites are available for exchange with various cations.
A study has been performed in which a single piece of 1200 EW
Nafion was successively placed in the H^+, Na^+, Cs^+, Mg^{2+}, and
Ca^{2+} forms; analyses were performed for each counterion after desorption (7). No difference could be found, within 1% relative
error, in the measured number of exchange sites for each counterion form. This result agrees with an infrared study of the conversion of hydrogen ion forms of Nafion samples of various
equivalent weights into univalent and divalent metal ion forms
(8). The degree of replacement of hydrogen ion was measured to
be 99-100% in all cases. However, Roche and co-workers (9)
estimated the extent of conversion of a H^+-form of Nafion to the
Na^+-form to be only 77%, as determined by flame photometry. No

apparent reason can be seen for this difference in findings.

Ion exchange selectivity coefficients for univalent ion and divalent ion-hydrogen ion exchange are given by the equations

$$
K_{H^+}^{M^+} = \frac{\chi_{M^+} \, C_{H^+} \, \gamma_{H^+}}{\chi_{H^+} \, C_{M^+} \, \gamma_{M^+}}
\qquad
K_{H^+}^{M^{2+}} = \left[\frac{\chi_{M^{2+}} \, C_{H^+}^2 \, \gamma_{H^+}^2}{\chi_{H^+}^2 \, C_{M^{2+}} \, \gamma_{M^{2+}}} \right]^{\frac{1}{2}}
$$

where χ represents the equivalent ionic fraction of ion in the polymer and C is solution molarity. Single ion activity coefficients, γ, approximately cancel for univalent ion exchange in dilute solution (0.01 M). However for divalent ion-hydrogen ion exchange they must be included even for dilute solution experiments, due to the asymmetry of the reaction. These selectivity coefficients are not constants in general, but depend upon the equivalent ionic fraction of exchanging ion. Average selectivity coefficients can be determined by measurement of values as a function of ionic fraction (ion exchange isotherm) followed by integration of the curve. A value of the selectivity coefficient determined at an equivalent ionic fraction of 0.5 for both ions is often a good approximation to the integrated result (2).

Univalent Ion-Hydrogen Ion Selectivities. Selectivity coefficient isotherms of 1200 EW Nafion for univalent cation-hydrogen ion exchange at 25°C are shown in Figure 1 (10, 11). These isotherms show a much less pronounced dependence of the selectivity coefficient on ionic fraction of metal ion than corresponding isotherms for cross-linked polystyrene sulfonates (2, 12-16). An exception is the large rise in the cesium ion selectivity coefficient at high metal ion content; indeed minima in selectivity isotherms are unusual even for polystyrene sulfonate exchangers. The water content of Nafion can be increased by boiling in water; the polymer reaches a constant value of water sorption which remains unchanged at 25°C for at least several weeks (7). Selectivity measurements were also performed for this expanded form, and results are shown in Figure 2 (10, 11).

Selectivity coefficients were calculated at an ionic fraction of metal ion equal to 0.5 by interpolation of the isotherms. Values are listed in Table I along with some values which were determined at 40°C. Included in Table I are the water to exchange site mole ratios for the polymer in each ionic form (7, 11). Both the normal and expanded forms of Nafion exhibit the expected order of selectivities of alkali metal ions for a sulfonate ion exchanger. Perhaps the most interesting feature of these values is their large spread compared to conventional sulfonate resins. For example, a 16% cross-linked polystyrene sulfonate has metal ion-hydrogen ion selectivity coefficients equal to 0.680 (Li^+), 1.61 (Na^+), 3.06 (K^+), 3.14 (Rb^+), and 3.17 (Cs^+) (12). The selectivity coefficients of Ag^+ and Tl^+ for Nafion are lower than those of polystyrene sulfonate resins of

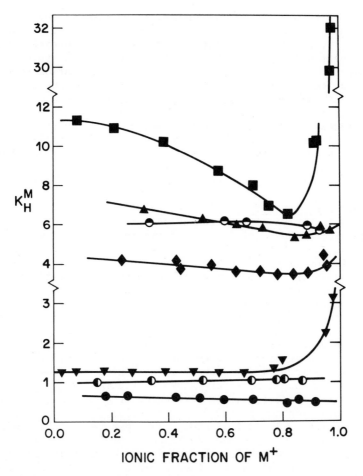

Figure 1. Ion exchange isotherms for 1200 EW Nafion at 25° C, 0.01 M ionic strength (10, 11). Key: ●, Li+; ▼, Na+; ◆, K+; ▲, Rb+; ■, Cs+; ◐, Ag+; ◓, T1+.

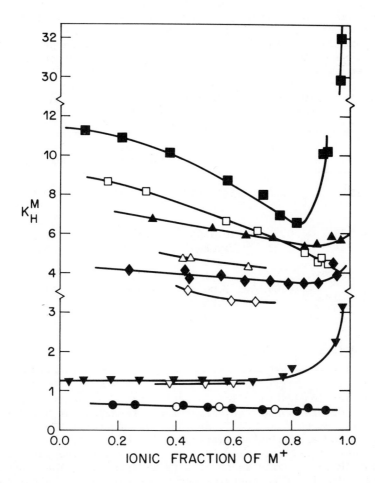

Figure 2. Ion exchange isotherms for 1200 EW Nafion at 25° C, 0.01 M ionic strength (10). Closed symbols: normal form; open symbols: expanded form of polymer. Symbols are the same as those in Figure 1.

various cross-linkings though, which indicates that for this
perfluorinated polymer with no aromatic content, nonelectrostatic
interactions between counterions and the polymer matrix are
reduced to a minimum. All selectivity coefficients are smaller
for the expanded form of Nafion, but the spread in alkali metal
ion values remains higher than for polystyrene sulfonate resins.
The water to exchange site mole ratios for the pure univalent
forms of Nafion are markedly dependent on the identity of the
counterion; thus these ion exchange reactions are accompanied by
significant changes in polymer water content. The expanded forms
of Nafion show the same trend, even though water contents average
50% higher than the as-received forms.

Table I.   Univalent Ion Selectivity Coefficients and
            Water Contents for 1200 EW Nafion.

Ion	$K_{H^+}^{M^+}$			mol H_2O / mol $-SO_3^-$	
	$25^\circ C$	$40^\circ C$	$25^\circ C$ (E)[a]	$25^\circ C$	$25^\circ C$ (E)[a]
H^+				16.7	22.3
Li^+	0.579	0.555	0.586	14.3	22.3
Na^+	1.22	1.31	1.18	11.9	18.4
K^+	3.97		3.48	8.8	13.3
Rb^+	6.26		4.71	7.7	11.8
Cs^+	9.11	9.04	7.06	6.6	11.3
Tl^+	6.12		3.83	8.0	11.7
Ag^+	1.07		0.90	12.2	17.6

[a] expanded form

Divalent Ion-Hydrogen Ion Selectivities. Selectivity coefficients determined at equivalent ionic fractions of 0.5 for
the alkaline earth ions, Co^{2+}, and Zn^{2+} are listed in Table II
along with corresponding polymer water contents (7). Again, the
normal order of selectivities is seen for the alkaline earth ions
for a low charge density exchange site environment. The order of
standard hydration free energies for these cations is $Zn^{2+} > Co^{2+}$
$> Mg^{2+} > Ca^{2+} > Sr^{2+} > Ba^{2+}$ which is the inverse order of exchange
selectivities. The spread in selectivities for the alkaline
earth ion series is much smaller compared to that of the alkali
metal ions however. The values for the normal form of Nafion in
Table II are similar to those of a 4% divinylbenzene cross-linked
polystyrene sulfonate resin: Mg^{2+}, 2.23; Ca^{2+}, 3.14; Sr^{2+}, 3.56;
Ba^{2+}, 5.66 (12). Resins of 8% and 16% cross-linking have a much
larger spread than do these values; thus the enhancement in

Table II. Divalent Ion Selectivity Coefficients and
Water Contents for 1200 EW Nafion.

Ion	$K_{H^+}^{M^{2+}}$			mol H_2O / mol $-SO_3^-$	
	$25°C$	$40°C$	$25°C$ (E)[a]	$25°C$	$25°C$ (E)[a]
Mg^{2+}	2.30	2.36	2.15	13.9	19.8
Ca^{2+}	3.60		2.87	12.9	17.5
Sr^{2+}	4.23		3.79	12.3	16.9
Ba^{2+}	5.55	5.27	4.61	11.6	14.9
Co^{2+}	1.24			13.7	20.0
Zn^{2+}	0.97			14.1	19.9

[a] expanded form

selectivity for univalent cations with Nafion is not present for
these divalent cations. This change is accompanied by a reduction in the dependence of polymer water content on the counterion's identity compared to the univalent ion case. In fact, the
water contents of sulfonate resins show a greater change from
Mg^{2+} to Ba^{2+} forms than does Nafion (12, 16). Finally, the expanded forms again show increases in water sorption which are
accompanied by selectivity losses.

Quaternary Ammonium Ions. In a recent study (17), 1200 EW
Nafion has been used to construct a membrane ion selective electrode. The electrode was placed in both the tetrabutylammonium
ion and cesium ion forms, and the response characteristics of
each form were measured. These electrodes show Nernstian
responses, and the tetrabutylammonium ion electrode has no interference from inorganic cations such as Na^+, K^+, and Ca^{2+}. However, this electrode shows a marked interference with decyltri-
methylammonium ion. In addition the cesium ion electrode
response is sensitive to the presence of tetrabutylammonium ion
and especially dodecyltrimethylammonium ion. Although membrane
electrode sensitivities are not in general proportional to
thermodynamic selectivity coefficients, the results do indicate
that these large, hydrophobic cations are preferred over smaller
inorganic cations by the polymer. The authors suggest that the
surfactant character of the two asymmetric tetraalkylammonium
ions may lead to non-electrostatic interactions with the
fluorocarbon regions of the polymer, which would enhance their
affinities (17).

The Nature of the Cation Exchange Process in Nafion

It is useful to compare the properties of the cation exchange
process for polystyrene sulfonate resins with those of Nafion in
order to develop insight into the differences in selectivity
coefficients. Boyd discusses hydrogen ion-metal ion exchange for
polystyrene sulfonates in terms of enthalpy, entropy, and volume
changes of the reactions (18). Enthalpy and entropy changes are
small and negative for alkali metal ion exchange in resins of low
cross-linking. No evidence for ion pairing is evident, and the
thermodynamic parameters are likely due only to differences in
ionic hydration in the solution phase between the two exchanging
ions. Ion pairs may form for higher cross-linked resins with
lower water contents, due to increased electrostatic interactions.
In these cases enthalpies and entropies of exchange become increasingly negative, and somewhat improved selectivities for
alkali metal ions result. In contrast, positive entropies of
exchange appear to control the magnitudes of selectivities for
divalent-univalent cation ion exchange reactions. A large
component of this entropy gain results from the gain in solution
entropy when a divalent ion is replaced by two univalent ions
with much smaller hydration entropies (18, 19). As divalent ions
exchange into the resin phase, larger amounts of water are released compared to univalent-univalent ion exchange (12, 16)
probably due to formation of solvent separated and contact ion
pairs (18). This release of water will also cause a positive
entropy change to the system, as will the statistical entropy
increase when divalent counterions distribute among twice as many
univalent exchange sites (19, 20). Because of the energy required to dehydrate a divalent cation as it enters the resin
phase, enthalpies of exchange also tend to be positive.
    Nafion would be expected to show differences in the thermo-
dynamics of cation exchange for at least two reasons. First, the
perfluorinated backbone should generate a much lower charge
density on the sulfonate exchange sites, reducing the possibility
of the formation of even solvent separated ion pairs. The use of
hydrogen ion forms of Nafion as "superacid" solid catalysts for
various organic reactions is a reflection of this low charge
density (21). Second, the lack of formal cross-links generates
a dynamic morphology for Nafion, in which the water content of
the polymer would depend to a larger extent on the hydration
characteristics of the counterion. In contrast, cross-linked
polystyrene sulfonates contain a relatively large fraction of
sorbed water in interstitial pores, especially for resins of low
cross-linking. Reichenberg (2) concludes that this water tends
to reduce selectivity differences among cations. For higher
cross-linkings, the rigid polymer matrix serves to dehydrate
counterions, especially polyvalent ones, in order to enter the
polymer phase.
    Both of these characteristics of Nafion would lead to

reduced enthalpies of ion exchange. This appears to be the case
for Li^+, Na^+, Cs^+, Mg^{2+} and Ca^{2+} exchange with hydrogen ion, as
inferred from the lack of temperature dependence of these
selectivity coefficients in Tables I and II, and the lack of any
trend in the temperature dependence. Small enthalpies of ion
exchange are difficult to detect in this way though, and calorimetry of these systems would be necessary to confirm this point.
It is clear though that increased heats of reaction are not
responsible for the wide spread in selectivity coefficients for
alkali metal ion-hydrogen ion exchange in Nafion.

The major factor which is involved in the ion exchange
selectivity behavior of Nafion appears to be the positive entropy
change associated with the replacement of hydrogen ion with a
metal ion. In all cases this exchange is accompanied by water
release and polymer contraction, which are entropy-producing
processes (19, 22). Rather large entropy increases occur when
water of hydration is released from ionic crystals, on the order
of 40 $J mol^{-1}K^{-1}$ (23, 24), or about 12 $kJ mol^{-1}$ at $25^O C$. Although much smaller increments in entropy would be expected for
Nafion, the large amounts of released water would be a compen-
sating factor. To test this possibility it is first assumed
that the enthalpies of exchange are negligible and that the
differences in selectivity are dominated by entropy production
associated with changes in water content of the polymer. A
relationship should then be found between the logarithm of
selectivity coefficient and the amount of water released per mole
of exchanging ions, $\Delta H_2 O_{EXCH}$:

$$-\Delta G^O = RT \ln K_H^M = -\Delta H^O + T\Delta S^O \simeq T\Delta S^O$$

$$\Delta S^O = f(\Delta S_{H_2 O})$$

$$\Delta S_{H_2 O} = (S_{H_2 O, SOLN} - S_{H_2 O, NAFION})(\Delta H_2 O_{EXCH})$$

Although other sources of entropy change are to be found in ΔS^O,
these would remain relatively constant for various metal ions
compared to $\Delta S_{H_2 O}$. This relationship is shown in Figure 3 for
normal and expanded forms of 1200 EW Nafion using both univalent
and divalent ion selectivity coefficients. Lines are least
squares fits for alkali metal ions and alkaline earth ions.

The linearity of these plots, coupled with the similarity of
slopes between univalent and divalent ion lines for each form of
Nafion, suggests that for both forms a constant increment in
entropy occurs per released water molecule. The slopes of these
lines are: 0.90 and 0.94 $kJ mol^{-1}$ at $25^O C$ for alkali ion and
alkaline earth ion plots, normal form; and 0.53 and 0.40 $kJ mol^{-1}$
for the expanded form of Nafion. These values can account for
the magnitudes of the selectivity coefficients, even though they
are less than 10% of the entropy increase for water release from

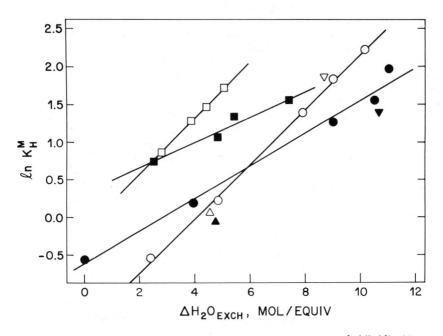

Analytical Chemistry

Figure 3. Logarithm of selectivity coefficients vs. change in water content for 1200 EW Nafion (7). Key: ◯, ●, alkali metal ions; ☐, ■, alkaline earth ions; △, ▲, Ag^+; ▽, ▼, Tl^+. Light symbols, normal form, dark symbols, expanded form.

ionic crystalline hydrates, when the amounts of released water are considered. The lower slopes for the expanded form are consistent with the expectation that ion clusters would be more aqueous-like, yielding smaller entropy increases per released water molecule, even though the amount of water released per equivalent of exchanged ions is generally larger compared to the normal form.

Thus it appears that the large spread in ion exchange selectivities for alkali metal ion exchange with hydrogen ion are caused by entropy increases upon desorption of water from Nafion. This effect is largely removed for divalent ion-hydrogen ion exchange. These ions have much larger hydration energies, and the dynamic character of the ion clusters permits large, hydrated metal ion species to enter the polymer phase. Also the low charge density of the exchange sites does not favor ion pair formation, a process which promotes water desorption. Thus selectivities for these ions are similar to those of lightly cross-linked polystyrene sulfonate resins.

Application of Nafion for Ion Exchange Chromatography

Although Nafion is generally considered for use only as a membrane material, its selectivity properties suggest that it would be possible to perform chromatographic separations of ions using the polymer as a stationary phase. The outstanding chemical stability of this perfluorinated material would be an advantage in situations where normal ion exchange resins may suffer degradation, such as in the handling of highly radioactive solutions. A granular form of 1200 EW Nafion was prepared by grinding membrane samples at liquid nitrogen temperatures, which generated a 40-60 mesh product (11). A chromatographic column was prepared using this material in the hydrogen ion form. Column dimensions were 0.8 cm diameter by 40 cm in length. Chromatographic separations at 25°C were performed for alkali metal ions and alkaline earth ions. Results are shown in Figures 4 and 5. A flow rate of 1.0 mL min^{-1} was used here, using increasing concentrations of HCl to effect the separation. Figure 4 demonstrates that the resolution of alkali metal ions is easily accomplished, even with a rather inefficient column made up of irregularly shaped particles of stationary phase. Higher flow rates generated asymmetric peaks and reduced separation. Since selectivity coefficients show a minimal temperature dependence, the same separation was successfully performed at 40°C at a flow rate of 1.5 mL min^{-1}, using eluent concentrations of 0.5 M (45 mL) followed by 1.5 M HCl (165 mL), with an improved separation time of 2 h. The separation of the alkaline earth ions is not as successful. However, with a properly designed column using smaller, more uniformly shaped particles, a complete resolution of all ions should be straightforward. Therefore Nafion does show promise as an ion exchange chromatographic phase for specialized applications.

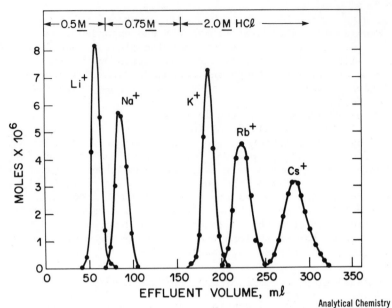

Analytical Chemistry

Figure 4. Chromatographic separation of alkali metal ions using 1200 EW Nafion at 25° C (11).

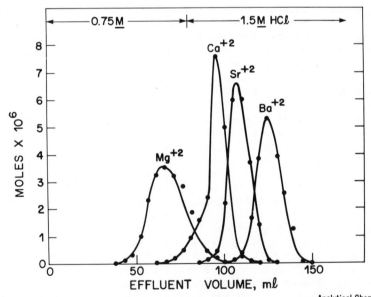

Analytical Chemistry

Figure 5. Chromatographic separation of alkaline earth ions using 1200 EW Nafion at 25° C (11).

Ion Exchange Properties in Methanol

Nafion is known to sorb large amounts of various nonaqueous solvents, especially those with hydrogen bonding protons (25). Selectivity isotherms for alkali metal ion-hydrogen ion exchange have been determined for anhydrous methanol solutions; these are shown in Figure 6 (7). The isotherms for K^+, Rb^+, and Cs^+ show rather large positive slopes. For these ions, the exchange process is accompanied by polymer volume decreases of up to 40%. It is suspected that this shrinkage may significantly alter the properties of the exchange site environment, and be responsible for these sloping isotherms. For Li^+ and Na^+, only minor volume changes are seen, and isotherms indicate ideal thermodynamic behavior.

These isotherms were integrated to produce overall selectivity coefficients. The results are: Li^+, 0.443; Na^+, 0.680; K^+, 4.68; Rb^+, 7.17; and Cs^+, 9.61. These values are very similar to those in aqueous environments, with the minor exception that sodium ion is now less preferred than hydrogen ion. It is interesting to note that Na^+ exchange is now accompanied by a slight increase in solvent sorption, which may account for this difference in terms of an entropy-driven process (10). The similarity of ion exchange behavior in the two cases suggests that increased electrostatic interactions due to the lower dielectric constant of methanol do not occur. In contrast, affinity reversals are common in non-aqueous solvent environments for polystyrene sulfonate resins. For example, the selectivity coefficients for hydrogen ion exchange in an 8% cross-linked sulfonate resin for methanol solutions are: Li^+, 0.335; Na^+, 3.23; K^+, 18; Cs^+, 10.0 (26). This change in order of selectivity can be ascribed to increased electrostatic interactions between exchange sites and alkali metal ions as ionic radius increases and solvation energy decreases (2, 3). Therefore the selectivity coefficients determined for Nafion in methanol are a further indication of the much lower charge density on the sulfonate exchange site compared to conventional sulfonate polymers.

Conclusion

The ion exchange properties of Nafion have not been extensively studied to date. However the results discussed here indicate that the polymer shows interesting and potentially useful properties for various applications in which ion exchange selectivity is required. These include not only the various configurations in which Nafion can be used in membrane form, but also its possible application as a chromatographic phase. The study of the ion exchange selectivity for ion clustered polymers of other chemical types is also suggested from these results.

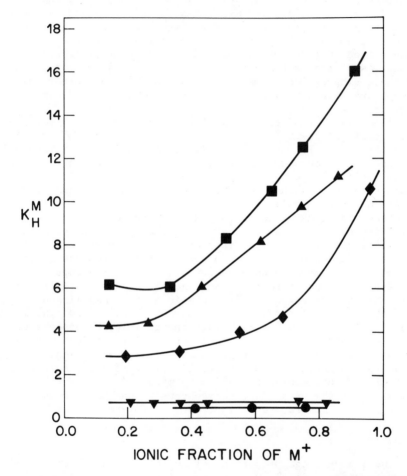

Figure 6. Ion exchange isotherms for 1200 EW Nafion in methanol at 25° C (7).
Key: ●, Li+; ▼, Na+; ◆, K+; ▲, Rb+; ■, Cs+.

The dynamic character of polymers such as Nafion in terms of solvent swelling can have a marked influence on the thermodynamics of ion exchange processes.

Literature Cited

1. Eisenman, G. Proc. 23rd Int. Congr. of Physiol. Sci.; Exerpta Medica 1965, 87, 489-506.
2. Richenberg, D. in "Ion Exchange", Marinsky, J.A., Ed.; Marcel Dekker: New York, 1966; Vol. I, Chapter 7.
3. Eisenman, G. in "Ion-Selective Electrodes", Durst, R.A., Ed.; Natl. Bur. Std. Spec. Publ. 1969, 314, Chapter 1.
4. Yeo, S.C.; Eisenberg, A. J. Appl. Polym. Sci. 1977, 21, 875-98.
5. Lopez, M.; Kipling, B.; Yeager, H.L. Anal. Chem. 1976, 48, 1120-22.
6. Yeager, H.L.; Kipling, B.; J. Phys. Chem. 1979, 83, 1836-39.
7. Steck, A.; Yeager, H.L. Anal. Chem. 1980, 52, 1215-18.
8. Peluso, S.L.; Tsatsas, A.T.; Risen, W.M., Jr. "Spectral Studies of Ions in a Perfluorocarbonsulfonate (Nafion) Ionomer", Report 1979, TR-79-01; Order No. AD-A080935, Avail. NTIS.
9. Roche, E.J.; Pineri, M.; Duplessix, R.; Levelat, A.M. J. Polym. Sci., Polym. Phys. Ed. 1981, 19, 1-11.
10. Steck, A. M.Sc. Thesis, The University of Calgary, Calgary, Alberta, Canada, 1979.
11. Yeager, H.L.; Steck, A. Anal. Chem. 1979, 51, 862-65.
12. Bonner, O.D.; Smith, L.L. J. Phys. Chem. 1957, 61, 326-29.
13. Bonner, O.D.; Argersinger, W.J.; Davidson, A.W. J. Am. Chem. Soc. 1952, 74, 1044-47.
14. Bonner, O.D.; Rhett, V. J. Phys. Chem. 1953, 57, 254-56.
15. Bonner, O.D.; Payne, W.H. J. Phys. Chem. 1954, 58, 183-85.
16. Bonner, O.D.; J. Phys. Chem. 1955, 59, 719-21.
17. Martin, C.R.; Freiser, H. Anal. Chem. in press.
18. Boyd, G.E. in "Charged Gels and Membranes". Selégny, E., Ed.; D. Reidel: Dordrecht, Holland, 1976, Volume I, pp. 73-89.
19. Boyd, G.E.; Vaslow, F.; Lindenbaum, S. J. Phys. Chem. 1967, 71, 2214-19.
20. Cruickshank, E.H.; Meares, P. Trans. Faraday Soc. 1957, 53, 1289-98.
21. Olah, G.A.; Parkash, G.K.S.; Sommer, J. Science 1979, 206, 13-20.
22. Gamalinda, I.; Schloemer, L.A.; Sherry, H.S.; Walton, H.F. J. Phys. Chem. 1967, 71, 1622-28.
23. Moore, W.J. "Physical Chemistry", 4th ed.; Prentice-Hall: Englewood Cliffs, N.J., 1972; Chapter 3.
24. Latimer, W.M. "Oxidation Potentials", Prentice-Hall: Englewood Cliffs, N.J., 1952; p. 364.
25. Yeo, R. Polymer 1980, 21, 432-35.
26. Fessler, R.G.; Strobel, H.A. J. Phys. Chem. 1963, 67, 2562-68.

RECEIVED August 7, 1981.

Transport Properties of Perfluorosulfonate Polymer Membranes

HOWARD L. YEAGER

Department of Chemistry, The University of Calgary, Calgary, Alberta T2N 1N4, Canada

Perfluorinated, high molecular weight sulfonate polymers, such as the Nafion materials (E.I. du Pont de Nemours and Co.), have the high chemical stability and strength to serve as ideal membranes in various separation applications. In addition, homogeneous and uniform membranes of large size can be produced by taking advantage of the thermoplastic characteristics of the polymer in the unhydrolyzed sulfonyl fluoride form (1). The capability of producing membranes of uniform composition and thickness is another important advantage for wide scale industrial application. Finally, these polymers sorb relatively large amounts of water (and other protic solvents) despite the fluorocarbon character of the polymer backbone. This latter feature is related to perhaps the most important characteristic of these materials: cations and water readily diffuse through the polymer, which enables electrolytic communication to be maintained through the membrane phase.

It is of course important to characterize the nature of transport processes in perfluorosulfonate polymer membranes in order to optimize their performance in separation systems. The ion-clustered morphology (2) of these polymers is unusual compared to conventional cross-linked sulfonate ion exchange resins, whose transport properties have been reasonably well studied. Therefore it is expected that differences in transport characteristics will be seen between the two types of polymers. These differences should lend insight into the nature of the ion clustering phenomenon. Dilute solution studies are of particular importance in this regard. Under these conditions, ion-containing polymers exhibit Donnan exclusion of anions, which prevents sorption of electrolytes from the solution (3). Only the cationic exchange counterions are then present in the membrane phase, which helps to simplify the interpretation of the material's transport properties. Experiments performed in concentrated solutions and at elevated temperatures are also necessary, because most applications of ion exchange membranes involve such conditions. It is also important to consider the driving force for transport of

0097-6156/82/0180-0041$05.75/0

species across the membrane. Gradients in concentration lead to diffusional processes, while electrical potential gradients generate ionic migration and electroosmotic effects. The combined effects of these forces yield the overall transport characteristics of the membrane.

Experimental results which yield insight into the nature of these processes in perfluorosulfonate membranes are emphasized in this chapter. This information, when correlated with structural studies and results of membrane performance in practical applications, should help to produce a unified understanding of this important new type of polymer membrane.

Diffusion in Nafion Perfluorosulfonate Membranes

Membrane Diffusion in Dilute Solution Environments. The measurement of ionic diffusion coefficients provides useful information about the nature of transport processes in polymer membranes. Using a radioactive tracer, diffusion of an ionic species can be measured while the membrane is in equilibrium with the external solution. This enables the determination of a self-diffusion coefficient for a polymer phase of uniform composition with no gradients in ion or water sorption. In addition, self-diffusion coefficients are more straightforward in their interpretation compared to those of electrolyte flux experiments, where cation and anion transport rates are coupled.

Tracer self-diffusion coefficients for sodium ion and cesium ion have been measured for 1200 equivalent weight Nafion membranes (4-7). Results obtained at 25°C are listed in Table I, along with

Table I.   Sodium Ion and Cesium Ion Self-Diffusion
Coefficients, 25°C

Medium	$\overline{D}$, cm^2 sec^{-1}		$\overline{D}_{Na^+}/\overline{D}_{Cs^+}$	E_{ACT}, kJ mol^{-1}	
	Na$^+$	Cs$^+$		Na$^+$	Cs$^+$
1200 EW Nafion	9.44×10^{-7}	5.20×10^{-8}	18	28.3[c]	66.1[c]
8.6% DVB-PSS[a]	9.44×10^{-7}	1.37×10^{-6}	0.69	27.1[d]	20.0[d]
H$_2$O[b]	1.33×10^{-5}	2.06×10^{-5}	0.65	19.1[c]	18.0[c]

[a]reference 9
[b]reference 10
[c]0-40°C
[d]0-25°C

similar results for an 8.6% divinylbenzene cross-linked polystyrene sulfonate resin and for aqueous solution. Cation diffusion coefficients in the sulfonate ion exchange resin are reduced by about a factor of ten compared to those in aqueous solution, due to increased tortuosity of the medium. The slightly increased activation energies of diffusion can be ascribed partly to the same cause and partly to electrostatic interactions with fixed anionic exchange sites. The ratio of Na^+ and Cs^+ diffusion coefficients remain the same in both environments, though. This suggests that effects such as ion pairing, which would be dependent on the cation's charge density, are not a significant factor in affecting diffusion in the sulfonate ion exchange resin. For 1200 EW Nafion, the sodium ion result is (coincidentally) identical to that in the sulfonate resin, but the cesium ion value is much lower, with an extremely large activation energy of diffusion. This Cs^+ activation energy is closer to that of Na^+ diffusion in NaCl crystal at about 600°C, 74 kJ mol^{-1} (8), than a value for a solution-like diffusional process. Ion-pairing of cesium ion to sulfonate exchange sites would not be suspected as the cause of this difference, for the exchange site charge density on the sulfonate group should be low due to the fluorocarbon content of the polymer.

In order to explore this anomaly, these diffusion coefficients were remeasured for the same sample of Nafion after dry storage for two years. A second sample of Nafion, of recent manufacture, was also studied (6,7). Results are shown in Table II and Figure 1. Sodium ion values are virtually identical for

Table II. Ionic Self-Diffusion Coefficients in 1200 EW Nafion (7)

Ionic Form	Membrane Sample	0°C	25°C	40°C	E_{ACT} (0–40°C) kJ mol^{-1}
Na^+	I	3.07	9.44	15.1	28.3
	I[a]	2.78	11.2	14.9	29.8
	II	3.18	9.83	14.8	27.3
Cs^+	I	0.038	0.520	1.58	66.1
	I[a]	0.520[b]	1.70	3.37	38.9[b]
	I[c]	0.446	2.38	3.01	35.9
	II	0.363	1.88	2.67	35.4

[a]after two years in as-received form
[b]value measured at 5°C
[c]mixed Na^+-Cs^+ form, ionic fraction of Cs^+ = 0.14

Journal of the Electrochemical Society

all three cases, but those of cesium ion have increased (with decreased activation energy of diffusion) in the aged membrane sample. Results for the second sample of Nafion show similar behavior to aged sample I for both cesium ion and sodium ion.

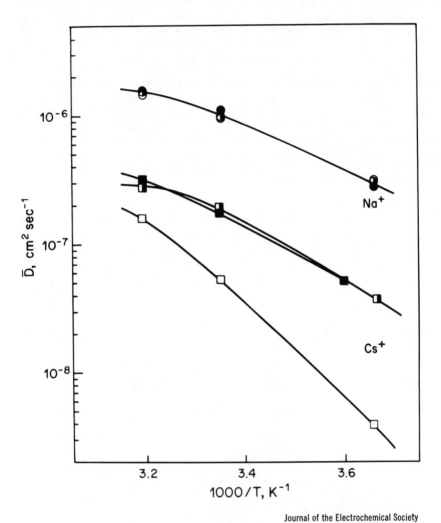

Journal of the Electrochemical Society

Figure 1. Logarithm of self-diffusion coefficient vs. reciprocal of absolute temperature for 1200 EW Nafion (11). Key: ○, □, Sample 1, 1978; ●, ■, Sample 1, 1980; ◐, ◪, Sample 2.

The water to exchange site mole ratio was measured for both ion
forms in the aged sample I and for sample II, and no change could
be found from the original values of 11.9 and 6.6 for Na^+ and Cs^+
forms, respectively.

Thus it appears that the aging process for the first sample
is accompanied by some kind of morphological change which affects
cesium ion diffusion but not sodium ion diffusion. The results
from the recently produced second sample of membrane may indicate
improved annealing of the polymer to yield more time independent
membrane behavior. The important feature of these results lies in
the difference in response of the two cations to whatever polymer
relaxation process that did occur. Originally, Cs^+ diffuses in a
different manner than Na^+, as concluded by comparison of diffusional activation energies. After aging, the Cs^+ activation
energy is closer to that expected of tortuous polymer diffusion,
but the Na^+/Cs^+ diffusion coefficient ratio, 6.6, is still far
from the expected value of 0.7. Thus the two cations appear to
have similar transport mechanisms with cesium ion encountering a
more tortuous diffusional pathway.

Of course, the water content of the polymer is a central
factor in the diffusional properties of a polymer. In
order to study water sorption as a variable, both as-received and
boiled forms of 1200 EW were used for diffusion experiments (7).
In addition, various Na^+-Cs^+ heteroionic forms of the polymer were
prepared and their water contents determined. Since the overall
water sorption decreases smoothly as the Cs^+ ionic fraction increases, both Na^+ and Cs^+ diffusion could be studied as a function
of polymer water content. Results are shown in Figure 2 (7).

The function $V_p/(1-V_p)$, where V_p is the volume fraction of
polymer in a water swollen material, is plotted as the abscissa in
Figure 2. The denominator of this term is therefore the volume
fraction of water in the membrane, calculated from sorption results. This function was developed by Yasuda and co-workers to
treat diffusion in various hydrophilic polymers (12). Their
equation:

$$\overline{D} = D^\circ \exp\left[-b\ V_p/(1-V_p\right. \tag{1}$$

describes the relationship between the aqueous and the polymer
diffusion coefficients of a species, D° and $\overline{D}$ respectively. The
equation is related to that derived by Cohen and Turnbull for the
diffusion coefficient of a molecule in a simple liquid

$$D = A \exp\left(-\gamma v^*/v_f\right) \tag{2}$$

where v^* is a characteristic volume for diffusion of the species,
v_f is the "free volume" per solvent molecule and γ and A are
constants (13). The constant b in Equation 1 is related to the
exponent in Equation 2, where now v_f would represent the free
volume of water. Equation 1 provides an excellent correlation of

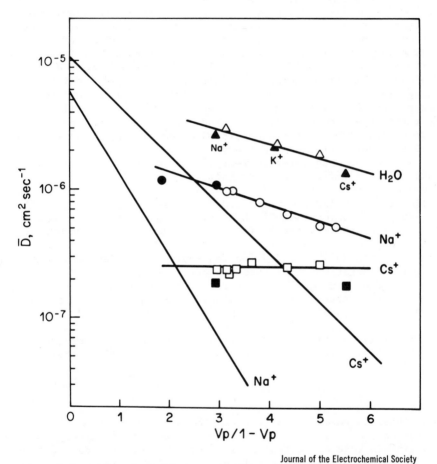

Journal of the Electrochemical Society

Figure 2. Logarithm of self-diffusion coefficient vs. polymer fraction function for 1200 EW Nafion, 25° C. Na+ and Cs+ lines without data points: polystyrene sulfonate behavior, (11); △, ○, □, denote heteroionic forms.

the binary diffusion coefficients for NaCl in a variety of water
swollen polymers, for $\overline{D}$ values ranging over five orders of magnitude (12). In addition, the equation successfully fits tracer
self-diffusion coefficients for Na^+ and Cs^+ in polystyrene sulfonate resins of varying cross-linking (and water content) (14).
The pre-exponential term in Equation 1 was found to depend on the
electrostatic attraction of the counterion to fixed charge sites
in addition to D° for these ion exchange polymers. Lines which
these authors found to describe self-diffusion coefficients for
Na^+ and Cs^+ are shown in Figure 2, for polymers where V_p varied
from 0.85 to 0.15 (14).

As seen in the Figure, diffusion results are very different
for Nafion compared to cross-linked polystyrene sulfonates.
Sodium ion has a much higher diffusion coefficient in Nafion for
a given water content, which supports the concept of ion clustering in Nafion as a morphology with considerable phase separation between fluorocarbon and hydrated cations and exchange
sites. Tortuosity would thus be reduced in Nafion, compared to a
cross-linked polymer with a more random distribution of exchange
sites. This result is also obtained for cesium ion in the as-
received homoionic form of Nafion. As water content is raised in
the polymer, by boiling or by partial exchange for Na^+, the Cs^+
diffusion coefficient remains constant however. Also, the
inverted order of magnitude for Cs^+ versus Na^+ diffusion coefficients is not removed at higher water contents, but actually
becomes more pronounced. The activation energy of Cs^+ in a sample
which is largely in the Na^+ form, in Table II, shows that the
mechanism of Cs^+ diffusion is largely independent of membrane
water content and counterion form.

The self-diffusion coefficient for iodide ion was also
measured at $25^\circ C$ in the first sample of 1200 EW Nafion, before
aging (4). The result, 9×10^{-8} cm^2 sec^{-1}, indicates that this
anion is more mobile than cesium ion, but less than a sodium ion.
This is also somewhat unusual. In general, co-ion diffusion coefficients are seen to be larger than those of counterions in ion
exchangers, because no electrostatic attractions to the polymer
phase exist. (The membrane concentration is only 5×10^{-3} mol L^{-1}
though, reflecting the effects of Donnan exclusion processes (4).)

Water self-diffusion coefficients have also been determined,
using tritium water tracer, for sample II (7). Results are given
in Table III and shown in Figure 2. Also shown in Figure 2 are
water diffusion coefficients for several Na^+-Cs^+ heteroionic forms
in order to test the effect of membrane water content. Values in
Table III for Na^+-form Nafion are similar to those obtained for
the H^+-form of 1155 equivalent weight Nafion, which were determined by measurement of the rate of water sorption (2,15). Water
diffusion coefficients are seen to follow a dependence on the V_p
function which is similar to that of sodium ion. The magnitudes
of the diffusion coefficients are quite large in relation to the
volume fraction of sorbed water. Also, activation energies of

Table III.   Self-Diffusion Coefficients of Water in
1200 EW Nafion (7).

Ionic Form	$\overline{D} \times 10^6$, cm^2 sec^{-1}			E_{ACT} (5-40°C) kJ mol^{-1}
	5°C	25°C	40°C	
Na$^+$	1.40	2.65	3.95	21.4
K$^+$	–	2.15	–	–
Cs$^+$	0.815	1.32	2.37	22.0

diffusion for the Na$^+$-form and Cs$^+$-form are only slightly larger
than the corresponding value in pure water, 17.8 kJ mol^{-1} (16).
The results indicate that there is a high degree of phase separation between fluorocarbon and ion-clustered regions, and that
water diffusion among clusters is facile. This is true even for
the Cs$^+$-form, where the water-exchange site mole ratio is only
6.6.

A Diffusional Model for Nafion. Several structural models of
Nafion have been proposed; these have been based on a variety of
transport and spectroscopic properties of the polymer (17-20).
The cluster-network model develops the concept of spherical ionic
regions separated by inter-connecting channels (17). These
channels are seen to have an important role in hydroxide ion
rejection in chlor-alkali cells. Rodmacq and co-workers propose
a three phase model in which fluorocarbon microcrystallites, ion
water clusters, and a second ionic region of lower water content
coexist (18). Falk sees evidence for two environments of sorbed
water in Nafion from infrared spectroscopic studies (19). The
first environment appears to be aqueous in nature, with the
strength of intermolecular hydrogen bonding reduced from that in
pure water. In the second environment, the water molecules are
not hydrogen bonded and appear to be exposed mainly to fluorocarbon. Extensive intrusions of fluorocarbon material into ion
clustered regions is inferred from these results (19). Lee and
Meisel have studied the microenvironment of the Ru(II)-2,2-
bipyridine complex in Nafion, and also find evidence for extensive
interaction of this cation with fluorocarbon phase (21).

A model of Nafion which is consistent with ionic diffusional
results and with the above observations has been proposed (7).
This approach also describes three regions in the polymer, as
shown in Figure 3. Region A consists of fluorocarbon backbone
material, some of which is in a microcrystalline form, as detected by Rodmacq and co-workers (18). Ion clusters form Region
C, in which the majority of sulfonate exchange sites, counterions,
and sorbed water exist. The interfacial Region B is seen as one

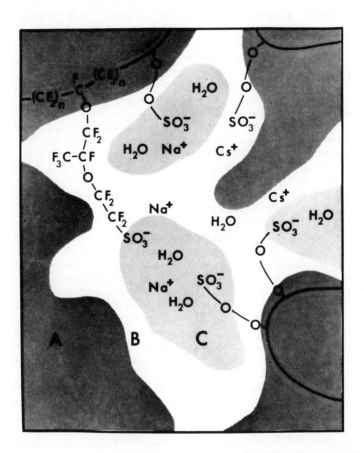

Journal of the Electrochemical Society
Figure 3. Three region structural model for Nafion: A, fluorocarbon; B, interfacial zone; C, ionic clusters (11).

of relatively large fractional void volume, containing pendant side chain material, a smaller amount of water, some sulfonate exchange sites which have not been incorporated into clusters, and a corresponding fraction of counterions. The relative numbers of ions in Regions B and C would depend on the size, charge density and hydration energy of the cation. Ions of low charge density or large size, such as Cs^+ or $Ru(bpy)_3^{2+}$, would prefer Region B, while those of larger charge density and hydration energy would localize in the more aqueous ionic clusters (within electroneutrality limitations).

In order to interpret the Na^+ and Cs^+ diffusional results in terms of this model, it is assumed that both cations would be able to diffuse readily in both the ionic clusters and interfacial regions. Cesium ion would experience a more tortuous diffusion path compared to sodium ion, and thus would have a smaller measured self-diffusion coefficient. The insensitivity of this diffusion coefficient to increasing water sorption may then be because most of this water serves to increase the size of ionic clusters, which would have a relatively minor overall effect on the diffusion path length.

As described earlier, aging of a sample of 1200 EW Nafion was accompanied by an increase of a factor of three in the diffusion coefficient of Cs^+, with a large decrease in activation energy of diffusion. Almost no change was seen in the corresponding values for Na^+. It is possible that portions of region B were not originally as well formed as shown in Figure 3, and contained diffusionally isolated portions. The aging process would then have consisted of a consolidation of aqueous and fluorocarbon phases. The originally isolated portions of the interfacial region would yield large activation energies of diffusion for counterions. Diffusion of cesium ion would be more sensitive to this incompleteness of phase separation compared to sodium ion. The study of heteroionic forms of the unaged sample I would have helped to resolve this point.

Thus the model in Figure 3 is consistent with spectroscopic and diffusional results, but is certainly an oversimplified picture nevertheless. Other approaches to the modeling of transport in Nafion, such as the recent application of percolation theory by Hsu and co-workers (22), may yield further insight into the problem.

Membrane Diffusion in Concentrated Solution Environments. Most of the current applications of perfluorosulfonate membranes involve electrochemical cells in which concentrated electrolyte solutions are employed, often at elevated temperatures. Relatively little diffusion data are available under these conditions, although a larger amount of membrane resistance and other operating data have been published. Sodium ion self-diffusion coefficients have been measured in various Nafion membranes in concentrated NaOH solutions at elevated temperatures (23). This electrolyte system is being studied because the emerging membrane chlor-alkali cell technology is the most important current application of perfluorinated ion exchange membranes. Sodium ion is the major current carrying species in the membrane phase for this application, and its mechanism of transport in Nafion under such conditions is of great interest.

Sodium ion self-diffusion coefficients in several Nafion membranes are plotted versus the reciprocal of absolute temperature in Figure 4 (23). In 9.5 M NaOH, considerable electrolyte sorption into the membrane phase occurs. In addition, polymer water sorption is reduced due to the dehydrating effect of the external solution. This results in a reduction in the Na^+ self-diffusion coefficient. For example, 1200 EW Nafion shows a value of 3.45 × 10^{-7} cm^2 sec^{-1} at 90°C, about three times smaller than the dilute solution, room temperature value. The 1150 EW material yields slightly larger diffusion coefficients than 1200 EW, due to the greater concentration of exchange sites and larger water sorption. The 1150 (EDA) membrane is also an 1150 EW polymer film, but one surface is treated with ethylenediamine while the film is in the sulfonyl fluoride precursor form. Upon hydrolysis, exchange sites in about a 0.04 mm thick layer are converted to sulfonamide groups. These weakly acidic exchange sites yield improved current efficiency in a chlor-alkali cell when the treated layer of the membrane faces the NaOH cathode solution. The effect of this layer is to increase the activation energy of diffusion for sodium ion, as seen in Figure 4. A fully converted membrane, labeled 'EDA', shows the effect more dramatically. The activation energies of Na^+ diffusion for these membranes are: 1150 EW, 10.5 kJ mol^{-1}; 1200 EW, 20.0 kJ mol^{-1}; 1150 (EDA), 28.9 kJ mol^{-1}; and EDA, 50.6 kJ mol^{-1}. The sulfonamide exchange sites produce a membrane with decreased sorbed water, and this appears to be the main factor for decreased Na^+ diffusion coeffients and increased activation energies of diffusion. Thus higher current efficiencies in operating cells are accompanied by higher membrane voltage drops as well for these types of membranes.

Diffusional Arrhenius plots for Nafion 295 at three NaOH concentrations are shown in Figure 5 (23). This membrane is similar to 1150 (EDA), but is backed with an open weave Teflon fabric for added strength. Sodium ion diffusion coefficients drop rapidly with increasing caustic strength and membrane dehydration. Indeed, at 25°C in NaOH solutions of 10 M or higher, these membranes are virtual nonconductors of ions (23). Activation energies of diffusion calculated for the 60–90°C temperature interval for the plots in Figure 5 are relatively constant at about 35 kJ mol^{-1}. Thus the mechanism of Na^+ diffusion appears to remain constant over this solution concentration range.

Another feature of the 295 membrane is seen in Figure 6 (23). Here Na^+ self-diffusion coefficients are plotted versus NaOH concentration from dilute to concentrated solutions. The rapid drop at high caustic strength is attributed to membrane

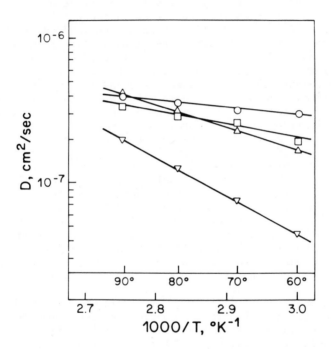

Journal of the Electrochemical Society
Figure 4. Arrhenius plots for Na+ diffusion in Nafion membranes, 9.5 M NaOH external solution (24). Key: ○, *1150 EW;* □, *1200 EW;* ▽, *1150 (EDA);* ▼, *EDA.*

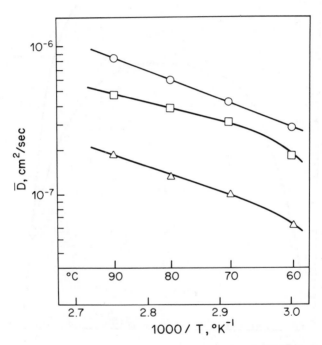

Journal of the Electrochemical Society

Figure 5. Arrhenius plots for Na+ diffusion in Nafion 295, NaOH external solution (24). Key: ○, 9.5 M; □, 11.0 M; △, 12.5 M.

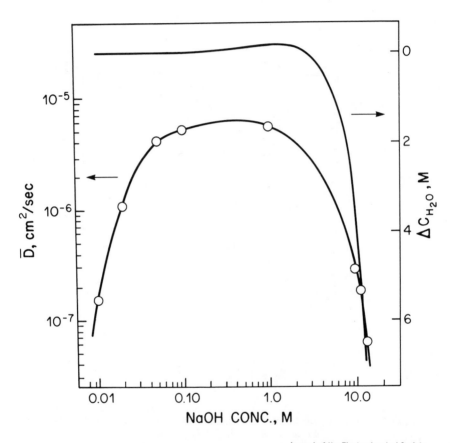

Journal of the Electrochemical Society

Figure 6. Na$^+$ diffusion coefficient in Nafion 295 at 60° C, NaOH external solution; and change in water molarity in NaOH solution (24).

dehydration. This effect is reflected in the rapid drop in NaOH solution water molarity in this concentration region, which is also shown in the Figure. A similar drop in membrane diffusion coefficient is seen in dilute solution as well. This is caused by protonation of the weakly acidic sulfonamide exchange sites. Thus membranes of this type can only be used in alkaline media. In addition, these results suggest that the pH gradient across such membranes in operating chlor-alkali cells will be an important factor in determining the membrane voltage drop.

The diffusion of molecular species has also been studied in concentrated solution environments (25,26). Yeo and McBreen measured the diffusion coefficients of H_2 and Cl_2 in 1200 EW Nafion membranes immersed in HCl solutions, and that of Br_2 in HCl and HBr solutions as a function of electrolyte concentration and temperature (25). In concentrated HCl solutions the order of diffusion coefficients is $H_2>Cl_2>Br$, as expected from molecular size. Activation energies of diffusion for H_2 and Cl_2 in 4.1 $\underline{M}$ HCl were found to be 21.6 and 23.3 kJ mol^{-1} respectively over the 25°-50°C temperature interval. These values are very similar to those for water diffusion in the same membrane in dilute solution, as seen in Table III. The authors utilize these results to estimate a coulombic loss of about 2% in a hydrogen-chlorine fuel cell, due mainly to chlorine migration through the membrane.

The interpretation of Br_2 diffusion was complicated by the formation of Br_3^- and possibly other anionic bromine species. Will (26) has also studied bromine diffusion, in 1200 EW and other Nafion membrane materials. The electrolytes used in these experiments were concentrated $ZnBr_2$ or NaBr solutions. The Br_3^- ion is the predominant bromine species in such media, although molecular Br_2 would appear to be responsible for transport across the membrane. Measured diffusion coefficients varied from 1×10^{-8} to 5×10^{-7} cm^2 sec^{-1} at room temperature. Values increased with decreasing equivalent weight of the polymer and with decreasing solution concentration. This again supports the view that membrane water content is an important factor in determining membrane diffusion coefficients, even for neutral diffusing species.

Membrane Diffusion in Nonaqueous Solvent Environments. Self-diffusion coefficients of Na$^+$ and Cs$^+$ for 1200 EW Nafion membranes in dilute methanol and acetonitrile solutions have been measured (5). Arrhenius plots of these results are shown in Figure 7 along with corresponding results for aqueous experiments; activation energies of diffusion are listed in Table IV. Diffusion coefficients of Na$^+$ in methanol and water-equilibrated membranes are very similar, and the activation energy of diffusion for the methanol system is only slightly higher than the respective value for Na$^+$ in pure methanol solvent, 12.9 kJ mol^{-1} (27). Thus a solution-like diffusion mechanism is inferred for both solvent systems. Cesium ion diffusion in the methanol equilibrated membrane is much slower than sodium ion diffusion; in fact the

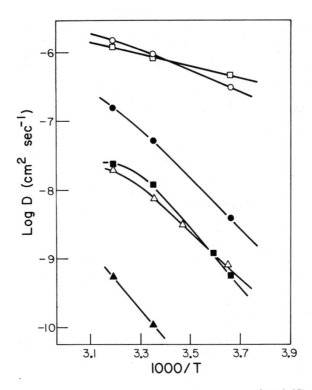

Journal of Physical Chemistry

Figure 7. Arrhenius plots for ion diffusion in 1200 EW Nafion (5). Na⁺: ○, H₂O;
□, CH₃OH; △, CH₃CN. Cs⁺: ●, H₂O; ■, CH₃OH; ▲, CH₃CN.

Table IV. Activation Energies of Diffusion in 1200 EW Nafion (5).

Ion	Solvent	Temp. Interval, °C	E_A kJ mol^{-1}
Na$^+$	H$_2$O	0–25	30.3
	CH$_3$OH	0–25	15.3
	CH$_3$CN	1–25	61.3
Cs$^+$	H$_2$O	0–25	70.3
	CH$_3$OH	0–25	83.1
	CH$_3$CN	25–40	83.3

difference is now more pronounced than for the aqueous case. Methanol sorption is far less with 1200 EW Nafion for Cs$^+$-form samples compared to the Na$^+$-form, which may form part of the reason (5). For membranes which have been equilibrated with anhydrous acetonitrile, diffusion coefficients for both ions are decreased further. The acetonitrile values are extremely sensitive to traces of water in the membrane (4). Only small residual amounts increase the membrane diffusion coefficients by several hundred percent. The very small values of cation diffusion coefficients suggest that either extensive ion pairing predominates for this aprotic solvent, or that communication among clusters is lost due to relatively small polymer swelling. The sensitivity of the diffusion coefficients to small amounts of water would suggest that the former may be responsible. The inability of the weak Lewis acid acetonitrile molecule to solvate exchange sites would promote sulfonate-cation ion pairs, a process which would be reversed by small amounts of sorbed water.

The diffusion of SO$_2$ and CH$_3$CN in acetonitrile-equilibrated forms of Nafion has also been reported (28). A similar sensitivity of diffusion rates to the presence of water was noted. Calculated diffusion coefficients were based on solution concentrations, and thus are not readily interpretable in terms of membrane properties.

Transport Properties under Conditions of Current Flow

Applications of perfluorosulfonate membranes commonly involve their use as separation materials in electrolytic cells, in which concentrated solutions are employed. A primary consideration in such applications is the conductivity of the membrane, because the ohmic loss due to membrane resistance can significantly increase energy consumption of the cell. The conductivities of common Nafion membranes have been investigated for several

industrially important electrolyte environments. For example,
Yeo and co-workers report membrane conductivities in concentrated
HCl (25) and in NaOH and KOH solutions (29) as a function of
temperature. For the latter case, Nafion shows larger conductivities when equilibrated with NaOH solutions than with KOH solutions of equal molarity; again a correlation is found between
membrane conductivity and water content.

   For the application of these membranes to the electrolytic
production of chlorine-caustic, other performance characteristics
in addition to membrane conductivity are of interest. The sodium
ion transport number, in moles Na^+ per Faraday of passed current,
establishes the cathode current efficiency of the membrane cell.
Also the water transport number, expressed as moles of water
transported to the NaOH catholyte per Faraday, affects the concentration of caustic produced in the cell. Sodium ion and
water transport numbers have been simultaneously determined for
several Nafion membranes in concentrated NaCl and NaOH solution
environments and elevated temperatures (30-32). Experiments were
conducted at high membrane current densities (2-4 kA m^{-2}) to
duplicate industrial conditions. Results of some of these experiments are shown in Figure 8, in which sodium ion transport
number is plotted vs NaOH catholyte concentration for 1100 EW,
1150 EW, and Nafion 295 membranes (30,31). For the first two
membranes, t_{Na^+} decreases with increasing NaOH concentration, as
would be expected due to increasing electrolyte sorption into the
polymer. It has been found that uptake of NaOH into these membranes does occur, but the relative amount of sorption remains
relatively constant as solution concentration increases (23,33).
Membrane water sorption decreases significantly over the same
concentration range however, and so the ratio of sodium ion to
water steadily increases. Mauritz and co-workers propose that a
tunneling process of the form

$$\underset{\delta^- \quad \delta^+}{Na^+\overset{\overset{\textstyle H}{|}}{O}\text{---}H} \quad OH^- \; \rightleftharpoons \; Na^+OH^- \quad H_2O$$

may enhance hydroxide ion transport in environments of decreased
water content due to increased polarization of the O-H bond (33).
This would explain the decrease of t_{Na^+} for 1100 EW and 1150 EW
membranes. The lower equivalent weight material shows a steeper
decrease presumably because of its larger concentration of sodium
ions.

   Nafion 295 shows a different type of dependence in t_{Na^+} with
increasing NaOH concentration. The shape of this curve is not
unique to this bilayer membrane, but has been seen in similar
Nafion products as well. Figure 9 shows a corresponding plot for
Nafion 227, which is a fabric-backed material composed of 1200 EW
polymer with the cathode surface converted to sulfonamide

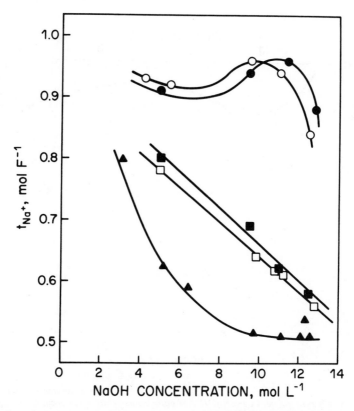

Figure 8. Sodium ion transport number vs. NaOH catholyte molarity for Nafion membranes, 80° C. Key: ○, ●, Nafion 295; □, ■, 1150 EW; ▲, 1100 EW. Anolyte solution is NaOH for light symbols and 5M NaCl for dark symbols.

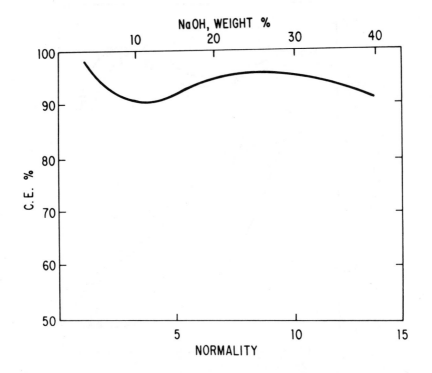

Figure 9. Current efficiency vs. NaOH catholyte concentration for Nafion 227 membrane in a chlor-alkali cell (34). Conditions: current density, 31 A/dm²; temperature, 85° C; anolyte concentration, 4.4 N NaCl; cell voltage, 4.6 V.

sites (34). The dependence of t_{Na^+} on NaOH concentration has
been recently reported for perfluorinated carboxylate membranes;
a minimum at intermediate concentration followed by a peak in
t_{Na^+} at higher concentration has been observed in these cases as
well (35-37). Several factors may be involved in this rather
complicated variation in t_{Na^+}. The common feature of both perfluorinated sulfonamide and carboxylate polymers is a lower inherent water content compared to sulfonate polymers. It is
possible that with increasing caustic strength, the associated
drop in water sorption would generate extensive ion pairing in
the membrane. This in turn would reduce O-H bond polarization in
remaining water molecules of hydration, and largely remove the
hydroxide ion tunneling mechanism of transport ($33,36$). Suhara
and Oda suggest instead that as the membrane contracts due to
increasing dehydration, the local concentration of exchange sites
in ionic clusters increases (35). This in turn re-establishes a
Donnan exclusion mechanism for hydroxide rejection.

In another approach, Kressman and Tye predict in general
terms that a minimum in t_{Na^+} can occur with increasing solution
concentration, if a sufficiently large electroosmotic effect is
present, due to a frictional interaction between water transport
and hydroxide ion migration (38). For experiments where anolyte
and catholyte are identical concentrations of NaOH, t_{H2O} values
decrease from about 3 mol F^{-1} to less than 1 mol F^{-1} for solution
concentrations of 5 M to 13 M for the 1150 EW membrane (30). For
Nafion 295, t_{H2O} varies from 5 to 1 mol F^{-1} under the same
conditions ($30,32$). For experiments in which 5 M NaCl is used as
anolyte, an osmotic component is also present to water transport.
For example, t_{H2O} remains constant at about 3 mol F^{-1} for NaOH
solution concentrations above 10 M for Nafion 295 (30). Smaller
increases in t_{H2O} are observed for the 1150 EW membrane. As seen
in Figure 8, this is accompanied by a shift in the t_{Na^+} peak to
higher NaOH concentration. Similar effects were observed with a
perfluorinated carboxylate membrane, although t_{H2O} values are
generally smaller (36). Therefore, while water transport is seen
to influence the ionic transport characteristics of these membranes, other factors must be considered to understand membrane
performance in these highly concentrated solutions.

Studies of these perfluorinated membranes in dilute and in
concentrated solution environments still leave many unanswered
questions about the nature of membrane transport properties.
However, the obvious importance of these polymers in membrane
separation applications, coupled with the fundamental significance of their ion clustered morphology, makes the continued
study of these materials a fruitful area of research for the
future.

Literature Cited

1. Price, E.H. "The Commercialization of Ion-Exchange Membranes

to Produce Chlorine and Caustic Soda", presented at the
152nd National Meeting, The Electrochemical Society, Atlanta,
Ga., Oct. 10-14, 1977.

2. Yeo, S.C.; Eisenberg, A. J. Appl. Polym. Sci. 1977, 21, 875-
    98.
3. Helfferich, F. "Ion Exchange", McGraw-Hill: New York, 1962;
    Chapter 5.
4. Lopez, M.; Kipling, B.; Yeager, H.L. Anal. Chem. 1977, 49,
    629-32.
5. Yeager, H.L.; Kipling, B. J. Phys. Chem. 1979, 83, 1836-39.
6. Yeager, H.L.; Steck, A. in "Proceedings of the Symposium on
    Ion Exchange"; Yeo, R.S.; Buck, R.P., Eds., The Electro-
    chemical Society: Pennington, N.J., 1981.
7. Yeager, H.L.; Steck, A. J. Electrochem. Soc. August 1981.
8. Mapother, D.; Crooks, H.N.; Maurer, R. J. Chem. Phys. 1950,
    18, 1231.
9. Boyd, G.E.; Soldano, B.A. J. Am. Chem. Soc. 1954, 75, 6091-
    99.
10. Calculated from limiting single ion conductivities in
    Robinson, R.A.; Stokes, R.H. "Electrolyte Solutions",
    Butterworths: London, 1959.
11. Reference 7, reprinted by permission of the publisher, The
    Electrochemical Society, Inc.
12. Yasuda, H.; Lamaze, C.E.; Ikenberry, L.D. Makromol. Chem.
    1968, 118, 19-35.
13. Cohen, M.H.; Turnbull, D. J. Chem. Phys. 1959, 31, 1164-69.
14. Fernandez-Prini, R.; Philipp, M. J. Phys. Chem. 1976, 80,
    2041-46.
15. Takamatsu, T.; Hashiyama, M.; Eisenberg, A. J. Appl. Polym.
    Sci. 1979, 24, 2199-220.
16. Tanaku, K. J. Chem. Soc., Faraday Trans. I 1975, 71, 1127-31.
17. Gierke, T.D. "Ionic Clustering in Nafion Perfluorosulfonic
    Acid Membranes and its Relationship to Hydroxyl Rejection
    and Chlor-Alkali Efficiency", presented at the 152nd National
    Meeting of the Electrochemical Society, Atlanta, Ga.,
    October, 1977.
18. Rodmacq, B.; Coey, J.M.; Escoubes, M.; Roche, E.; Duplessix,
    R.; Eisenberg, A.; Pineri, M. in "Water in Polymers", S.P.
    Rowland, Ed.; ACS Symposium Series, No. 127, American
    Chemical Society: Washington, D.C., 1980; Chapter 29.
19. Falk, M. Can. J. Chem. 1980, 58, 1495-1501.
20. Mauritz, K.A.; Hora, C.J.; Hopfinger, A.J. in "Ions in
    Polymers", A. Eisenberg, Ed.; ACS Advances in Chemistry
    Series, No. 187, American Chemical Society: Washington,
    D.C., 1980, Chapter 8.
21. Lee, P.C.; Meisel, D. J. Am. Chem. Soc. 1980, 102, 5477-81.
22. Hsu, W.Y.; Barkley, J.R.; Meakin, P. Macromolecules 1980,
    13, 198-200.
23. Yeager, H.L.; Kipling, B.; Dotson, R.L. J. Electrochem. Soc.
    1980, 127, 303-07.

24. Reference 23, reprinted by permission of the publisher, The Electrochemical Society, Inc.
25. Yeo, R.S.; McBreen, J. J. Electrochem. Soc. 1979, 126, 1682-87.
26. Will, F.G. J. Electrochem. Soc. 1979, 126, 36-42.
27. Calculated from limiting single ion conductivities in Kay, R.L. J. Am. Chem. Soc. 1960, 82, 2099-105; and Vidulich, G.P.; Cunningham, G.P.; Kay, R.L. J. Solution Chem. 1973, 2, 23-35.
28. Kimmerle, F.M.; Breault, R. Can. J. Chem. 1980, 58, 2225-29.
29. Yeo, R.S.; McBreen, J.; Kissel, G.; Kulesa, F.; Srinivasan, S. J. Appl. Electrochem. 1980, 10, 741-47.
30. Yeager, H.L.; O'Dell, B.; Twardowski, Z. J. Electrochem. Soc. 1981, in press.
31. Yeager, H.L.; unpublished results.
32. Dotson, R.L.; Lynch, R.W.; Hilliard, G.E. in "Proceedings of the Symposium on Ion Exchange"; Yeo, R.S.; Buck, R.P., Eds., The Electrochemical Society: Pennington, N.J., 1981.
33. Mauritz, K.A.; Branchick, K.J.; Gray, C.L.; Lowry, S.R. Polym. Prepr., Am. Chem. Soc., Div. Polym. Chem. 1980, 20, 122-23.
34. Hora, C.J.; Maloney, D.E. "Nafion Membranes Structured for High Efficiency Chlor-Alkali Cells", presented at the 152nd National Meeting of The Electrochemical Society, Inc., Atlanta, Ga., October 10-14, 1977.
35. Suhara, M.; Oda, Y. in "Proceedings of the Symposium on Ion Exchange"; Yeo, R.S.; Buck, R.P., Eds., The Electrochemical Society: Pennington, N.J., 1981.
36. Yeager, H.L.; Twardowski, Z. "Measurement of Ionic and Water Transport Numbers in a Membrane Chlor-Alkali Cell", presented at the 159th National Meeting, The Electrochemical Society, Minneapolis, Minn., May 10-15, 1981.
37. Seko, M. "Membrane for Chlor-Alkali Electrolysis", presented at the 159th National Meeting, The Electrochemical Society, Minneapolis, Minn., May 10-15, 1981.
38. Kressman, T.R.E.; Tye, F.L. Trans. Faraday Soc. 1959, 55 1441-50.

RECEIVED August 7, 1981.

Solubility Parameter of Perfluorosulfonated Polymer

RICHARD S. YEO

The Continental Group, Incorporated, Energy Systems Laboratory, 10432 North Tantau Avenue, Cupertino, CA 95014

The solubility parameter, δ, defined as the square root of the cohesive energy density is used considerably in the field of polymer science (1,2). The cohesive energy density is the ratio of the molar energy of vaporization minus RT, the work of expansion on vaporization, to the molar volume. It is not defined in this manner for polymers which cannot be vaporized. However, for various theoretical reasons, the cohesive energy density of materials which cannot be vaporized equals that of vaporizable solvents in which they dissolve athermally. The solubility parameter of a material is a measure of the inter-molecular forces in a given substance and is a fundamental property of all matter. A knowledge of intermolecular forces in polymers would enable a better understanding of their physical and chemical properties on a molecular basis. It is convenient to express the cohesive energy density in cel/cc units and to refer to the solubility parameter by the symbol Hb (Hildebrand).

The solubility parameter concept has been used to correlate many physical phenomena. Miscibility of solvents with polymers, diffusion of solvents within polymers, effects of intermolecular forces on the glass transition temperature and interfacial in-teractions within copolymer materials would be included, just to mention a few examples. In many cases, meaningful interpreta-tion of results was facilitated with the use of the solubility parameter.

The solubility parameter of Nafion membranes has been determined experimentally in a recent study (3). The samples which have been studied have an equivalent weight (EW) of either 1100 or 1200 (weight of polymer per sulfonic acid group). Since the samples are not soluble, the solubility parameter of the polymer can be determined only from the swelling technique (4).

Determination of δ_2 from Swelling Measurement

The degree of swelling of the membrane in solvent is related to the closeness between the solubility parameters of

0097-6156/82/0180-0065$05.00/0

TABLE I
SOLUBILITY PARAMETERS AND MOLAR VOLUMES OF SOLVENTS
AND SOLVENT UPTAKE BY NAFION (3).

Solvent	$\left(\frac{\delta}{cal/cm^3}\right)^{1/2}$	$\frac{V_1}{cm^3/mole}$	% Increase in weight	
			1100^a	1200^a
Triethyl Amine	7.4	139.4	22	24
Diethyl Amine	8.0	103.2	21	40
2-Ethyl Hexanol	9.5	158.0	--	77
n-Amyl Alcohol	10.9	109.0	73	59
Cyclohexanol	11.4	106.0	--	64
n-Butanol	11.4	91.5	74	65
2-Propanol	11.5	76.8	58	50
1-Propanol	11.9	75.2	55	40
Ethanol	12.7	58.5	50	32
Methanol	14.5	40.7	54	37
Ethylene Glycol	14.6	55.8	66	44
Glycerol	16.5	73.3	56	40
Formamide	19.2	39.8	56	37
Water	23.4	18.0	21	17

a. Equivalent Weight

the polymer and the solvent (4), as well as to the hydrogen
bonding capability of the solvent (2).

Swelling in Pure Solvents. Table I shows the solvent
uptake by Nafion-H. Figure 1, a plot of the solvent uptake for
the 1200 EW sample against the solvent solubility parameter, δ_1,
exhibits two distinct swelling envelopes. Apart from the
presence of a sharp peak at 9.5 Hb, a broad peak spans from 12.8
to 23.4 Hb. They are denoted as envelopes I and II and have been
tentatively ascribed to the organic backbone and the ion clusters
of the membrane, respectively (3,5).

While one of the δ_2 values is equal to the peak position
of envelope I, the other δ_2 value cannot be determined accurately
from Figure 1 due to the uncertainty of the peak position and low
intensity of the broad peak (envelope II). Yeo has thus
calculated (3) the interaction (χ) parameter based on the modified Flory-Rehner equation (6):

$$\ln (1 - \nu_2) + \nu_2 + \chi\nu_2^2 = \frac{V_1\rho}{M_c} (1/2\ \nu_2 - \nu_2 1/2) \qquad [1]$$

where ρ and M_c are the density and EW of the polymer, respectively. V_1 is the molal volume of the solvents. Figures 2
and 3 show the $(1 - \nu_2)$ and χ against δ_1, respectively.

The parameter χ is a measure of the interaction between
the solvent and the polymer. It is related to the closeness
between the solubility parameters of the polymer (δ_2) and the
solvent (δ_1), giving the following equation:

$$\chi = \beta + \frac{V_1}{RT} (\delta_1 - \delta_2)^2 \qquad [2]$$

where β is a constant $(0.1-0.4)$.

After rearrangement, equation [2] yields the following
equation:

$$\frac{\delta_1^2}{RT} - \frac{\chi}{V_1} = \left(\frac{2\delta_2}{RT} \right) \delta_1 - \frac{\delta_2^2}{RT} - \frac{\beta}{V_1} \qquad [3]$$

Figure 4 represents a plot of $(\delta_1^2/RT - \chi/V_1)$ versus
δ_1 for Nafion of 1200 EW, exhibiting two straight lines.
The lines are denoted as lines I and II, which correspond to envelopes I and II of Figure 1. It is evident that the value of
δ_2 can be accurately calculated from the intercept and the slope
the straight lines. Table II summarizes the results.

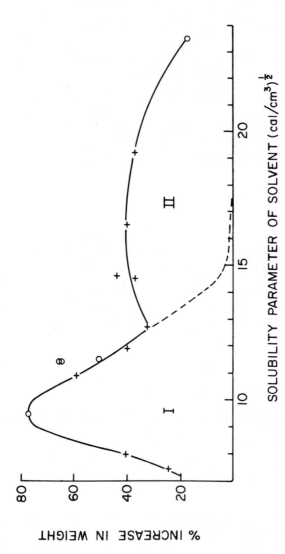

Figure 1. *Solvent uptake of Nafion 120 vs. solubility parameters of solvents* (3).

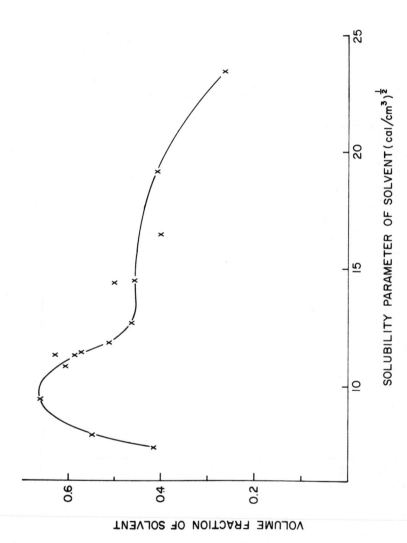

Figure 2. Volume fraction of the solvent vs. solubility parameter of solvents.

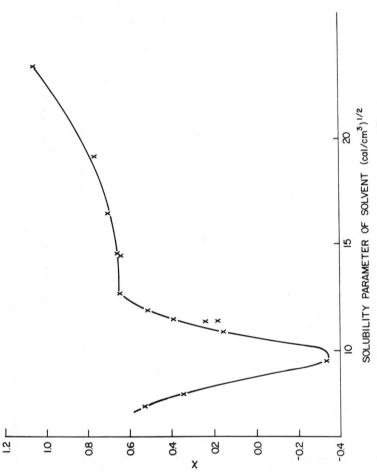

Figure 3. Interaction (χ) parameter of the polymer with solvents vs. solubility parameter of solvents.

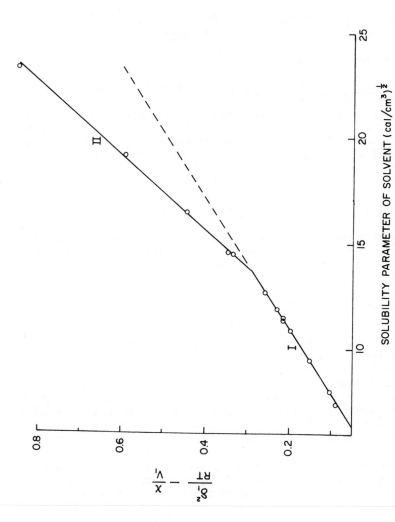

Figure 4. Plot of $(\delta_1{}^2/\mathrm{RT} - x/\mathrm{V}_1)$ *vs. solubility parameter of solvents* (3).

Table II

Solubility Parameters (Hb) of Nafion Determined from Eq. [3]

	Line I		Line II	
	1100^a	1200^a	1100^a	1200^a
Calculated from intercept	10.04	9.61	16.86	17.37
Calculated from slope	10.08	9.68	16.71	17.27

a. Equivalent Weight

Swelling in Mixed Solvents. The swelling of Nafion in mixed solvents is greater than in pure solvents (7), presumably because of the larger difference in chemical potentials between the membrane and the solvent. Figure 5 shows the solvent uptake of Nafion in three different solvent/water mixtures (3). The peak of the swelling envelopes appears around 16-17 Hb and is possibly related to the swelling envelope II for the case of pure solvents.

Calculation of δ_2 from the Structural Formulas

Small's Method. This method (8) is based on the assumption that the contributions of the individual atoms or groups to the overall value of the solubility parameter of the molecule are additive:

$$\delta_2 = \rho \sum_i G / M \qquad [4]$$

where G is the molar-attraction constant, ρ and M are the density and molecular weight of the polymer, respectively. As shown in Table III, δ_2 values calculated for Nafion are less than the values obtained experimentally by the maximum swelling method. The effects of the ionic bonding of the polymer and the hydrogen bonding of the solvents used in the swelling experiments account for the discrepancy because Small's method assumes no hydrogen or ionic bonding (8).

Hayes Method. A relationship between polymer structure, glass transition temperature and molar cohesive energy (cohesive energy density multiplied by the molar volume) was found by Hayes (9):

$$\delta_2^{\,2} = \frac{n}{V_1} \left(\frac{RTg}{2} - 25 \right) \qquad [5]$$

where V_1 is the molar volume, R is the gas constant, and n is a number related to the degree of freedom of certain atoms and groups.

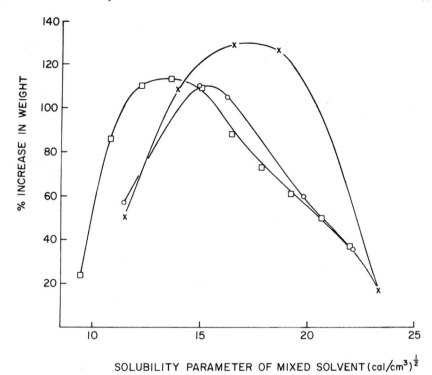

Figure 5. Solvent uptake vs. solubility parameter of mixed solvent (3). *Key:* ×, *iso-propanol/water;* ○, *cellosolve/water;* □, *diglyme/water.*

Table III

Solubility Parameter of Nafion Calculated from Small's and
Hayes' Methods (3)

Methods	Sample	Solubility Parameter (Hb)
Small	Nafion-H (1100 EW)	7.60
	Nafion-H (1500 EW)	7.20
Hayes	Nafion-precursor (1200 EW)	9.28
	Nafion-H (1200 EW)	10.93
	Nafion-Cs (1200 EW)	12.39

The increase in glass transition temperature of polymers by ion incorporation is related to the concentration of the ionic groups and to the strength of the interaction between anion and cation. According to Hayes' equation, the δ_2 values for the ionic materials would be higher than that for the non-ionic materials. Table III shows the trend in δ_2 values in going from Nafion precursor (10) to Nafion-H (11) to Nafion-Cs (11). It is interesting to note that the calculated δ_2 for the Nafion precursor, a non-ionic material, is similar to the δ_2 (line I) determined experimentally, while the calculated δ_2 for Nafion-Cs is close to the average of the two δ_2 values obtained from lines I and II.

The Nature of the Dual Solubility Parameter

It is obvious that Nafion, in contrast to many other polymers, exhibits two solubility parameter values. This feature is uncommon for any other material whose solubility parameter has been reported (2). However, in a recent study (5) on radiation grafted membranes, it is reported that membranes consisting of sulfonic acid groups also show two solubility parameter values.

Effect of Equivalent Weight. Although the effect of equivalent weight on the δ_2 of Nafion was not studied in detail, some trend can be seen. The separation of the two δ_2 values is smaller for the lower EW samples. It appears that as the EW decreases further, the two δ_2 values will likely merge into one and finally reach a value close to that of CF_3SO_3H, a low carbon compound resembling Nafion. Samples of EW below 1000 may be soluble because of high ionic content. As a matter of fact, soluble materials always exhibit one single δ_2 value.

Effect of Neutralization. Figure 6 shows the solvent uptake for Nafion in both H and Li form (5). It is clear that the swelling behavior of Nafion in Li form is rather similar to that of the un-neutralized sample. The magnitude of the swelling envelope II decreases as the membrane is converted to Li form. In contrast to this, the swelling peak at low δ_1 of the radiation-grafted sulfonic acid membranes disappears upon neutralization (5). Yeo, et al, have proposed that the disappearance of the dual solubility parameter is related to the decrease of the organic character of the material.

Relation between Membrane Swelling and Conductivity. The membrane conductivity of Nafion in various solvents has been measured in a very recent study (12). For high δ_1 solvents, the transport of ions, like Li^+, in the Nafion membrane is related to the swelling of the membrane in the solvent, as shown in Figure 7. In contrast to this, for solvents of less than 12 Hb,

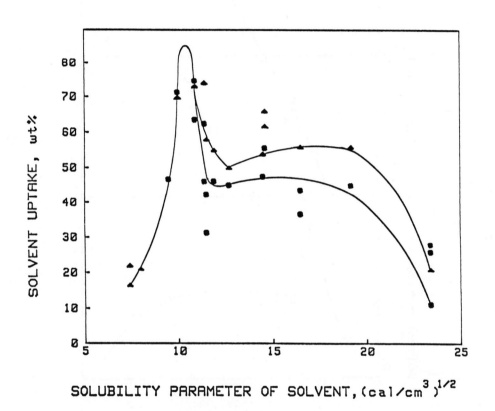

*Figure 6. Solvent uptake of Nafion vs. solubility parameter of solvents (5). ▲, H form;
● , Li form.*

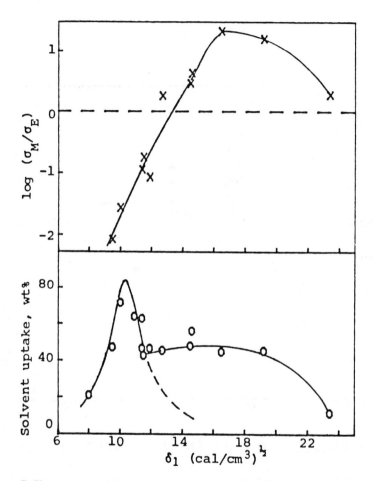

Figure 7. Upper part: conductivity ratio (σ_M/σ_E) of Nafion in solvent $+$ 0.05 N LiOH vs. solubility parameter of solvent. σ_M, membrane conductivity; σ_E, electrolyte conductivity. Lower part: solvent uptake of Nafion–Li vs. solubility parameter of solvent (12).

the membrane conductivity is far less than the conductivity of the electrolyte, in spite of the extensive swelling of the membrane in these solvents.

Conclusion and Suggested Future Works

The majority of studies on the structure and properties of Nafion membranes are very often performed on the dry or humidified samples while many important applications of these materials are in the "wet" form. The knowledge pertaining to the interaction between the solvents and the polymer by the use of the solubility parameter should facilitate the understanding of the structure-property-performance relationship. Investigations of ionic transport (13), spectroscopic properties (14) and dielectric loss tangent (12) of the membrane in light of the solubility parameter could prove to be an interesting and perhaps profitable line of inquiry.

Literature Cited

1. Hildebrand, J.H.; Scott, R.L.; "The Solubility of Non-Electrolytes," 3rd Ed., Dover, New York, 1949.
2. Burrell, H.; in "Polymer Handbook." Brandrup, I; Immergut, E.H., Eds.; 2nd Ed.; Wiley-Interscience, N.Y.; 1975, pg. IV-337-59.
3. Yeo, R.S.; POLYMER, 1980, 21, 432.
4. Gee, G.; Trans. Faraday Soc., 1942, 38, 418.
5. Yeo, R.S.; Chan, S.F.; Lee, J.; J. Membrane Sci., 1981, 9.
6. Flory, P.J.; Rehner, J., Jr.; J. Chem Phys., 1943, 11, 521.
7. Grot, W.G.F.; Chem Eng. Technol., 1972, 44, 167.
8. Small, P.A.; J. Appl. Chem., 1953, 3, 71.
9. Hayes, R.A.; J. Appl. Polym. Sci., 1961, 5, 318.
10. Hodge, I.M.; Eisenberg, A.; Macromolecules, 1978, 11, 289.
11. Yeo, R.S.; Eisenberg, A.; J. Appl. Polym. Sci., 1977, 21, 875.
12. Yeo, R.S.; in "Proceedings of the Symposium on Ion Exchange," Yeo, R.S.; Buck, R.P., Eds.; the Electrochemical Soc., Pennington, N.J., 1981, pg. 235.
13. Lopez, M.; Kipling, B.; Yeager, H.L.; Anal. Chem., 1977, 49, 629.
14. Heitner-Wirguin, C.; Bauminger, E.R.; Levy, A.; Labensky de Kanter, F.; Ofer, S.; POLYMER, 1980, 21, 1327.

RECEIVED August 3, 1981.

Mechanical Relaxations in Perfluorosulfonate Ionomer Membranes

THEIN KYU and ADI EISENBERG

Department of Chemistry, McGill University, Montreal, Quebec H3A 2K6, Canada

This review article describes the relaxation phenomena of Nafions as determined by regular (dry state) and under-water stress relaxation and dynamic mechanical (torsion pendulum and vibrating reed) studies. The thermal stability and the glass transition temperatures of the membranes, as examined by differential scanning calorimetry, linear thermal expansion and density studies are also reviewed. Dry state stress relaxation studies show a movement of the primary relaxation curve to higher temperatures upon ionization of the precursor and neutralization of the acid. This movement results from the strong interactions within the ionic domains as deduced from results of the under-water stress relaxation studies. In dynamic mechanical studies, three relaxation peaks termed α, β, and γ, in descending order of temperature, are observed. The original assignments of the mechanisms of these relaxations are reanalysed in connection with recent results of the water sensitivity and the structure of the ionic aggregates in the material. A review is also presented on the mechanical relaxations of the precursor as investigated by dynamic mechanical methods, and the relaxation peaks are discussed in conjunction with the dielectric studies. In addition, the effects of various parameters, such as the effects of degree of neutralization, of the type of counterion, and of the degree of crystallinity on the mechanical relaxations are described.

It is well known that a wide range of physical properties of polymers can be modified profoundly through ion incorporation (1,2). In many materials, the polymer matrix is effectively crosslinked through the association of these ionic groups which form small aggregates termed multiplets and larger aggregates termed clusters. The ionic aggregates can relax thermally, with the temperature of the relaxation depending on a range of molecular parameters which influence the structure of the ionic domains. The relaxation of the ionic aggregates yields a new peak in addition to that of the glass transition of the matrix; this new peak

0097-6156/82/0180-0079$08.00/0
© 1982 American Chemical Society

usually appears at a higher temperature than that of the matrix. This is a feature that is not encountered in neutral polymers but is quite common for ionomers. Moreover, the glass transition of the polymer matrix is raised to significantly higher temperatures as a result of the strong ionic interactions.

Many studies (3-11) have been devoted to the elucidation of the structure of the ionic domains and their effect on various properties. The rheological properties of the polymer are strongly influenced by varying the parameters of the ion such as the nature of the ionic comonomer, the ion concentration, the counterion type, the degree of neutralization and so on (12-17). Since a sizable amount of original literature (3-17), review papers (18-21), books (1,2) and proceedings of symposia (22-25) relevant to ion-containing polymers are available, no general survey on the progress of the field will be given here.

Several ionomer systems are under extensive investigation. One of these, which has recently become increasingly important in electrochemical applications, and which is one of the materials discussed extensively in this volume, is the Nafion ionomer family. These materials were developed by the duPont company, and consist of hydrophobic fluorocarbon backbone chains, with hydrophilic perfluorinated ether side chains terminated by sulfonic acid groups or corresponding alkali salts. The Nafions possess many exceptional properties which are not encountered in other ionomer systems, particularly the high water permeability (26,27), permselectivity with regard to ion transport (28-30), durability in strong alkali (26), thermal stability (26,31), and others. Many of these properties are discussed in detail in this volume.

The understanding of the structures of the ionic aggregates and of the physical properties of the Nafions are of crucial importance for the improvement and diversification of their industrial utility. Various experimental techniques, such as small angle X-ray scattering (31-33), neutron scattering (33), Mössbauer spectroscopy (34,35), nuclear magnetic resonance (36) and infrared spectroscopy (37,38) have been employed for the elucidation of the structure of the ionic aggregates. Details of the experimental methods and the results of these studies are also presented in various chapters of this book. The common conclusions reached by these and the preceeding studies are that the Nafions are semi-crystalline polymers with a low degree of crystallinity, which is related to the equivalent weight. The hydrophobic fluorocarbon and hydrophilic ionic regions in these materials are phase separated, the Bragg distance of the hydrated ionic domains being of the order of 5 nm. Based on these structural results, some theoretical models have been proposed (32,36,39).

The primary objective of this paper is to review the mechanical behavior of dry and hydrated Nafions with an emphasis on the mechanical relaxations. The tentative assignments of the relaxation mechanisms underlying the three mechanical relaxations are discussed in connection with the structure of the ionic aggregates

and the influence of water absorption. In addition, various
effects such as that of the degree of neutralization, the kind of
counterion and the degree of crystallinity on the mechanical
relaxations are also discussed.

Thermal Studies

Prior to discussion of the mechanical behavior of the Nafions,
some thermal properties should be noted. A differential scanning
calorimetry study of Nafion acid and of amorphous Nafion-Na at a
heating rate of 20°C/min was conducted by Kyu and Eisenberg (40)
and the results are shown in Figure 1. A pronounced transition,
probably associated with the α peak (tan δ) of the torsion pendulum studies (31), was observed in the DSC curves in the temperature region of about 70°C for the undried acid and of about 140°C
for the undried salt. Successive DSC runs, after progressively
severe heat treatments, showed a movement of this transition to
higher temperatures. This indicates that the glass transition
region of Nafion (matrix or the ionic regions) increases markedly
upon drying although the exact transition temperature is difficult to determine from these DSC curves. The exact assignment of
this glass transition is still in some doubt; it will therefore
be discussed in considerable detail in the overview section at the
end of this review.

It should be emphasized that in the Nafions, as in other
polymers, and especially ionomers, the glass transition temperature can be strongly influenced by the thermal history and the
moisture content of the polymer. Furthermore, in the present
case, some decomposition can be seen at ca. 190°C in the acid samples, which show considerably lower thermal stability than is
observed in the salts. These results are consistent with those
reported earlier by Yeo and Eisenberg (31), based on weight loss
in thermogravimetric studies. This feature appears to be a common
phenomenon in sulfonated systems; for example, in the sulfonated
polysulfones, improved thermal stability is also observed in the
neutralized materials (2).

The linear thermal expansion study by Takamatsu and Eisenberg
(41) is relevant to the present discussion. Their measurements
showed considerable complexity due to the dual effects of the drying procedure and mechanical anisotropy of the sample on the
results. For example, the linear thermal expansion of an anisotropic Nafion acid film at a heating rate of ca. 1°C/min shows
appreciable differences between successive heating runs. The
thermal expansion coefficient is reproducible in the temperature
range below 80°C; however, for wet or anisotropic samples, the
slope changes drastically above that temperature, depending on the
number of previous heating runs. The shifting of the inflection
temperature of the linear thermal expansion curve in successive
heating runs to higher temperature parallels the results of the
DSC runs as mentioned before. This confirms that the glass

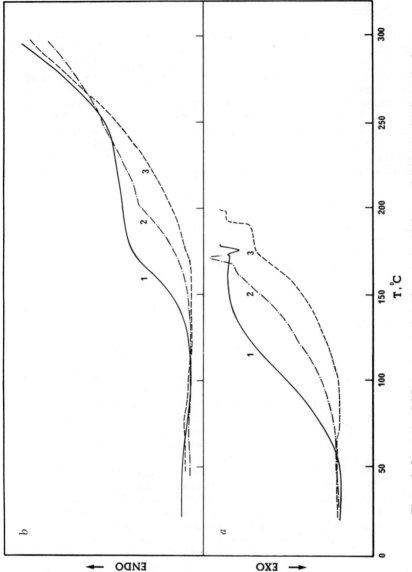

Figure 1. Successive DSC runs scanned at 20°/min for Nafion acid (a): 1, undried sample; 2, preheated to 105° C for 1 h; and 3, preheated to 150° C for 1 h; amorphous Nafion–Na (b): 1, undried sample; 2, preheated to 220° C for 10 min; and 3, preheated to 300° C for 10 min.

transition is strongly affected by the drying procedure of the sample. This is seen even more clearly in the examination of the expansion coefficient of an annealed salt sample which is made wet, in that the effect of residual water is extremely complex. In successive runs, as the sample dries, the values of the linear thermal expansion coefficient approach those of the annealed samples.

In the case of the annealed salt samples, two distinct breaks in the linear thermal expansion curve are observed. The upper break was originally thought to be related to the glass-transition of the material, while the lower one was ascribed to the T_g of the ionic region as it corresponds in temperature to the mechanical β relaxation. As mentioned before, some doubt exists concerning these assignments, and the topic will be discussed more extensively later. As the linear thermal expansion behavior of the annealed samples can be correlated with fundamental properties of the material such as the glass transition, among others, the inflection temperatures and the linear thermal expansion coefficients are noteworthy. They are listed in Table 1 for both the acid and the salts.

The densities of the Nafions were also investigated (41). The experimental results, in spite of considerable scatter, show clearly that the densities decrease with increasing moisture content. However, they seem to be independent of the equivalent weight. The data on the decrease in density with increasing water content are explainable on the basis of simple volume additivity. The large scatter of the experimental densities (standard deviations of ±4%) is attributed to the possibility that the sample is subject either to a microphase separation process which is strongly influenced by the sample history and which affects the densities of the material very strongly, or to partial crystallization (or recrystallization) during annealing, or both.

Stress Relaxation Studies

The stress relaxation behavior of the Nafion system exhibits some unusual characteristics; the relaxation master curves of the precursor as well as of Nafion in its acid and salt forms are very broad, and are characterized by a wide distribution of relaxation times. Figures 2, 3, and 4 show the individual stress relaxation curves and the master curves, with the reference temperatures indicated in the captions, for the precursor (42), Nafion acid and Nafion-K (31), respectively. Figure 3 also shows the master curves for styrene and two sytrene ionomers for the sake of comparison. Time-temperature superposition of stress relaxation data appears to be valid in the precursor and in the dry Nafion acid, at least over the time scale of the experiments. In the case of Nafion-K, time-temperature superposition is not valid, leading to a breakdown at low temperatures, but is reestablished at high temperatures (above 180°C). Similar behavior was also observed

TABLE I

Mean Linear Thermal Expansion Coefficients ($\overline{\alpha_i}$) and Inflection Temperatures T_i of Nafion Membranes

NAFION SPECIMEN	$\overline{\alpha_I}$ (below T_I)	T_I (°C)	$\overline{\alpha_{II}}$ (between T_I & T_{II})	T_{II} (°C)	$\overline{\alpha_{III}}$ (above T_{II})
Acid		Room Temp.	1.47 ± 0.04	115 ± 5	2.9 ± 0.25
Li salt	1.23 ± 0.13	98 ± 9	1.98 ± 0.29	200 ± 5	3.3*
Na salt	0.95 ± 0.15	132 ± 9	1.40 ± 0.21	206 ± 12	2.8*
K salt	1.23 ± 0.16	104 ± 11	1.44 ± 0.22	213 ± 14	2.5*
Cs salt	0.94 ± 0.14	125 ± 3	1.20 ± 0.17	199 ± 4	2.6*

* Only one determination

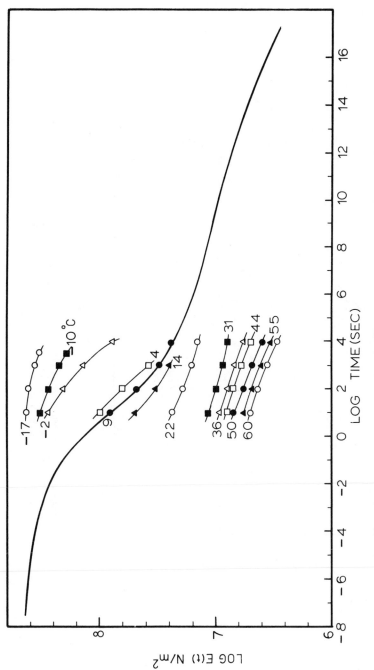

Figure 2. The individual stress relaxation curves and master curve reduced to the glass transition temperature for the precursor (42).

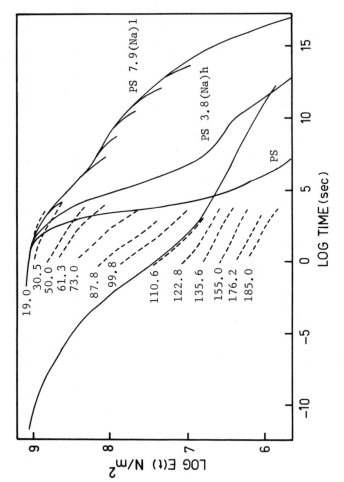

Figure 3. The individual stress relaxation curves and master curve reduced to the α glass transition temperature for Nafion acid, compared with those of polystyrene and two styrene ionomers reduced to their glass transition temperatures (31).

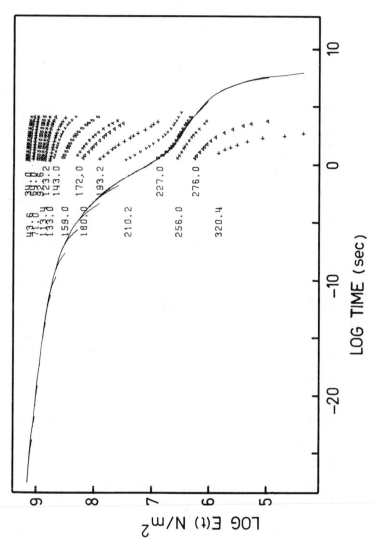

Figure 4. The individual stress relaxation curves and master curve reduced to the α glass transition temperature for Nafion–K (31).

for a low molecular weight ($\sim$5 x 10^4) styrene ionomer (2). The
addition of small amounts of water to the Nafion acid also leads
to a breakdown in time-temperature superposition. Nevertheless,
the pseudo-master curve definitely shows that stress relaxation is
faster in the wet state than is observed in the dry samples. This
implies that the introduction of water accelerates the rate of
stress relaxation due to the plasticization effect of the water
within the ionic domains.

The relaxation curves (with the reference temperatures at the
glass transitions as they were assigned in the original paper) are
broad for the precursor and the ionized Nafions. Since the precursor is a semicrystalline polymer, the breadth of the spectrum
may be mainly attributable to the presence of crystals. However,
in the case of ionized Nafions, this great breadth must be
ascribed not only to the influence of crystallinity but also to the
effect of strong ionic aggregation. No discussion is presented in
the original publication concerning the significance of the shift
factors obtained by time-temperature superposition. It should be
noted that in view of the problem in assigning the T_g's of these
polymers, the position of the relaxation curves of the Nafion acid
and salts should not be taken as absolute. Additional horizontal
shifting may be required. Only the shapes of the curves are
significant here.

In Figure 5, the ten-second tensile modulus, E(10 sec), versus
temperature plots for the precursor, the Nafion acid and the
Nafion-K salt are compared with those of   styrene and its ionomer (42). In each case, the value of the modulus at the glass
transition temperature, as it was assigned at that time, is of the
order of $\sim$10^7 N/m^2, about one to two orders of magnitude lower
than the values for styrene or its ionomers. The primary relaxation temperature of the precursor, which is about 10°C, rises
dramatically upon ionization, i.e. to 110°C for the acid and to
220°C for the salt, if the high-temperature relaxations are taken
as the glass transitions (but see below for more extensive discussion of α and β peak assignments). This dramatic rise in the
glass transition is undoubtedly related to the effects of ion
aggregation, which will be discussed in detail in a subsequent
section.

Under-water Stress Relaxation Studies

In order to elucidate the effect of ionic aggregation on the
primary relaxation process, it seems useful to reduce the ionic
interaction by the introduction of water into the ionic regions.
Since water is largely incompatible with the fluorocarbon matrix,
although some of it is closely associated with the CF_2 groups due
to the small size of the aggregates (38), the water-plasticization
can be expected to occur preferentially in the ionic domains
rather than in the matrix. As will be shown later, however, the
proximity of the water does influence the matrix T_g appreciably

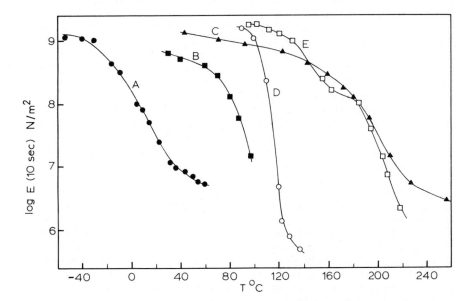

Figure 5. The comparison of E(10 s) *vs. temperature curves for the precursor, Nafion acid and Nafion–K; A, precursor; B, Nafion acid; C, Nafion–K; D, polystyrene; and E, styrene — 9% Na MA (42).*

also (38). The contribution of ionic aggregation to the primary
relaxation process can then be explored directly as a result of
changes in the strength or hardness of the ionic aggregates in the
dry and wet states. These considerations suggest the utility of
under-water stress relaxation studies (40) in which the sample can
be stretched in a water environment. This technique permits the
study of stress relaxation of ionic materials in which the strong
ionic interactions have been reduced considerably by ion hydration.

Under-water stress relaxation of the Nafion acid and Nafion-
Na samples with an equivalent weight of 1200 was carried out by
Kyu and Eisenberg (40). The degree of neutralization of the
Nafion-Na was measured to be ca. 80%. Time-temperature superposed
master curves of the two systems, reduced to a reference temperature of 50°C, are shown in Figure 6. The under-water stress relaxation behavior of Nafion acid resembles that of Nafion-Na, except
for the fact that the elastic modulus is somewhat lower in the
acid. This latter feature may be due to the difference in the
degree of water absorption of the acid and salt samples (26,31).
The swelling of Nafion acid is greater than that of Nafion-Na,
which yields a material of lower modulus.

The most striking feature of this under-water stress relaxation is perhaps the great similarity of the superposed relaxation
master curves of the acid and the salt samples. In contrast, the
corresponding conventional stress relaxation curves in the dry
state are significantly different from one another (31). The
tendency is more evident in the comparison of the ten-second tensile modulus E(10 sec) versus temperature curves of Nafion acid
and Nafion-Na in the regular (dry-state) and under-water stress
relaxation studies, as shown in Figure 7. The temperature depen-
dencies of the moduli of the dry samples differ appreciably due to
the great difference in their glass transitions. The under-water
stress relaxation study does not cover a wide range of temperatures due to instrumental limitations; however, over the range
investigated, the rate of stress relaxation appears to be the same
in the acid and the salt.

The resemblance of the under-water stress relaxation curves
and the dissimilarity of the stress relaxation behavior of the
Nafion acid and the salt in the dry state may be explained as
follows. The phase separated hydrophilic regions are expected to
contain a substantial fraction of the ether side chains which are
anchored in the ionic domains by their polar end groups. In the
dry state, the coulombic interactions within the ionic aggregates
are so strong that these domains probably serve as effective cross-
links. This would not only reduce the mobility of molecules within
the domains but would also control the mobility of the fluorocarbon
matrix through the side chains; this, in turn, leads to the rise in
the primary relaxation temperature.

When the samples are immersed in water, the ionic domains must
swell due to their hydrophilic nature. According to Mauritz et al.,
(36,37) the direct interaction of the bound anion and the unbound

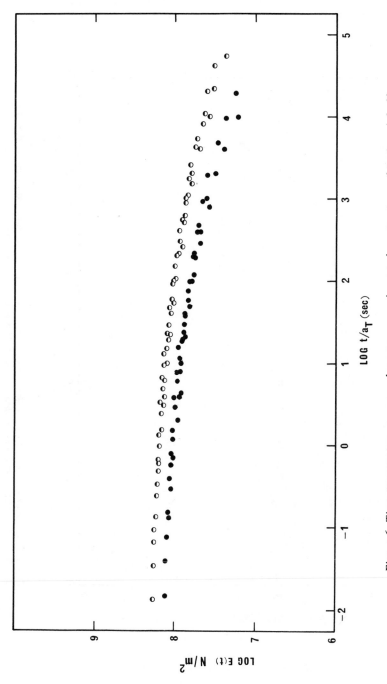

Figure 6. Time–temperature superposed master curves from underwater stress relaxation of the Nafion acid (●) and Nafion–Na (◉), reduced to a reference temperature of 50° C.

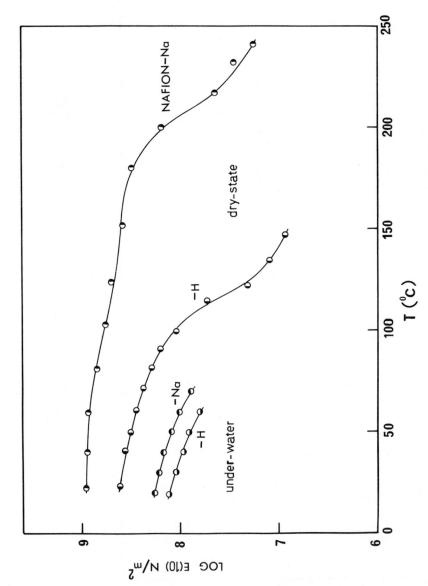

Figure 7. Comparison of E(10 s) *vs. temperature curves of the regular (dry state) and underwater stress relaxation studies for Nafion acid and Nafion–Na.*

cation is reduced upon hydration and the cation becomes highly
mobile with increasing degree of hydration. This fact suggests
that the strength of the ionic association and thereby also its
effect on the glass transition must be diminished by hydration.
This is analogous to the well explored ionic domain plasticization
effect (43), as a result of which the stress relaxation is accelerated in the plasticized condition over that in the dry state.
The great similarity in the shapes of the under-water relaxation
curves for the two systems suggests that the relaxation process
may be independent of the kind of counter cation in the sample.
In other words, the counter cation may be isolated from the bound
anion by water shielding, resulting in a weakening of the interaction between the cation and anion. Hence, the effect of the
counterion on the mechanical relaxation in the under-water state
is not significant. It is obvious, therefore, that the enhancement of the primary relaxation upon neutralization of Nafion membranes in the dry state is mainly due to the strong ionic interactions.

Dynamic Mechanical Studies

      The dynamic mechanical behavior of Nafion acid and various
alkali salts (EW = 1365) was investigated by Yeo and Eisenberg (31)
by means of a torsion pendulum in a temperature range of -150 to
250°C at a frequency of ca. 1 Hz. The temperature dependence of
the storage shear modulus, G', and the loss tangent, tan δ, for
the Nafion acid and Nafion-Cs are shown in Figure 8. Three dispersion regions, labeled α, β, and γ in descending order of temperature are evident in each case. Of the three mechanical
      dispersions, the α peak shows the highest intensity, the β
peak appearing as a shoulder of lower intensity. Similar behavior
was observed for Li, Na, and K salts. Except for the Li^+ salt,
the position of the α peaks for Na^+, K^+ and Cs^+ salts is in qualitative agreement with the $T_g \propto cq/a$ relation (where c stands for
ion concentration, q for electrical charge and a for distance
between centers of charge of anion and cation), which has been
found operative in many ion containing polymers (2). The comparison of the α peak positions of different salts is by no means
straightforward since the position can be influenced by the degree
of neutralization as well as by the degree of crystallinity, as
will be discussed later. In addition to the above mentioned
effects, it is conceivable that the effect of residual water, even
at these high temperatures, may by responsible for the failure of
the Li^+ ion to fit the relation. The effect of water on the position of the α peak could not be studied since the salts lose water
at a very rapid rate at elevated temperatures.
      The mechanical β peak in the dried acid occurs at ∿20°C at
ca. 1 Hz and in the dried salt it appears at ∿150°C. The water
sensitivity of the β peak position is very remarkable and is depicted in Figure 9. With increasing water content, the β peak

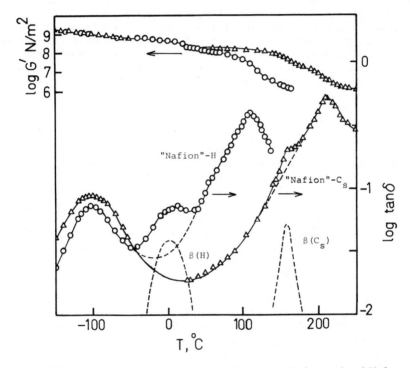

Figure 8. Variations of G′ *and tan* δ *with temperature for Nafion acid and Nafion–Cs at ca. 1 Hz.*

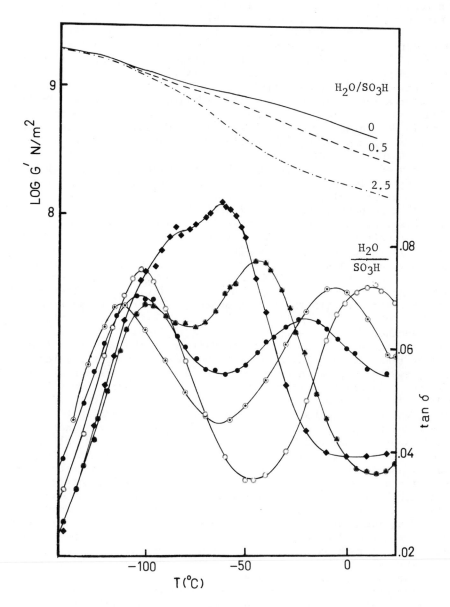

Figure 9. Variations of G′ and tan δ with temperature for Nafion acid with varying water content at ca. 1 Hz (31).   Key: ——, G′ = 0; - - -, G′ = 0.5; — · —, G′ = 2.5; ○, tan δ = 0.0; ✪, tan δ = 0.1; ●, tan δ = 0.5; ▲, tan δ = 0.9; ◆, tan δ = 2.5.

migrates to lower temperature and eventually merges with the γ
peak. Similar behavior was observed for the salts, the decrease
in temperature of β peak position with increasing water content
being perhaps even more drastic. This remarkable sensitivity to
water led originally (31) to the suggestion that the β peak may be
associated with molecular motions within the ionic domains. The
depression of the β peak temperature with moisture content is also
found in the dielectric studies on the Nafion acid (31).

While the water sensitivity of the β peak is generally agreed
on, contradictory evidence exists in regard to the α peak. Preliminary evidence from Yeo and Eisenberg (31) suggests that the α
peak in Nafion acid might be insensitive to moisture content since
water is, by and large, incompatible with the fluorocarbon backbone. However, a very recent study (40), as mentioned before,
presents evidence of a definite shift to lower temperature of the
primary transitions of Nafion acid and its sodium salt on immersion in water, relative to those obtained in the dry state.

At this point, the original assignments on the relaxation
mechanisms of the α and β peaks, which have been attributed to the
glass transitions of the fluorocarbon matrix and the ionic regions,
respectively, need to be reexamined. The arguments in favour of
the above assignments (31) are strongly based on the preliminary
findings that the β peak is highly sensitive to water content,
while the α peak show little or no sensitivity. This latter observation contrasts with the recent under-water stress relaxation
results, which show a definite shift of the primary transition.

Another feature is the qualitative similarity of the tan δ
versus temperature behavior of the Nafions to that of other ionomer systems (2,44). In the hydrocarbon based ionomers studied to
date, two tan δ peaks have been observed in the glass transition
region. The one occurring at the lower temperature is usually
identified with the matrix T_g while the one at the higher temperature, which usually has the greater intensity, is attributed to
the T_g of the ionic regions. As an example one may consider the
studies of Rigdahl and Eisenberg (44), who investigated polystyrene sulfonated to varying extents and who showed that the 2-9
mole % samples exhibit the ionic T_g peak at higher temperatures
and with a greater intensity than that of the styrene matrix.
Many other examples exist (2). If the Nafions are not an exceptional case for the ionomer systems, the original assignments
regarding the α and β peaks could be reversed. In subsequent sections, additional evidence bearing on this problem will be introduced, and a more extensive discussion presented.

The mechanical γ peak occurs at ca. -100 C at ca. 1 Hz for
both the acid and the salts and corresponds to the γ peak found in
poly(tetrafluoroethylene) at the same temperature (45). In contrast to the β and α peaks, the γ peak position remains unaffected
by variations in counterion size and degree of neutralization, as
will be demonstrated in a later section. However, the height
(tan δ) of the γ peak is inversely proportional to the counterion

radius, except for the acid, for which it is anomalously low.
Introduction of water into both the acid and the salts does not
appreciably affect the position of the γ peak, at least at the low
water levels investigated.

The dissimilarity of the effect of water on the behavior of
the γ peak on the one hand, and on that of the α and β peaks on
the other, suggests that the molecular mechanism of the γ peak
must be very different from those of the α and β regions. It
seems evident that the water plasticization effect which dominates
the α and β regions is no longer operative in the γ region. On
the basis of the good correspondence of both the γ peak position
and the activation energy with that of the γ peak in PTFE (45),
the γ dispersion may tentatively be assigned to local short range
motions of the fluorocarbon chains, as it has been in PTFE (45).

A study of the dynamic mechanical relaxations of the precursor was performed by Hodge and Eisenberg (42). The major techniques employed in their work utilized vibrating reed, torsion
pendulum and dielectric relaxation. Three peaks are observed in
the loss tangent or loss modulus versus temperature curves from
the vibrating reed studies. These relaxations are termed as α, β,
and γ in descending order of temperature and are shown in Figure
10. Only two peaks, corresponding to the α and β regions, are
seen in the torsion pendulum studies, as illustrated in Figure 11.
The β peak is found to be due to two different processes which are
labeled β' and β". The conclusions relevant to the assignment of
the relaxation mechanisms may be summarized as follows.

A low-temperature γ relaxation at ca. -160°C at ca. 300 Hz
has an activation energy of 17 kJ/mol as determined dielectrically.
It is mechanically very weak ($E''_{max} \sim 5 \times 10^7$ N/m^2, tan δ $\sim 10^{-2}$),
but it is dielectrically most intense ($\varepsilon'' \sim 10^{-2}$). It is due to
the -SO$_2$F group and is probably caused by either the rotation
about the C-S bond or by a localized wagging motion of the SO$_2$F
group itself. This γ relaxation is not detected in the frequency-
temperature range of the torsion pendulum studies and is not found
in the dielectric studies of the ionized Nafion membranes.

A β relaxation occurs in the temperature range from -100 to
-20°C. At low frequencies, ca. 0.3 Hz, it can be resolved into
two components at ca. -130 and -95°C, labeled β' and β", respectively. The activation energy for the dielectric β relaxation process is ca. 45 kJ/mole. The high temperature β" process primarily
reflects the dielectric β relaxation process which likely originates from motions of the fluorinated ether side chains. On the
other hand, the low temperature β' process, which corresponds to
the mechanical γ relaxation in the ionized Nafions, as well as in
PTFE, is attributed to local short-range motions of the fluorocarbon backbone. The mechanism of the β" process should also be
observable in the case of ionized Nafion; however, this is hidden
by the pronounced water-dependent β peak.

An α relaxation is unambiguously assigned to the glass transition of the precursor matrix and has an apparent activation

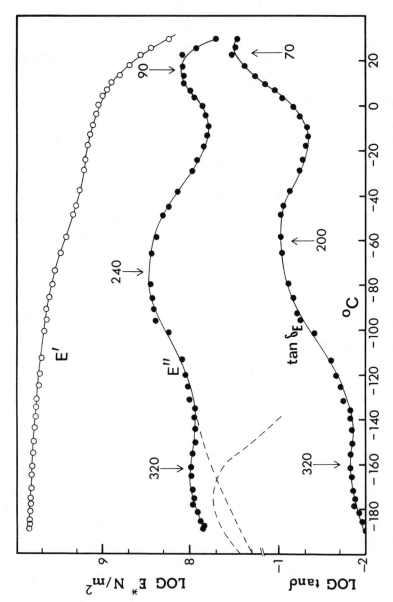

Figure 10. The mechanical loss tangent, storage and loss moduli vs. temperature for the precursor from vibrating reed studies. The frequencies in Hz at which the maxima occur are also shown (42).

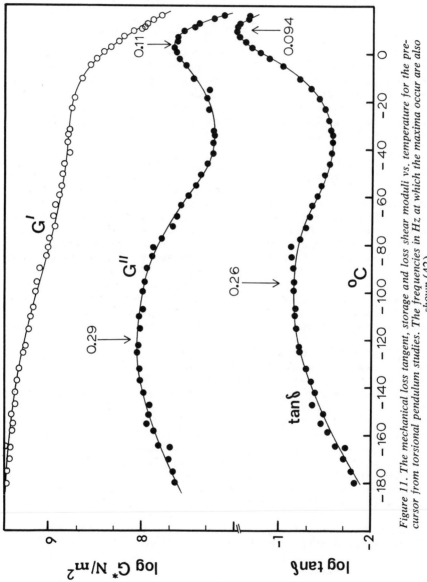

Figure 11. The mechanical loss tangent, storage and loss shear moduli vs. temperature for the precursor from torsional pendulum studies. The frequencies in Hz at which the maxima occur are also shown (42)

energy of ca. 310 kJ/mole at the T_g. The molecular mechanism of
this transition is analogous to that of the α process of ionized
Nafion except for the fact that the α transition of ionized Nafion
is influenced by the dual effects of crystallinity and ionic
aggregation, while the α relaxation of the precursor reflects
mainly backbone mobility, possibly affected by crystallinity.

Effects of Various Parameters on Mechanical Relaxations

The presence of water has been demonstrated to affect the
relaxation behavior of the Nafions profoundly. It is therefore of
interest to inquire about the effect of various other variables,
notably structural ones, on the relaxations of this material.
This section will deal with various effects such as the degree of
neutralization, the type of counterions and the degree of crystallinity on the mechanical relaxations.

Effect of Degree of Neutralization. It is evident that a
wide range of properties, particularly the rheological properties,
are altered by the introduction of ions into precursor to form
Nafion. Furthermore, since the properties of the acid differ
greatly from those of the salts, there is no doubt that these properties must also depend on the degree of neutralization. The
effect of the degree of neutralization on the mechanical relaxations was explored by Hashiyama and Eisenberg (46) by means of a
torsion pendulum. The samples were neutralized by immersing the
acid (EW = 1155) in an appropriate CsOH solution for varying
periods. The degrees of neutralization of the various Nafion-Cs
samples thus prepared were 39%, 50%, 76% and 90%. The experiments covered a temperature range of -150 to 250°C at a frequency
of ca. 1 Hz.

The storage shear modulus, G', and loss tangent, tan δ, versus temperature for the Nafion-Cs with different degrees of neutralization are shown in Figure 12. It is evident that the relaxation region (the region in which the logarithmic modulus curve
exhibits the sharpest drop, not identical with, but close to, T_g)
shifts systematically to higher temperatures with increasing
degree of neutralization of the Cs-salts. This suggests that an
increase in the degree of neutralization increases progressively
the degree of ionic aggregation or the strength of the aggregates,
thereby increasing the relaxation temperature region or the glass
transition temperature. At low temperatures, the drop in the
modulus in the γ region is quite clear, while in the β region it
is somewhat less so, although still discernible.

As might be expected, the loss tangent versus temperature plot
reveals three distinct peaks, corresponding to the α, β, and γ
relaxations of the ionized Nafions, in descending order of temperature. The γ region occurs at ca. -90°C, regardless of the variation in the degree of neutralization of the sample, which is consistent with the previous study (31). In contrast to its effect

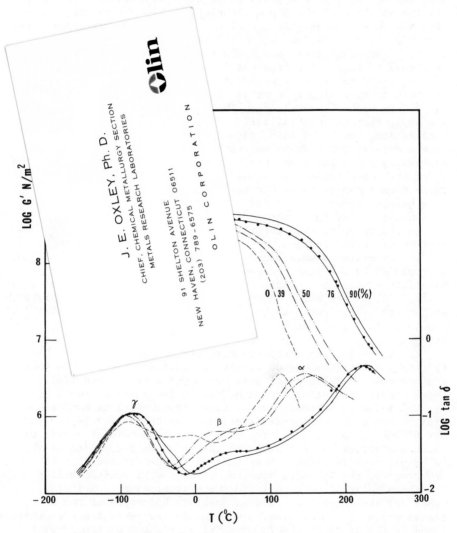

Figure 12. Variation of G' and tan δ with temperature for Nafion–Cs with varying degrees of neutralization.

on the α and β peaks, the effect of the neutralization on the
position of the γ peak appears to be extremely small.

The position of the β peak, except for the 39% sample, exhibits a tendency of shifting to higher temperatures with increasing
degree of neutralization, while the peak intensity decreases markedly. This behavior is analogous to that of the lower temperature peak of many other ionomers (2,44), in which it shows a gradual movement to higher temperature with increasing ion content,
accompanied by a slight decrease in intensity. This peak can be
identified with the matrix T_g. This behavior is explainable on
the basis that the neutralization enhances the strength of ionic
interaction, which, in turn, reduces the mobility of the fluorocarbon matrix, and thereby increases the peak temperature while
depressing the relaxation intensity. If the β relaxation were the
T_g of the ionic region, one would not expect such depression of
relaxation intensity with increasing degree of neutralization.
As mentioned before, the α and β peak assignments will be discussed more fully in a subsequent section.

In analogy with the temperature dependence of G' with various
degrees of neutralization, the position of the α peak moves to
higher temperatures with increasing degree of neutralization,
accompanied by a slight increase in intensity. This feature
parallels that of the higher temperature peak of other ionomer
systems (2,44), in which it shows a marked counterion sensitivity,
and shifts dramatically to higher temperatures with increasing ion
content. This peak can be identified with the T_g of the ionic
region.

In Figure 13, the α peak temperature is plotted against the
Cs^+ ion content (the degree of neutralization). It is seen that
the increase of T_α is not linear but appears to be sigmoidal.
This behavior cannot be accounted for on the basis of the
$T_g \propto \frac{cq}{a}$ relation because the ion concentration (c) is not essentially changed by the neutralization; the charge (q) is the same
for H^+ and Cs^+ and the distance (a) between centers of charge of
anion and cation for Cs^+ would be larger since the radius of the
Cs^+ counterion is greater than that of the unhydrated H^+. In
other words, the T_g would have to decrease with neutralization if
the above relation were valid in the present case. Obviously,
other effects intervene, such as possible hydration of the H^+ or
the strength of the dipoles. The behavior is reminiscent of that
found in the study of the T_g of ethyl acrylate - sodium acrylate
copolymers (17) where the onset of the sigmoid has been associated
with the onset of clustering of the ions, as opposed to simple
multiplet formation. However, this concept is not operative in
the present case because both the acid and the salt originally
possess the phase separated ionic domains as revealed by the SAXS
(31,32,33), SANS (33) and IR (38) studies.

Effect of Counterion Charge. Since the effect of neutralization by monovalent cations on the mechanical relaxations presents

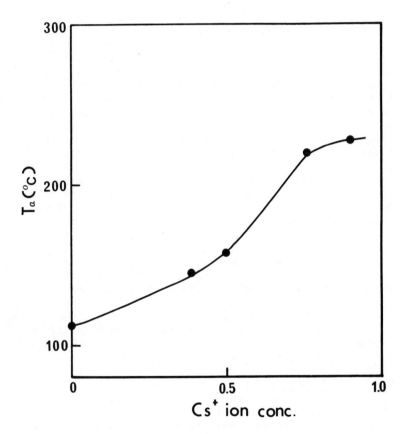

Figure 13. Variation of the glass transition peak Tα with the degree of neutralization (Cs+ ion concentration).

interesting features, it is also of interest to explore the effect
of divalent cations. A preliminary study of this kind was performed by Hashiyama and Eisenberg (46). Nafion acid of 1155 EW
was neutralized with an appropriate $BaCl_2$ solution to prepare a
Ba-salt sample. The torsion pendulum results in the form of G',
G" and tan δ versus temperature are depicted in Figure 14. A peak
is evident at ca. -90°C in tan δ and G" curves. Judging from the
peak temperature, this γ relaxation is probably caused by the same
mechanism as in the acid and the monovalent salt samples described
before. The β peak occurs at ca. -20°C and thus overlaps slightly
with the γ peak. The observation of the β region at such a low
temperature may be due to the presence of some residual water
which would act as a plasticizer within the ionic domains. Other
factors may also be present to depress the peak position.

A clear drop in G' and the corresponding peak in G" are evident at about 275°C. This peak is possibly due to the crystal
melting process, as it corresponds to the crystal melting temperature found in the DSC (40) and the WAXD studies (47) of the monovalent Nafion salts. It is also conceivable that the glass transition temperature of the material has been raised to that level,
and that the peak in this case is due to two mechanisms; i.e.
crystal melting and T_g of the ionic region. The α peak is not
observed in the tan δ plot, at least up to ca. 320°C. Comparing
the temperature dependence of G' of the Ba-salt with those of monovalent systems, it is seen that the T_g is higher than those found
for monovalent cations. This phenomenon is expected from the q/a
effect on the T_g, which has been found operative in many ion-containing polymers (2). This feature supports the idea that the α
peak is most likely attributed to the glass transition of the
ionic regions because it is hard to conceive the matrix T_g to be
higher than the crystal melting temperature. Hence, the present
authors favor the concept that the α peak should be assigned to
the T_g of the ionic regions, while the β peak should be associated
with the glass transition of the Nafion matrix.

Effect of Crystallinity. While neutralization is known to
exert a strong influence on mechanical relaxations, it is conceivable that the degree of crystallinity is also significant. To
explore this aspect, an amorphous Nafion salt sample was prepared
by rapid quenching from the melt (40). The original Nafion salt
(EW = 1200) had a degree of crystallinity of ca. 7% as revealed by
a wide angle X-ray diffraction study (40). The quenched sample
was also found to be completely amorphous by the absence of the
crystalline diffraction peak.

Figure 15 illustrates the comparison of ten-second tensile
modulus E(10 sec) versus temperature of the amorphous and semicrystalline Nafion-Na as observed by dry state and underwater
stress relaxation studies. It is evident that the rate of stress
relaxation is faster and the relaxation temperature is lower in
amorphous Nafion, relative to those of semicrystalline Nafion.

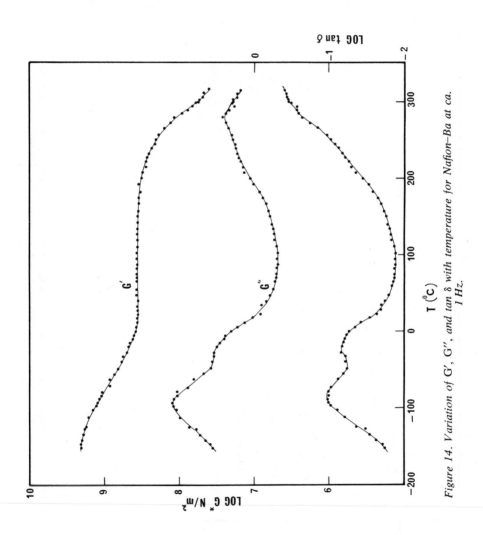

Figure 14. Variation of G', G'', and tan δ with temperature for Nafion–Ba at ca. 1 Hz.

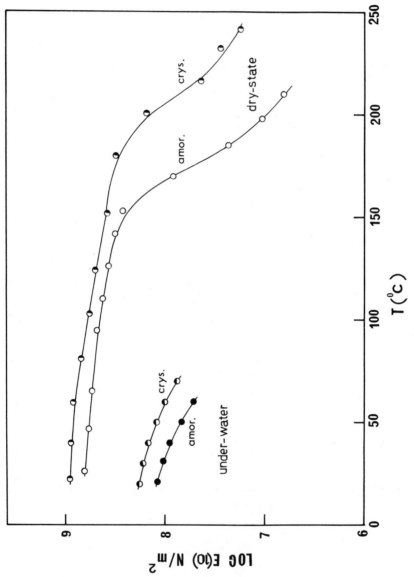

Figure 15. The comparison of E(10 s) vs. temperature curves of the dry state and underwater stress relaxation studies for amorphous and semicrystalline Nafion–Na.

This fact suggests that the presence of crystals hinders the relaxation process by restricting the molecular motion of the amorphous chains such that the relaxation times become longer.

Below the primary relaxation region, chain mobility is hindered both by the presence of crystals which immobilize backbone segments, and by the ionic aggregates, which tie down the side chains through strong coulombic interactions. In the primary relaxation region, the relaxation times for chain motion are of the same order as the residence times of the segments in the two immobilizing species, i.e. the crystallites or the ionic aggregates. The disappearance of the crystallites removes one of these retarding species, which naturally accelerates the chain relaxation process. This aspect could be further explored by a study, through a range of equivalent weights, of the relaxation behavior of Nafions with varying degrees of crystallinity and of amorphous Nafions.

An Overview Of The Relaxation Mechanisms

The recent observation of the water sensitivity of the primary relaxation of both Nafion acid and its salts has led to conflicting ideas regarding the assignment of the relaxation mechanisms of the α and β peaks. At this time, it should be of interest to reanalyse the phenomena based on the overall evidence available to date. The phenomena which make this assignment so difficult may be summerized as follows.

The most important feature that needs to be borne in mind to appreciate the problems of assignment of tan δ peaks to the various glass transitions (matrix or ionic regions) is the fact that the sizes of the phase separated regions are very small. Thus, whatever happens in one of the phases, would be expected to have a strong effect on the other phase because of the intimate contact between the phases coupled with their small size. The other feature is that since the matrix and the ionic regions are in such close proximity, the water molecules can be accommodated very close to the fluorocarbon backbone (38), which in turn, suggests that they would have the ability to interact with or plasticize the matrix, if only by creating a "soft" surface near the fluorocarbon regions. Since a majority of the ether side chains are anchored in the ionic regions through their polar end groups, the mobility of the backbone chains may be controlled, to some extent, by these side chains. The plasticization by water of the ionic domains would thus influence not only the molecular motion within the ionic domains, but also the fluorocarbon backbone chains.

For this reason, the water sensitivity of both the α and β peaks are quite reasonable; thus, their unambiguous assignment to the T_g's of particular phases is difficult. The thermal expansion coefficient studies show two inflections corresponding to the β and α relaxations, but again one cannot assign them to specific T_g's. In the DSC studies, one transition is observed at ca. 140°C

in the undried salt sample, which, upon drying, shows not a typical transition but a movement to higher temperatures with successive heat treatments. This transition is believed to correspond
to the α relaxation. The primary relaxation observed in the
stress relaxation studies of dry Nafions corresponds to the α
region. The rise of the primary relaxation to higher temperatures
upon ionization of the precursor and neutralization of the acid
suggests that the process is due to the strong ionic interactions
within the ionic aggregates. This indicates that the α peak
would be associated with the T_g of the ionic regions. This is
further confirmed in the study of the divalent salts.

The most instructive evidence, although by no means conclu-
sive, which would suggest that the α peak is due to the T_g of the
ionic regions is the qualitative similarity of the tan δ behavior
of the Nafions to that of many other ionomers (2,44). In the
hydrocarbon based ionomers studied to date, two tan δ peaks have
been observed in the glass transition region. The one occurring
at lower temperature exhibits a gradual movement to higher temperatures with increasing ion content, accompanied by a slight
decrease in intensity. This has been identified with the matrix
T_g. The higher temperature peak, which is found ca. 50 - 150 °C
above the T_g of the pure matrix material (for ion contents in the
2 - 20 mol % range), shows a marked counterion sensitivity, and
shifts dramatically to higher temperatures with increasing ion
content. These features are also observed in Nafions.

Hence, it seems reasonable to reverse the original assignments (31) of the α and β peaks. By doing so, some unanswered
problems can be explained. The problem of low modulus
($E \simeq 10^7$ N/m^2) at the α peak temperature is no longer a problem if
one assign the β peak to the glass transition of the matrix. That
is, the modulus at the β peak temperature of the Nafions, which is
the order of $\sim 7 \times 10^8$ N/m^2, is comparable to that of the precursor at its glass transition. Since the observation of the glass
transition peak of the ionic regions above the matrix T_g, the
Nafions can be classed with the other ionomers as far as the
relaxation behavior is concerned. Another anomaly which is eliminated by the reassignment concerns the position of the master
curves of the Nafions in Figures 3 and 4. If the curves are
replotted with the β peak temperature as the reference temperature, then the position of the acid master curve (Fig. 3) would
coincide with the original 19°C curve, and would thus be very
close that of the 7.9% styrene ionomer. In Fig. 4, the new position would be located between the 143 and 159°C curves. This
would make the relative positions of the time-temperature sur-
posed master curves of the precursor, the Nafion acid and the
potassium salt (42) much more reasonable. In other words, when
these master curves are reduced to the α peak temperature, the
relaxation time of the precursor appears to be the longest, while
that of the potassium salt is even shorter than that of the acid.
This feature certainly contradicts the comparison of the

ten-second moduli versus temperature curves of these polymers shown in Figure 5 as well as the torsion pendulum results. This anomaly can be eliminated if time-temperature superposition is performed with the α peak temperatures as the matrix glass-transition temperatures for the Nafion acid and the potassium salt.

Acknowledgements

It is a pleasure to acknowledge partial support from the U.S. Army Research Office, as well as from the Chemicals Group, Olin Corporation.

Literature Cited

1. Holliday, L. "Ionic Polymers"; Wiley, New York, 1975.
2. Eisenberg, A.; King, M. "Ion-containing Polymers"; Academic Press, New York, 1977.
3. Longworth, R; Vaughan, D.J. Nature (London) 1968, 2, 85.
4. Otocka, E.P.; Eirich, F.R. J. Polym. Sci., Part A-2, 1968, 6, 921.
5. Delf, B.W.; MacKnight, W.J. Macromolecules, 1969, 2, 309.
6. Roe, R.J. J. Phys. Chem., 1972, 76, 1311.
7. Marx, C.L.; Caulfield, D.F.; Cooper, S.L. Macromolecules, 1973, 6, 344.
8. Eisenberg, A.; Navratil, M. Macromolecules, 1974, 7, 90.
9. MacKnight, W.J.; Taggart, W.P.; Stein, R.S. J. Polym. Sci. Polym. Symp., 1974, 45, 113.
10. Pineri, M.; Meyer, C.; Levelut, A.M.; Lambert, M. J. Polym. Sci. Polym. Phys. Ed., 1974, 12, 115.
11. Roche, E.J.; Stein, R.S.; Russell, T.P.; MacKnight, W.J. J. Polym. Sci. Polym. Phys. Ed., 1980, 18, 1497.
12. Fitzgerald, W.E.; Nielson, L.E. Proc. R. Soc. (London) 1964, A282, 137.
13. Ward, T.C.; Tobolosky, A.V. J. Appl. Polym. Sci. 1967, 11, 2403.
14. MacKnight, W.J.; McKenna, L.W.; Read, B.E. J. Appl. Phys., 1967, 38, 4208.
15. Bonotto, S.; Bonner, B.F. Macromolecules, 1968, 1, 510.
16. Eisenberg, A. Macromolecules, 1970, 3, 147.
17. Matsuura, H.; and Eisenberg, A. J. Polym. Sci. Polym. Phys. Ed., 1976, 14, 773.
18. Eisenberg, A. Adv. Polym. Sci., 1967, 5, 59.
19. Otocka, E.P. J. Macromol. Sci. Revs. Macromol. Chem., 1971, C5, 275.
20. MacKnight, W.J.; Earnest, T.R. Jr. Macromol. Rev., in press.
21. Bazuin, C.G.; Eisenberg, A. Ind. Eng. Chem. - Prod. Res. and Dev., 1981, 20, 271.

22. Birkales, N.M. ed., "Water-soluble Polymers"; Plenum Press, New York, 1973.
23. Eisenberg, A. Ed., "Ion-containing Polymers"; Wiley, New York, 1974.
24. Rembaum, A. and Sélégny, E. Eds., "Polyelectrolytes and Their Applications"; Reidel, Dordrecht, 1975.
25. Eisenberg, A. Ed., "Ions in Polymers"; ACS Symposium Series #187, Washington D.C., 1980.
26. Grot, W.G.; Munn, G.E.; Walmsley, P.N., 141st Meeting of the Electrochemical Society, Houston, Texas, 1972.
27. Yeager, H.L.; Kipling, B. J. Phys. Chem., 1979, 83, 1836.
28. Burkhardt, S.F. 152nd Meeting of the Electrochemical Society, Atlanta, Georgia, 1977.
29. Yeo, R.S.; McBreen, J. J. Electrochem. Soc., 1979, 126, 1682.
30. Yeager, H.L.; Kipling, B; Dotson, R.S. J. Electrochem. Soc., 1980, 127, 303.
31. Yeo, S.C. and Eisenberg, A. J. Appl. Polym. Sci., 1977, 21, 875.
32. Gierke, T.D. 152nd Meeting of the Electrochemical Society, Atlanta, Georgia, 1977.
33. Duplessix, R.; Escoubes, M.; Rodmacq, B.; Volino, F.; Roche, E.; Eisenberg, A; Pineri, M. in "Water in Polymers", ACS Symposium Series #127, S.P. Roland Ed., 1980, p. 469, ibid. 1980, p. 478.
34. Rodmacq, B.; Pineri, M. Rev. Phys. Appl., 1980, 15, 1179.
35. Rodmacq, B.; Pineri, M; Coey, J.M.D.; Meagher, A. in preparation.
36. Komoroski, R.A.; Mauritz, K.A. J. Am. Chem. Soc., 1978, 100, 7487.
37. Lowry, S.R.; Mauritz, K.A. J. Am. Chem. Soc., 1980, 102, 4665.
38. Falk, M. Can. J. Chem., 1980, 58, 1495.
39. Mauritz, K.A.; Hora, C.J.; Hopfinger, A.J. in "Ions in Polymers", ACS Symposium Series #187, Eisenberg, A., Ed., 1980, 123.
40. Kyu, T.; Eisenberg, A. in preparation.
41. Takamatsu, T.; Eisenberg, A. J. Appl. Polym. Sci., 1979, 24, 222.
42. Hodge, I.M.; Eisenberg, A. Macromolecules, 1978, 11, 289.
43. Lundberg, R.D.; Makowski, H.S.; Westerman, L. in "Ions in Polymers", ACS Symposium Series #187, Eisenberg, A. Ed., 1980, p. 67.
44. Rigdhal, M; Eisenberg, A. J. Polym. Sci. Polym. Phys. Ed., in press.
45. McCrum, N.G. J. Polym. Sci., 1959, 34, 355.
46. Hashiyama, M.; Eisenberg, A. in preparation.
47. Fujimura, M.; Hashimoto, T.; Kawai, H. submitted to Macromolecules.

RECEIVED September 16, 1981.

SPECTROSCOPIC, STRUCTURAL, AND MODELING STUDIES

Nuclear Magnetic Resonance Studies and the Theory of Ion Pairing in Perfluorosulfonate Ionomers

R. A. KOMOROSKI

B. F. Goodrich Company Research and Development Center, 9921 Brecksville Road, Brecksville, OH 44141

K. A. MAURITZ

Diamond Shamrock Corporation, T. R. Evans Research Center, P. O. Box 348, Painesville, OH 44077

Nafion (a registered trademark of E. I. du Pont de Nemours and Co.) and other perfluorinated ion exchange membranes have received much recent consideration as electrolytic separators in electrochemical applications, particularly chlor-alkali cell technology. The systems of current commercial interest, as well as descriptions of many physical property investigations, are reported throughout this volume.

To be sure, most of the exceptional properties exhibited by these materials, in applications reserved for ion exchange membranes, must implicate the ionic cluster morphology that exists on about a 50 Å microstructural level. An unambiguous under standing of the ionic-hydrate molecular architecture within the confines of a cluster must precede a rational understanding of the action of these membranes in selective ion transport. Even at equilibrium, the structure within these encapsulated micro-solutions must be considered as dynamic. In addition to investigating the molecular mobility of the ionic and water components, a study of the dynamic conformational aspects of the perfluorinated polymer backbone and comparison with similar conventional hydrocarbon materials, is clearly needed. These studies, of course, would logically be performed with polymer equivalent weight, ion exchange functionality, counterion type, internal water content, and temperature, as independent variables.

This contribution to the series is a report of an initial use of nuclear magnetic resonance spectroscopy in an effort to begin to answer some of these structural questions. Also included is an outline of a molecular-based theoretical model of sidechain-counterion interactions, as suggested by the experimental results. The content of this work is concerned with Nafion in the monovalent cationic salt form, i.e., $\{OCF_2CF\}_m O(CF_2)_2 SO_3^- X^+$, with no co-ions being present.
$\phantom{\{OCF_2}CF_3$

Characterization by Nuclear Magnetic Resonance Techniques

Background. Nuclear magnetic resonance (NMR) is a powerful technique for the study of the molecular structure and morphology of polymeric materials. Standard high resolution techniques (1, 2) have provided information on the stereochemistry and defect structures of homopolymers and the composition and sequence distribution of copolymers. High resolution studies usually demand that the polymer be soluble in an appropriate solvent or exist as a low viscosity liquid. It is under these circumstances that the strong dipolar and other anisotropic interactions characteristic of solids are suitably averaged. High power pulsed and wide-line NMR techniques (3, 4) have been used to obtain morphological information concerning crystalline, glassy, and rubbery polymers in bulk. However, the utility of these techniques has been limited by difficulties in relating observed relaxation processes to specific motions of chemical entities of the polymer.

Nafion, like other fluoropolymers, is not soluble in any known solvent at temperatures below at least 200°C. Hence, it is not amenable to standard high resolution techniques using ^{19}F or ^{13}C nuclei. To our knowledge, no pulsed NMR study similar to that done on polytetrafluoroethylene (PTFE) has been published for Nafion (5).

Much of the interest in Nafion materials stems from its use as a membrane separator for chlor-alkali production and other electrochemical applications (6). Hence, it is relevant to study the material in the presence of water of swelling, with or without sorbed electrolytes. The presence of water and counterions gives us an NMR handle with which to study Nafion. Solvent and ionic mobility is rapid enough in the aqueous regions of swollen Nafion to yield high resolution NMR signals.

Counterions. 1. Sodium-23: Alkali metal NMR is a sensitive probe of the immediate chemical environment and mobility of alkali metal ions in aqueous and nonaqueous solvents (7, 8). The chemical shifts of alkali metal nuclei will respond to electronic changes only in the immediate environment of the cation since alkali metals rarely participate in covalent bonding (7). All alkali metal nuclei have spins greater than 1/2 and hence have quadrupole moments. The interaction of these moments with electric field gradients, produced by asymmetries in the electronic environment, is modulated by translation and rotational diffusive motions in the liquid. It is via this relaxation mechanism that the resonance line width is a sensitive probe of ionic mobility.

The first published NMR studies of Nafion examined the mobility and hydrative properties of sodium counterions using ^{23}Na
NMR ($\underline{9}$, $\underline{10}$). A single ^{23}Na resonance was observed for the sodium
form of Nafion in the presence of water but with no excess in-
ternal electrolyte (Na^+:SO_3^- = 1:1). This resonance shifted upfield and broadened significantly with decreasing water content.

Figure 1 depicts the behavior of the chemical shift with
%H_2O. The shift change occurs over a range of about 130 ppm,
rather large for the sodium ion ($\underline{7}$). The behavior of the linewidth is qualitatively similar to the shift behavior in Figure 1.
For saturated Nafion, the linewidth is about 220 Hz at 30°C, and
increases to thousands of Hz for very low (1-2%) water contents.
Linewidth and chemical shift changes with decreasing water
content are reversed with increasing temperature ($\underline{9}$). The above
chemical shift and linewidth behavior of the sodium form of
Nafion parallels that of the model salt CF_3SO_3Na as a function of
concentration in aqueous solution ($\underline{9}$, $\underline{10}$). Such a comparison can
only be made within the solubility limits of the model salt.
However, since the directions and magnitudes of the changes are
the same, we can conclude that the interaction responsible for
the changes is the same in both cases. This is the cation-anion
electrostatic attraction.

At saturation, the ^{23}Na chemical shift is close to that of
the model salt at comparable H_2O/Na^+ molar ratio, suggesting that
the immediate electronic environment of the two are similar.
This is circumstantial evidence for the existence of hydrated
ionic clusters in Nafion. Isolated, hydrated ion pairs might be
expected to have a significantly different chemical shift than
the comparable electrolyte solution due to the closer proximity
of the surrounding fluorocarbon phase.

Unlike the chemical shift, the ^{23}Na linewidth of saturated
Nafion is not close to that of the model salt. The 220 Hz linewidth of saturated Nafion is an order of magnitude larger than
that of CF_3SO_3Na. This suggests that some or all of the sodium
ions in saturated Nafion are restricted relative to the electrolyte solution, even though the chemical shift is not substantially affected. This result can influence the calculation of
the fractions of bound cations in saturated Nafion, as shown in
subsequent discussion.

The large shift changes at low water contents are strong
evidence for the formation of contact ion pairs ($\underline{9}$, $\underline{10}$). The
amount of water present at these levels is not sufficient to

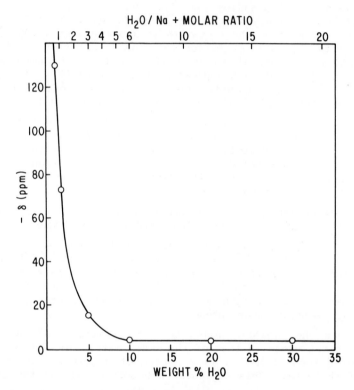

Figure 1. Plot of the ^{23}Na chemical shift of Nafion (1100 EW) vs. % H_2O and the H_2O/Na^+ molar ratio (9, 10).

provide effectively for a large concentration of solvent-sepa rated ion pairs. The sensitivity of the ^{23}Na chemical shift to local changes implies that solvent-separated ion pairs should not differ substantially in chemical shift from the "isolated" cation (or anion) in dilute solution.

It is possible to quantitatively interpret the ^{23}Na results in terms of a two-state dissociation equilibrium between free and bound species. The observed chemical shift (δ_{obs}) and linewidth ($\Delta\nu_{obs}$) are the weighted averages of the values in the free and bound states:

$$\delta_{obs} = P_f \, \delta_f + P_b \, \delta_b$$

$$\Delta\nu_{obs} = P_f \, \Delta\nu_f + P_b \, \Delta\nu_b$$

$$P_b = 1 - P_f$$

P_b and P_f are the mole fractions of bound and free cations, respectively. To estimate the number of cations bound at a given water content, it is necessary to know the chemical shifts and linewidths in the free and bound states. The values for the free state can be estimated from the corresponding values for the model salt (9, 10), although, as pointed out earlier, a linewidth of about $2\overline{2}0$ Hz for saturated Nafion calls this approximation into question for that parameter. It is difficult to estimate δ_b and $\Delta\nu_b$, since we cannot make measurements on the totally bound species. Here, we will use values 50% higher than those observed at 1% H_2O:

$$\delta_f = -1.5 \text{ ppm} \qquad\qquad \delta_b = 195 \text{ ppm}$$

$$\Delta\nu_f = 220 \text{ Hz} \qquad\qquad \Delta\nu_b = 6000 \text{ Hz}$$

In Table I are the values calculated using δ for percent sodium ions bound at 30°C.

Table I

Fraction of Sodium Ions Bound in Nafion at 30°C*

% H_2O	% Ions Bound
30	0.4
20	0.5
10	1
5	7
2	37
1	66

*Calculated from chemical shift data in references 9 and 10.

Nafion equivalent weight = 1100.

The ^{23}Na NMR parameters of Nafion are not substantially affected by equivalent weight in the range of water content where a valid comparison can be made (10).

Arrhenius plots of the linewidth for various equivalent weights are, for the most part, approximate straight lines (10). An activation energy of 3.2 kcal/mole was obtained from the plot for the sample of 1100 equivalent weight. It is difficult to assess the relative importance of the specific dynamic processes that give rise to the measured activation energy. Clearly, the motion of solvent molecules in and out of the ionic hydration sphere and the binding of the cation to an anionic site will be involved. Calculations of the energetics involved in the dissociation equilibrium of bound and unbound cations yield activation barriers in reasonable agreement with this number (11). Previous NMR studies (12) on pure water yielded a value of 3.5 kcal/mole between 40° and 100°C. Using ^{23}Na NMR, a somewhat lower value (2.5 kcal/mole) was obtained for the reorientational process of water molecules in the hydration shell of Na^+ in aqueous NaCl (12).

2. Other Alkali Metals: It is possible to observe the NMR signals from other alkali metal forms of Nafion (10). Lithium-6

and 7 and cesium-133 have been observed for Nafion samples. Rubidium-85, which yields a broad line in an aqueous electrolyte, could not be observed under high resolution conditions (10). Apparently, potassium-39 has not been attempted, to date.

Figures 2 and 3 show the behavior of the ^{7}Li and ^{133}Cs resonances of Nafion, respectively, with variation of water content. The resonances of the water-saturated materials are broadened relative to that found in a corresponding aqueous electrolyte. This is evidence of restricted Li$^+$ and Cs$^+$ (and Rb$^+$) ion mobility in the presence of the polymer. The ^{7}Li linewidth changes gradually with water content while the ^{133}Cs linewidth and chemical shift change very rapidly at the lowest water contents. The ^{7}Li chemical shift change is small over the entire range.

The amount of broadening experienced by alkali metal ions upon binding to a polymer such as Nafion will depend on several factors including nuclear spin, electric quadrupole moment, stability of the hydrated ion, and electronic polarizability. This last factor relates to the ease with which electric field gradients can be produced by binding to an anionic site. Based on the relative change in linewidth upon going from saturation to low water contents, the ions studied so far are related such that Li$^+$ > Na$^+$ > Cs$^+$. Similar changes are seen in the -SO$_3^-$ symmetric stretch frequency in the IR spectra of Nafion with various counterions (13). The above order also holds for the gradualness of the NMR linewidth changes with change in water content. This behavior can be rationalized on the basis of the hydration sphere sizes for the various cations.

Sorbed Water. Because Nafion is a perfluorinated material, it is straight-forward to observe the NMR signal of water hydrogens free of interference. Proton (^{1}H) NMR is sensitive to hydrogen bonding in liquids and electrolyte solutions (14, 15). Formation of a hydrogen bond causes the resonance of the bonded proton to move downfield. Hence, an upfield shift can be correlated with the breaking of hydrogen bonds. This has led to the characterization of ions in terms of their effect on water structure (15).

The state of sorbed water has been studied by NMR for Nafion in the acid (16) and sodium (10) forms. In both cases, the ^{1}H resonance broadens and shifts upfield with decreasing H$_2$O content. The water resonance of the acid form (16) is broader than that of the sodium form (10). This is probably due, in part, to bulk susceptibility contributions to the linewidth of the acid

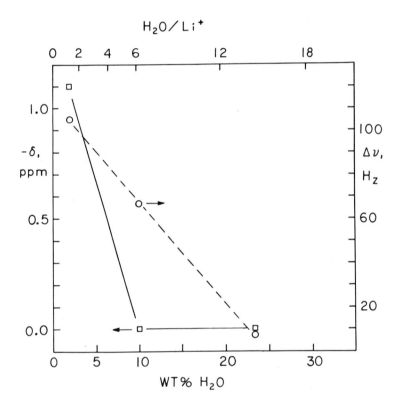

Figure 2. Plots of the ⁷Li chemical shift and linewidth dependences on water content.

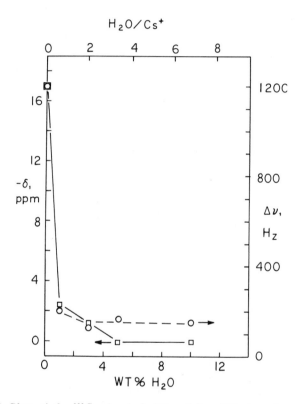

Figure 3. Plots of the ^{133}Cs chemical shift and linewidth dependences on water content.

form, which was probably examined in a granular state (10, 16).
The above chemical shift results suggest the breakup of water
hydrogen bonding with decreasing water content.

Both the resonance linewidth and the spin lattice relaxation
time (T_1) of the water protons have been studied as a function of
temperature at different water contents (16). For samples with
low water content (approximately 2.7%), the linewidth decreases
with increasing temperature from 230 to 300°K. For samples of
higher water content, the linewidth decreases up to about 250°K,
then remains constant. The T_1 data (Figure 4) exhibits a minimum
that shifts to higher temperatures with decreasing water content.
This indicates restricted water mobility at low water content.
The curves for higher (5-20%) levels of water show two minima,
suggesting a complex motion or possibly two components.

Fluorocarbon Backbone.   No NMR study of the Nafion fluorocarbon regions has yet been reported. Preliminary ^{19}F pulsed
NMR results have been obtained by Stengle (17). It was found
that the ^{19}F second moment of Nafion is reduced by about 1 G^2
upon going from the dry to the water-swollen state for both the
acid and potassium forms. This result is suggestive of a more
mobile polymer chain in the presence of water, on the average.
This sensitivity of chain mobility in the fluorocarbon regions to
the presence of water may relate to the observation of a minimum
in the ^{19}F T_1 curve near room temperature.   Clearly, further
studies are needed to elucidate the nature of the fluorocarbon-
water interaction.

Ion Pairing in Ionomeric Membranes

Background. The sidechain-counterion association-dissociation equilibrium in polyelectrolyte gels has been recognized and
is, in fact, a major ingredient in the Harris and Rice molecular-
based theoretical model of equilibrium thermodynamic states
(18-21). Certainly, examples of ion pairs in electrolyte solutions, particularly at high concentrations, are legion.   One
might expect this phenomenon to be enhanced within the domain of
solid polyelectrolyte systems for a number of reasons. Firstly,
there exists, within the interior, a large concentration of ion
exchange groups as well as a reduced availability of water to
provide for hydration of the internal ions, compared to a "dilute" external electrolyte solution.   Secondly, the low dielectric constant of the organic polymeric matrix would increase the
anion-cation coulombic attraction over that as would exist in a
strict aqueous solution of equal ionic molarity.   Lastly, it
would be reasonable to assume that cations and anions would be

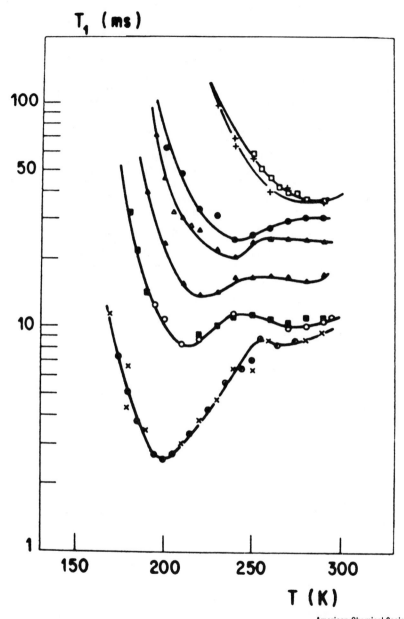

Figure 4. T_1 vs. temperature curves for various fixed water contents (16). *Key:* $+$, 2.7%; □, 2.8%; ●, 4.4%; △, 5%; ▲, 5.5%; ■, 7.7%; ○, 8.6%; ●, ×, 20%.

brought into mutual proximity in being confined to finite interstitial or hydrophilic cluster volumes. In other words, the
overall ionic mobility is reduced owing to the boundaries imposed
by the space preempted by the macromolecular architecture. In
this way, the dissociation equilibrium would shift to that corresponding to a greater population of cations being engaged in
some state of association with the charged sidechain.

     It would be appropriate, at least from a mechanistic or
"first principles" standpoint, to possess a theoretical molecular-based analysis of this equilibrium, including internal water
content hydration effects. The remaining discussion of this
topic is concerned with the efforts of Mauritz, et al., in providing a fundamental molecular representation of the phenomenon
of ion-pairing and subsequently relate this model to the swelling
of membranes in specific cation-exchanged salt forms.

     As discussed by Falk elsewhere in this volume, Fourier
transform infrared, as well as ^{23}Na NMR spectroscopy has been
used to monitor cation interactions with the sulfonate moeities
in Nafion membranes having a cation/sidechain molar ratio of
unity (13). In summary, the broadening and shift to higher
frequencies, with decreasing water content, of the $-SO_3^-$
symmetric stretching mode has been also attributed to an
increasing population of counterions that are strongly
interacting with the sulfonate group, the interactive magnitude
diminishing with increasing bare ioni (Li^+, Na^+, K^+, Rb^+) size.

The onset of both IR and NMR spectra changes occur at membrane
water contents that just barely provide enough water molecules
for complete ionic hydration.

     Molecular Model of Counterion Dissociation Equilibrium. The
following molecular concept is supported, or suggested, by both
these spectroscopic observations and past ultrasonic investigations of simple aqueous electrolytes. In particular, a four-
state model reminiscent of the multistep ionic dissociation
mechanism of Eigen et al., (22, 23) was adopted (24). With
regard to Figure 5, the states are classified as: 1) completely
dissociated hydrated ion pairs, 2) ion pairs at the contact of
undisturbed primary hydration shells, and 3) outer and 4) inner
sphere complexes. The relative populations of these states, (P_i;
i = 1-4), have been determined from the use of Boltzman statistics utilizing specific solvation and interionic energetic for-
mulations. In addition to the direct quantitative incorporation
of hydration energetics, the model differs from most classical
approaches that are based on the theory of Bjerrum (25) insofar

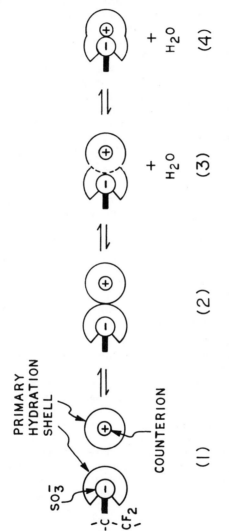

Figure 5. Four-state model of hydration-mediated equilibrium between unbound and sidechain-associated counterions in ionomeric membranes (9, 10, 13).

as a discrete set of well-specified states of ion-water "complexes" is considered.

The desolvation energetics encountered in passing between states 2 and 3, and 3 and 4, are computed by the utilization of a simple model in which the energy change accompanying the formation of a given ionic association is proportional to the change in the fractional degree of volume overlap between the primary hydration shells of the ionic sidegroups and counterions. The requisite number of ejected water molecules accompanying the overlap is then taken to be the product of this volume ratio and the original molecular population of the penetrated hydration shell. On the average, each displaced water molecule results in a hydrative energy loss of H_+/n_+, where H_+ and n_+ are the experimentally-determined hydration energies and numbers, respectively, of the given (positive or negative) ion. State 3 is actually considered as a hybridization, that is, energetically-weighted mixture of the two alternate modes of hydration shell overlap depicted in Figure 6. The cation $-SO_3^-$ interaction, presently, is represented by a simple modification of Coulomb's law for two point charges of opposite sign that are at the energetically-favorable separation for the given ionic-hydrate complex. Also, a "molecular" dielectric constant, K, that varies linearly with n, the average number of water molecules available to a cation $-SO_3^-$ pair, as shown in Figure 7, is utilized in the formulation.

Of course, Van der Waals and other secondary forces can be included in more sophisticated computations. Thus, as the water content drops below a level at which completely populated hydration shells can exist, the electrostatic shielding effect of the water becomes less effective resulting in enhanced cation $-SO_3^-$ attraction. Thus the ionic dissocation equilibrium model is dependent on water content for $n < n_+ + n_-$.

To be sure, the specific primary hydration structures of single ions, viewed within a dynamic framework, or otherwise, as well as "hydration numbers" and "hydration energies," deriving from numerous experimental procedures, are still under debate. Furthermore, equating the hydration properties of ions in polyelectrolyte gels, or ionomers, with those of the appropriate electrolyte analogues, may be questioned, in a strict sense. Nonetheless, the concept of primary hydration shells, or ionic "co-spheres," as they were originally dubbed by Gurney, (26) has been useful through years of ionic solution theory and has received considerable attention in the area of biopolymers, as well. Also, if one maintains consistency in utilizing n_+'s and H_+'s, in that the numbers from determinations using different techniques are not intermixed, and if the particular context in which the numbers are to be used, e.g., transport calculations,

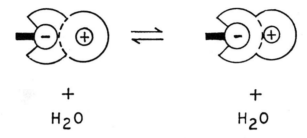

+

H_2O

+

H_2O

Figure 6. Two possible modes of hydration shell penetration in the conceptual outer sphere complex formation (24).

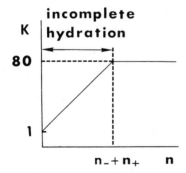

Figure 7. Linear variation of dielectric parameter, K, with n at low water contents (24).

equilibrium hydrative structure calculations, are matched to the experimental conditions that most reasonably simulate actual systems, the comparison of computational results for a series of ions is of significance, at least qualitatively.

Finally, the relative populations of the various states can be derived from the corresponding energetics by:

$$P_i = Q^{-1}\exp(-E_i/RT), \quad i = 1,2,3,4, \tag{1}$$

where:

$$Q = \sum_{j=1}^{4} \exp(-E_j/RT). \tag{2}$$

E_i is the combined hydration and coulombic energy per given ion pair for the state i and T the Kelvin temperature. P_1 is obviously the theoretical degree of dissociation and the overall bound percent of cations is $(1-P_1) \times 100\%$.

The calculated shift in the spectrum of states with decreasing water content for the Na^+ form is shown in Figure 8. P_1 steadily drops after a value of n, corresponding to an approximate combined hydration number of a $Na^+SO_3^-$ pair, is reached. State 3, the outer sphere complex, becomes increasingly populated and becomes physically identical to State 2 after the water content has dropped to about 4 moles of water per equivalent of resin. The inner sphere complex, or contact ion pair, while possessing the greatest electrostatic binding, has a population density of practically zero, however, because of the difficulty in removing the final interposed water molecules that must interact with both the Na^+ ion and SO_3^- moeity.

Plotted in Figure 9 are (a) the bound percent of Na^+ ions and (b) B, the overall population percentage of Na^+ ions having insufficient thermal kinetic energy needed to overcome electrostatic binding to the sulfonate group. The situation depicted in Figure 9, i.e., a progressively greater proportion of intimate cation-sulfonate interactions with dimishing internal water content, is at least qualitatively consistent with the physical model arising from spectroscopic interpretation.

It is not possible to directly compare the ^{23}Na NMR experiments and the results of the theoretical calculations of the

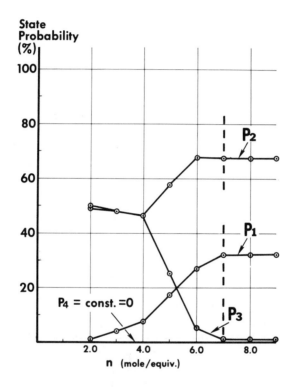

Figure 8. Dependence of the distribution of states of ionic association on n *for the Na⁺ form (24).*

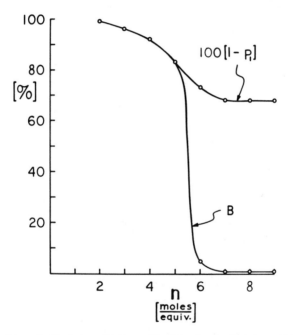

Figure 9. Variation of 1-P₁ and B with n for the Na⁺ form (24).

ionic dissociation equilibrium. Rapid multistage exchange processes cannot be distinguished from simple two-site rapid exchange. A comparison with the theory could be made if the chemical shift and linewidth in each state were known.

Although a direct comparison is not possible, the ^{23}Na experiments still provide considerable insight into a theoretical framework. First, the population of contact ion pairs, P_4, as calculated using a two stage model, is small at all except the lowest water contents, in agreement with theory. The fact that water-saturated Nafion exhibits a ^{23}Na linewidth larger than that of the model salt indicates restricted ionic mobility for some or all of the cations in this case. The theory predicts a sizeable population for P_2 at relatively high water contents. Ions in State 2 would be expected to have an increased linewidth due to their proximity to anions. Of course, rapid exchange between States 1 and 2 would produce the single, broadened line seen for water-saturated Nafion.

A Theory of Membrane Internal Water Activity. From a thermodynamic standpoint, the (water-swollen) equilibrium membrane structure must depend, in part, upon the internal osmotic pressure which is determined by the water activity, $\bar{a}_w$, within the microscopic cluster regions. $\bar{a}_w$, in turn, should be a function of the relative population of unpaired ions and free water molecules in the cluster solution.

If the resin "solution" were ideal, in a thermodynamic sense, then the internal water activity would simply be the mole fraction of sorbed water, i.e.,

$$\bar{a}_w = n/(n+1), \qquad (3)$$

where the unity term in the denominator refers to one equivalent of resin. However, since osmotic effects are primarily determined by the mixing of free water with free ions, $\bar{a}_w$, computed by equation (3), must be incorrect because: 1) there exists a finite population, $P_2+P_3+P_4$, of associated or bound cations; 2) of the n water molecules per ion exchange site, a certain portion will be constrained in a state of hydrative binding. This hypothetical division of the internal electrolyte into its osmotically "active" and "inactive" constituents is conceptualized in Figure 10.

Adopting the more realistic definition of the internal water activity as being the ratio of the number of moles of free water

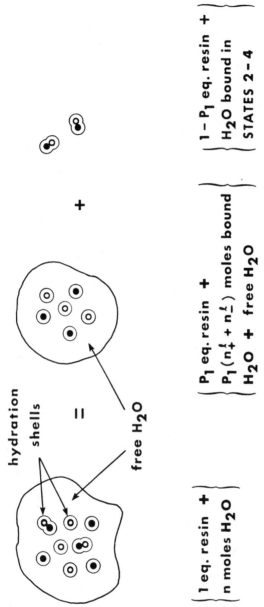

Figure 10. *Hypothetical ionic cluster solution as partitioned into a fraction containing all the dissociated (and hydrated) ions and free water, and a fraction containing all the bound (+)(−) pairs with their requisite water of hydration (24).*

to the number of moles of free water plus moles of free cation-
anion pairs per equivalent of resin, equation (3) becomes:

$$\bar{a}_w = n_{free}/(n_{free}+P_1),$$ (4)

where:

$$n_{free} = n - (n_+ + n_-)(P_1 + P_2) - P_3 \langle N_3 \rangle - P_4 \langle N_4 \rangle.$$ (5)

In addition to n_+ and n_-, which are the primary hydration numbers
of cation and anion, respectively, $\langle N_3 \rangle$ and $\langle N_4 \rangle$, the average
number of water molecules bound in the hydrative associations of
States 3 and 4, are also computed using the model for the association-dissociation equilibrium between bound and unbound cations
described previously.

A comparison of theoretically-determined internal water
activities for membranes in the monovalent cation forms over a
range of water contents is provided in Figure 11. For increasingly large n, differences between cationic forms becomes negligible and $\bar{a}_w$ asymptotically approaches unity (dilution effect).
At low n, however, a systematic differentiation is apparent
wherein the water activities increase in the following progression: $Rb^+ > K^+ > Na^+ > Li^+$. The primary cause of this behavior,
as well as the overall decrease from the pictured ideal solution
activity, is the ability of the smaller cations to engage a
larger number of water molecules in a state of hydration and
therefore effectively prevent them from participating in osmotic
swelling effects. The positive deviation of $\bar{a}_w$ from that of the
ideal solution, at large n, is due to incomplete ionic dissociation. Since, in the process of equilibration with pure water ($\bar{a}_w$
= 1), the gradient of water activity across the membrane/liquid
interface increases in the reverse ionic progression, the internal osmotic pressure will be greatest for Li^+ and lowest for
the Rb^+ form. It is, in fact, experimentally observed that the
equilibrium water uptake of Nafion membranes follows the order:
$Li^+ > Na^+ > K^+ > Rb^+$. In the establishment of chemical equilibrium for small cationic forms, additional water must necessarily
enter the membrane and result in greater swelling as compared to
large cationic forms. Of course, the absolute quantitative

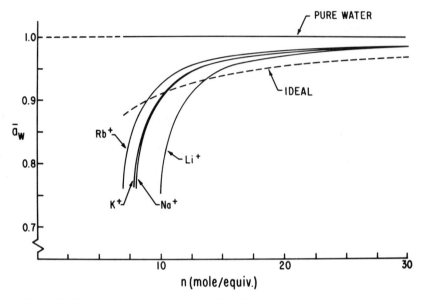

Figure 11. Theoretical $\bar{a}_W$ vs. n for Li^+, Na^+, K^+, and Rb^+. Also shown for comparison over the range of n is the ideal solution activity (24).

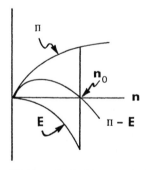

Figure 12. Hypothetical increase in the internal osmotic swelling pressure, Π, and counteractive polymer matrix elasticity (reflected in an increasing cohesive energy density, E), with increasing water content, n. n_0 is the number of moles of water, per ionic sidechain, for which Π-$E = 0$ (24).

prediction of water uptake also depends on the limit of polymer "network" deformation.

The approach to equilibrium for an initially-dry ion exchange membrane (in a given counterion salt form and containing no co-ions) that is subsequently immersed in pure water can be visualized in the following way: Although the interaction between the organic polymer backbone and water is endothermic and may influence the rate of swelling, the strong exothermic tendency of the counterions and ionogenic sidechains to hydrate results in having the initially-arrived water molecules strongly bound in ionic solvation shells resulting in little or no volume expansion of the network. In the truly dry state, the counterions are strongly bound by electrostatic forces in contact ion pairs. Further uptake of water beyond that which is barely required for maximum occupancy of all the hydration shells results in moving the association-dissociation equilibrium between bound and unbound counterions toward increased counterion mobility. The driving force for swelling is the tendency for the water to dilute the polymer network. Stated in thermodynamic terms, the difference between the water activity in the interior and exterior of the membrane gives rise to an internal osmotic pressure that serves as the driving force for polymer chain network deformation.

As the water uptake proceeds, the increased sidechain-counterion dissociation allows for more complete ionic hydration. The deformation of the polymer chain network upon further incorporation of water molecules also proceeds by a shift in the distribution of rotational isomers to higher energy conformations and changes in other intramolecular, as well as intermolecular interactions. Consequently, the increased overall energy state, for a given membrane water content of n moles, per equivalent of resin, is manifested by polymer chain retractive forces that resist expansion of the network. Accordingly, the configurational entropy decreases as less conformations become available within the matrix. Eventually, an equilibrium water content, n_o, is reached, at which the osmotic pressure, Π, is balanced by the cohesive energy density, E (see Figure 12). It should be stressed that there presently is no satisfactory theoretical molecular representation of the retractive polymeric response, E, particularly in the case of hydrophilic/hydrophobic phase-separated systems.

A Look to the Future

Several technical advances in the NMR field appear promising for future studies of Nafion. All of these techniques have the acquisition of high resolution spectra of the solid polymer as their goal. These techniques are briefly mentioned here to indicate the direction the authors believe NMR studies will go.

Special NMR probes that can operate at temperatures up to 400°C have been developed (27, 28). With such a probe, high resolution ^{19}F FT NMR spectra have been obtained for fluoropolymers both in solution and in the melt. The spectra were of sufficient quality to yield compositional and microstructural information (28). High temperature ^{19}F NMR might yield information concerning comonomer sequence distribution in Nafion or like materials.

It is now well-established that ^{13}C NMR spectra in which resolution approaches that of the liquid state can be obtained for solid hydrocarbon polymers by employing a combination of esolution (^{1}H dipolar decoupling and magic angle spinning) and sensitivity (^{13}C - ^{1}H cross polarization) enhancement techniques (29, 30). This work has recently been extended to the fluoropolymers PTFE and poly(chlorotrifluoroethylene) (PCTFE) using ^{19}F dipolar decoupling and ^{13}C - ^{19}F cross polarization (31). Narrow lines, on the order of 15 to 70Hz, were obtained for both polymers. The two nonequivalent carbons of PCTFE were resolved. These exciting results suggest that such techniques could be quite powerful for the study of Nafion. In addition to microstructural information such as copolymer sequence distribution, information on polymer mobility should be forthcoming as for hydrocarbon polymers (29). The high resolution technique would allow both the polymer backbone and the sidechains to be independently probed as a function of environmental conditions. Molecular information concerning the fluorocarbon/aqueous interface might be forthcoming.

The above technique is applicable to nuclei in low natural abundance (e.g., ^{13}C, ^{15}N) in solids. Progress has also been made toward the observation of high resolution spectra of abundant spins in solids (32). High resolution spectra of ^{1}H or ^{19}F in solids can be observed by application of multiple-pulse sequences specially designed to eliminate the strong homonuclear dipolar interactions responsible for most of the line broadening. Multiple-pulse dipolar averaging NMR has been applied to PTFE and poly(TFE-hexafluoropropylene) (33, 34). Individual ^{19}F resonances arising from the crystalline and amorphous regions of PTFE were resolved. Hence, it proved possible to study independently polymer chain motions in each region and to obtain a direct measurement of the degree of crystallinity. Applications to Nafion systems are obvious.

Conclusions

Preliminary results of the NMR investigation of the molecular nature of the ionic cluster regions of Nafion perfluorosulfonate membranes have been presented. An advantage of the NMR probe is the ability to monitor short-range dynamic aspects of the ionic-hydrate molecular architecture for equilibrium states. It has also been demonstrated how the resultant concepts can be translated into a molecular-based model that may be significant in providing a base for mathematical representations of various macroscopic physical properties.

On the other hand, it seems that, aside from the overall clustering aspect, little has been discovered, to date, regarding structural features of the polymer itself, particularly from NMR studies. What, for example, can be learned of comonomer sequence distribution, or backbone and sidechain conformations with the accompanying dynamic effects? How is conformation affected by counterion type, water content, and temperature, to name a few? What can we say about the fluorocarbon/aqueous interface at the molecular level? Future studies similiar to those mentioned in the last section may resolve these remaining problems.

Literature Cited

1. Bovey, F. A. "High Resolution NMR of Macromolecules;" Academic Press: New York, 1972.
2. Randall, J. C. "Polymer Sequence Determinations;" Academic Press: New York, 1977.
3. McCall, D. W. Acc. Chem. Res. 1971, 4, 223.
4. McBrierty, V. J. Polymer 1974, 15, 503.
5. McCall, D. W.; Douglass, D. C.; Falcone, D. R. J. Phys. Chem. 1967, 71, 998.
6. Chapter _____, this work.
7. Popov, A. I. Pure Appl. Chem. 1975, 41, 275.
8. Templeman, G. J.; Van Geet, A. L. J. Am. Chem. Soc. 1972, 94, 5578.
9. Komoroski, R. A.; Mauritz, K. A. J. Am. Chem. Soc. 1978, 100, 7487.
10. Komoroski, R. A. Advan. Chem. Ser. 1980, 187, 155.
11. Mauritz, K. A.; Hora, C. J.; Hopfinger, A. J. Advan. Chem. Ser. 1980, 187, 123.
12. Hertz, H. G. Prog. Nucl. Magn. Reson. Spectrosc. 1965, 3, 159.
13. Lowry, S. R.; Mauritz, K. A. J. Am. Chem. Soc. 1980, 102, 4665.
14. Burgess, J.; Symons, M. C. R. Q. Rev. Chem. Soc. 1968, 22, 276.

15. von Goldhammer, E. in "Modern Aspects of Electrochemistry," Vol. X, 1; Bockris, J. O'M., et al., Eds.; Plenum: New York, 1975.
16. Duplessix, R.; Escoubes, M.; Rodmacq, B.; Volino, F.; Roche, E.; Eisenberg, A.; Pinieri, M. ACS Symp. Ser. 1980, 127, 469.
17. Stengle, T. R., private communication.
18. Harris, F. E.; Rice, S. A. J. Chem. Phys. 1956, 24, 1258.
19. Rice, S. A.; Harris, F. E. Z. Phys. Chem. (Frankfurt) 1956, 8, 207.
20. Helfferich, F. "Ion Exchange;" McGraw Hill: New York, 1962; p. 118.
21. Rice, S. A.; Nagasawa, M. "Polyelectrolyte Solutions;" Academic Press: New York, 1961; p. 461.
22. Diebler, H.; Eigen, M. Z. Phys. Chem. (Frankfurt) 1959, 20, 299.
23. Eigen, M.; Tamm, K. Z. Elektrochem. 1962, 66: 93, 107.
24. Mauritz, K. A.; Hopfinger, A. J. "Structural Properties of Membrane Ionomers," in "Modern Aspects of Electrochemistry," Vol. XIV; Bockris, J. O'M., et al., Eds.; Plenum: New York, to appear; also, detailed theory submitted for publication by Mauritz, K. A.; Lowry, S. R.; Komoroski, R. A.
25. Bjerrum, N. Mat. Fys. Medd - K. Dan. Vidensk. Selsk 1926, 7, No. 9.
26. Gurney, R. W. "Ionic Processes in Solution;" McGraw-Hill: New York, 1953; p. 4.
27. Bruker Report 1/2/77; Bruker-Physik AG Product Literature; p. 18.
28. English, A. D.; Garza, O.T. Macromolecules 1979, 12, 351.
29. Schaefer, J.; Stejskal, E. O.; Buchdahl, R. Macromolecules 1977, 10, 384.
30. Earl, W. L.; VanderHart, D. L. Macromolecules 1979, 12, 762.
31. Fleming, W. W.; Fyfe, C. A.; Lyerla, J. R.; Vanni, H.; Yannoni, C. S. Macromolecules 1980, 13, 460.
32. Haeberlen, U. "High Resolution NMR in Solids-Selective Averaging;" Academic Press: New York, 1976.
33. English, A. D.; Vega, A. J. Macromolecules 1979, 12, 353.
34. Vega, A. J.; English, A. D. Macromolecules 1980, 13, 1635.

RECEIVED August 11, 1981.

Infrared Spectra of Perfluorosulfonated Polymer and of Water in Perfluorosulfonated Polymer

MICHAEL FALK

Atlantic Research Laboratory, National Research Council of Canada, 1411 Oxford Street, Halifax, Nova Scotia B3H 3Z1, Canada

Infrared spectroscopy provides information on the microscopic structure of hydrated polymers which is not easily available by other means. Nafion is a new material and so far only four infrared studies on it have been reported (1-4). The scope of these studies is summarized in Table I; they are of preliminary nature and do not exhaust the possibilities of the infrared technique. In the present chapter the structural information which has been so far derived from infrared studies of Nafion will be collected, and some additional results from the author's laboratory will be presented.

Control of Water Content of Nafion Membranes in Infrared Studies

The first infrared measurements on Nafion were carried out without controlling its water content (1,2). Because membranes exposed to the air tend to lose water by evaporation, especially when exposed to the heat of the infrared beam, the spectra of ref. 1 and 2 correspond to rather low water contents, probably below one H_2O molecule per sulfonate group.

Control of water content was introduced by Lowry and Mauritz (3) who observed that membranes pressed tightly between flat plates did not lose water during the recording of the spectrum. This enabled them to record Attenuated Total Reflectance (ATR) spectra (5) at different stages of water loss, starting with a thoroughly soaked membrane and allowing some of the water to evaporate between consecutive measurements. They estimated the water content by quickly weighing the membrane before and after each spectrum and again after thorough drying (3).

In the author's laboratory, two experimental techniques have been developed for controlling the water content. In the vapor equilibrium technique the membrane is suspended in a hygrostatic

NRCC No. 19534

TABLE I

Previously Reported Infrared Spectra of Nafion

Authors	Nafion code[a]	Cation form	Technique	State of hydration	Spectral range (cm^{-1})
Lopez, Kipling & Yeager [1]	125	H^+, Na^+, Cs^+	Transmission	Low	4000–1350
Heitner-Wirguin [2]	125, 152	H^+, Na^+, Co^{++}, Ni^{++}, Cu^{++}, Fe^{+3}	Transmission, ATR	Low	4000–200
Lowry & Mauritz [3]	113	Li^+, Na^+, K^+, Rb^+	ATR	Variable	1100–900
Falk (1980)[4]	125, 142	Na^+	Transmission	Low	Water bands only
Present work	125, 142	Mainly Na^+	Transmission, ATR	Variable	4000–200

[a]The first two digits of the numerical code indicate the equivalent weight in units of 100, while the third digit indicates thickness in mils (1 mil = 25.4 μm). Thus Nafion 142 membrane has equivalent weight 1400 and thickness 2 mils = 51 μm.

cell in which moist air of controlled humidity and H/D ratio is
circulated. This technique, introduced in ref. 4, yields
reproducible water contents. However, the determination of water
by weighing the membrane becomes inappropriate because during the
recording of the spectrum the portion of the membrane which is
subjected to the infrared radiation is heated to an estimated
10°C above ambient temperature, causing temporary local
diminution of the water content. When water contents of the
suspended membranes were determined from the absorbance of the
H_2O band at 1620 cm^{-1} (as described in the following section) it
was found that the highest levels of hydration attained with this
technique were only about 2 $H_2O/-SO_3^-$ and not 6 $H_2O/-SO_3^-$ as had
been calculated from weighing measurements (4). We have since
learned to increase the water content of the membranes in the
vapor equilibrium technique to about 5 $H_2O/-SO_3^-$ by procedures
minimizing the effect of beam heating, but basically this
technique is limited to the study of incompletely hydrated
membranes.

     More recently, a complementary sandwiched film technique was
developed in which spectra of Nafion membranes containing 1 to 14
H_2O per sulfonate were routinely obtained. The top level of
hydration can be taken to represent a membrane completely
saturated with water under our experimental conditions. This
technique is an adaptation of the method of Lowry and Mauritz of
sealing a membrane of a given water content between two flat
plates (3). The procedure is equally suitable for ATR
measurements (in which the membrane is sandwiched between the
internal reflectance crystal and the back plate, as in ref. 3) or
for transmittance measurements (in which the membrane is
sandwiched between two CaF_2 or AgCl plates). The disadvantage of
the sandwiched film technique is that when the membrane is
tightly pressed a certain amount of liquid water can be observed
to be squeezed out and trapped between the membrane and the
confining plates. Thus the transmittance spectrum, and even more
so the ATR spectrum may contain a contribution from a thin layer
of pure water, the more so the higher the original water content.
This does not interfere unduly with the bands of Nafion, but
falsifies to some extent, especially at high water contents, the
spectral band shapes of water in Nafion.

     As of the time of writing, an ideal method of recording
infrared spectra of Nafion membranes at all levels of hydration
is still being sought.

Determination of Water Content from Absorbance at 1620 cm^{-1}

     In what follows, absorbance is defined as $A = -\log_{10} T$, where
T is transmittance. The use of integrated absorbance of the
1620-cm^{-1} band (B_{1620}) or its peak absorbance (A_{1620}) to
determine the membrane water content represents a considerable
improvement over weighing techniques, being direct and reflecting

the actual water content of the part of the membrane within the
infrared beam. The preference for the use of the H_2O bending
fundamental at 1620 cm^{-1} is based on its insensitivity to changes
in molecular environment. This is demonstrated by the nearly
coincident values measured in three different laboratories for
the integrated absorptivity of this fundamental in liquid water,
water vapor, and solution of water in acetone: 5.9×10^6 cm/mol
(6), 6.4×10^6 cm/mol (7), and 5.8×10^6 cm/mol (8),
respectively. The standard deviation of these values is 5%,
about the expected experimental error. Using the mean value,
6.1×10^6 cm/mol, and physical constants appropriate to Nafion
142 membrane* (thickness 0.0051 cm, equivalent weight 1400 g/mol
SO_3^-, dry density 1.98 g/cm^3) and allowing for the difference
between natural logarithms (used in refs. 6–8) and decadic
logarithms, we obtain the following relation for the ratio R of
H_2O molecules to sulfonate groups:

$$R = \frac{B_{1620}[cm^{-1}] \times 1400[g/mol\ -SO_3^-] \times 2.303}{0.0051[cm] \times 6.1 \times 10^6[cm/mol\ H_2O] \times 1.98[g/cm^3]} = 0.052 \times B_{1620}$$

where B_{1620} is the measured band area (abscissa: cm^{-1}; ordinate:
decadic absorbance) of the bending fundamental of H_2O in Nafion.
Or, since the halfwidth (i.e. the full width at half height) of
this band is about 50 cm^{-1}, and assuming that the band area
equals the product of peak absorbance by the halfwidth, we obtain
the alternate expression:

$$R = 2.6 \times A_{1620}$$

where A_{1620} is the peak absorbance of the H_2O bending
fundamental. For Nafion 125 a parallel calculation yields
$R = 0.017 \times B_{1620}$ or $R = 0.83 \times A_{1620}$. Whenever the absorbance
of the water band at 1620 cm^{-1} was too high to be measured
accurately, the absorbance at 1700 cm^{-1}, A_{1700}, away from the
band center could be measured instead, as it has been found that,
conveniently, $A_{1700} = 0.10 \times A_{1620}$. These relations have been
used to calculate H_2O contents in our recent measurements on
Nafion; they are estimated to be accurate to 15%.

It may be noted that the absorbance of the OH stretching
fundamental has also been used to estimate the water content of
Nafion (1), but it is much less appropriate for this purpose
because of the very large changes accompanying any changes in
hydrogen bonding. For example, the integrated absorptivity of

---

*The explanation of the numerical codes is in footnote to Table I.
Density and other physical constants of Nafion membranes are
tabulated in Dupont Product Information Bulletins.

the two stretching fundamentals of H_2O increases by a factor of 18 going from vapor to liquid, and by a factor of 2 going from liquid to ice (6).

Principal Features of the Spectra

Sodium salt of Nafion. Figure 1 shows the infrared transmission spectrum of the thinnest commercially available Nafion membrane, Dupont's Nafion 142, whose thickness is about 51 μm. Films of this thickness are suitable for studies of the less intense infrared absorption bands by ordinary transmission techniques. However, the more intense bands of Nafion, such as those in the region of 1340-1100 cm^{-1} and the water bands in the region of 3650-3150 cm^{-1} for samples of high water content, absorb completely or almost completely at such thicknesses. Preparation of thinner films is difficult but films of any thickness may be studied by the technique of Attenuated Total Reflectance (ATR) (2,3). Figure 2 shows the ATR spectrum of the same Nafion film as in Figure 1. One minor disadvantage of the ATR technique is the common occurrence of spurious peaks due to uncompensated absorption by atmospheric H_2O and CO_2 and various surface contaminants. A more serious disadvantage of ATR is that it is a surface technique with a depth of penetration of the order of wavelength, i.e. 3-10 μm (5) and the possibility is always present that the spectra observed are not representative of the bulk sample (3). It is therefore advisable to verify findings from ATR experiments by transmission spectroscopy as far as possible. We have observed that bands due to Nafion in ATR spectra correspond closely to the corresponding transmittance spectra, but that this is not generally true of bands due to water in Nafion.

Table II lists the positions and relative intensities of the main absorption bands of the sodium salt of Nafion, together with the best available assignments to vibrational modes of the structural components of Nafion: the fluorinated hydrocarbon main chain, the ether-linked fluorinated side-chains, the ionic end groups $-SO_3^-Na^+$ and water of hydration.

The most intense absorptions in the spectrum are those due to the fluorocarbon main chain. The spectrum of Nafion therefore strongly resembles that of polytetrafluoroethylene (PTFE; Teflon), all of the major bands of PTFE being also observed in Nafion at very nearly the same wavenumbers. Especially in the region of symmetric and antisymmetric CF_2 stretching (1350 to 1100 cm^{-1}) these intense PTFE-like bands obscure all other absorptions of Nafion. The CF_2 and CF units in the side-chains have no distinctive absorptions but the ether linkages give rise to the well-separated band at 980 cm^{-1}, which has been said to originate in the C-O-C symmetric stretch (2). The $-SO_3^-Na^+$ end groups give rise to only one distinctive absorption band

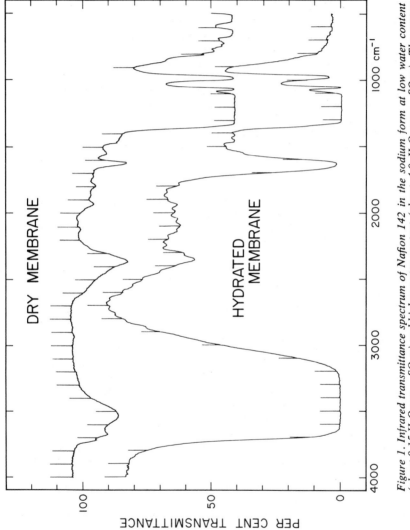

Figure 1. Infrared transmittance spectrum of Nafion 142 in the sodium form at low water content (about 0.15 H_2O per $-SO_3^-$) and higher water content (about 4.0 H_2O per $-SO_3^-$). The upper spectrum has been moved up the transmittance scale by 40%.

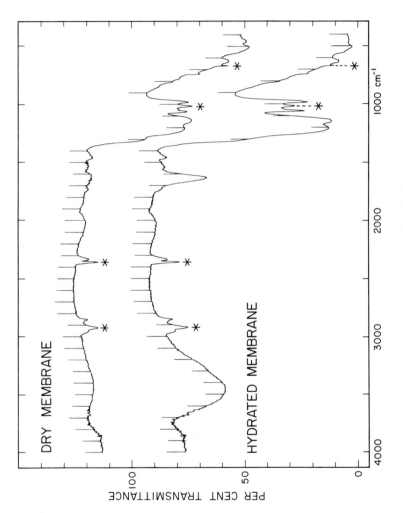

Figure 2. ATR spectra of the same Nafion samples as in Figure 1. The upper spectrum has been moved up the transmittance scale by 40%. Asterisks indicate spurious bands, artifacts of the ATR technique.

TABLE II

Infrared Absorption Bands of Nafion in the Sodium Form [a]

Band Position	Relative Intensity	Origin	Assignment or comments
3715[b]	variable	water	OH stretch ($CF_2 \cdots$H-O-H$\cdots CF_2$)
3670[b]	variable	water	OH stretch (O-H$\cdots CF_2$)
3530[b]	variable	water	OH stretch (O-H$\cdots$O)
3250[b]	variable	water	2x1620 = 3240
2930	medium	$-CF_2-CF_2-$	cf. 2925 in PTFE(9)
2360	very weak	$-CF_2-CF_2-$	cf. 2367 in PTFE (1152 + 1213 = 2365)(9)
2115	weak	$-CF_2-CF_2-$	
1930	weak	$-CF_2-CF_2-$	cf. 1935 in PTFE (553 + 1380 = 1933)(9)
1860	very weak	$-CF_2-CF_2-$	cf. 1859 in PTFE(9)
1790	weak	$-CF_2-CF_2-$	cf. 1792 in PTFE (638 + 1152 = 1790)(9)
1720	very weak	$-CF_2-CF_2-$	
1700	very weak	$-CF_2-CF_2-$	cf. 1703 in PTFE (1152 + 553 = 1705)(9)
1620[b]	variable	water	HOH bend
1545	weak	$-CF_2-CF_2-$	cf. 1545 in PTFE(9)
1470	very weak	$-CF_2-CF_2-$	cf. 1451 in PTFE (203 + 1242 = 1445)(9)
1450	very weak	$-CF_2-CF_2-$	cf. 1420 in PTFE (203 + 1213 = 1416)(9)
1420	very weak	$-CF_2-CF_2-$	
1210[c]	very strong	$-CF_2-CF_2-$ $-SO_3^-$	antisym. CF_2 stretch; cf. 1242 and 1213 in PTFE(9) deg. $-SO_3^-$ stretch; cf. doublet at 1200 in sodium polystyrene sulfonate(10)

cm^{-1} [a]	intensity	group	notes
1140^c	very strong	$-CF_2-CF_2-$ / $-SO_3^-$	sym. CF_2 stretch; cf. 1152 in PTFE(9)
$1060^{b,d}$	strong	$-SO_3^-$	$-SO_3^-$ sym. stretch; cf. 1058 in NH_2SO_3Na(11); 1034 in sodium polystyrene sulfonate(10); 1048 in sodium polyethylene sulfonate(12)
$980^{b,d}$, $960^{b,d}$	e		
960 (cont.)	strong	$-C-O-C-$	COC sym. stretch(2)(?)
850	weak	$-CF_2-CF_2-$	cf. 850 in PTFE(9)
805	weak	$-CF_2-CF_2-$	
775	weak	$-CF_2-CF_2-$	cf. 778 in PTFE(9)
740	weak	$-CF_2-CF_2-$	cf. 738 in PTFE(9)
720	weak	$-CF_2-CF_2-$	cf. 718 in PTFE(9)
630^d	very strong	$-CF_2-CF_2-$	Probably CF_2 scissor(13),f; cf. doublet at 638 and 625 in PTFE(9); 637 in perfluorocyclohexane(13)
530^d	very strong	$-CF_2-CF_2-$	cf. 516 in PTFE(9)^g
465	weak	$-CF_2-CF_2-$	
380	weak	$-CF_2-CF_2-$	cf. 384 in PTFE(14)
345	weak	$-CF_2-CF_2-$	
320	weak	$-CF_2-CF_2-$	cf. 321 in PTFE(14)
295	weak	$-CF_2-CF_2-$	
270	weak	$-CF_2-CF_2-$	cf. 277 in PTFE(14)
235	weak	$-CF_2-CF_2-$	
205	medium	$-CF_2-CF_2-$	cf. 203 in PTFE(14)

a Units = cm^{-1}.
b Band position depends on the counter ion and on the degree of hydration.
c From ATR spectra of ref. 2 (verified in the present study).
d From ATR spectra of ref. 3 (verified in the present study).
e Not always resolved.
f This intense absorption may also contain contributions from C-S stretch (2).
g This intense absorption may also contain contributions from S-O deformation modes (2).

unobscured by the fluorocarbon bands: The $-SO_3^-$ symmetric
stretch at about 1060 cm^{-1}.

Infrared spectra of Nafion, except those of very thoroughly
dried specimens, contain prominent bands due to the stretching
and bending fundamentals of water of hydration, the relative
intensity of these bands increasing with the water content of the
specimen. The OH stretching fundamental occurs in the region of
3750-3200 cm^{-1}, devoid of absorptions by other groups of Nafion,
so that the spectrum here is due only to water molecules. This
region contains information concerning hydrogen bonding of water.
The HOH bending fundamental at about 1620 cm^{-1} also lies clear of
major Nafion absorptions and contains relatively little
structural information about water in Nafion but provides a
convenient measurement of the water content. The H_2O librational
fundamentals absorb in the 800-500 cm^{-1} region too strongly
obscured by Nafion absorptions to be structurally useful. The OD
stretching band of D_2O occurs in the region of 2750-2350 cm^{-1},
and contains similar information to the OH stretching fundmental,
though it suffers from being superposed on an overtone of CF
stretching vibrations centered at 2360 cm^{-1}. On the reasonable
assumption that the band shape of this overtone is independent of
the state of hydration or deuteration of Nafion, the overtone
absorption may be compensated by a matching film of dry,
undeuterated Nafion in the reference beam of the spectrometer (4).

Other Salts of Nafion. Spectra of Nafion with other
counterions differ somewhat from those of the sodium salt. The
most interesting among such differences involve bands due to
water. These are presently under study in the author's
laboratory (15). There are also observable differences in the
Nafion bands at 1060 and 980 cm^{-1}. These will be discussed in
later sections.

Acid Form of Nafion. The infrared spectrum of the acid form
of Nafion is distinct from the spectra of its salts. Figure 3
shows the transmission spectrum of Nafion 142 in the acid form at
three water contents. It has been noted by Lopez et al. (1) and
by Heitner-Wirguin (2) that Nafion membranes in the acid form
absorb almost completely below 3700 cm^{-1}. As Figure 3 shows,
this is true only at high water contents. Dry or nearly dry
membranes yield spectra characteristic of the acid group $-SO_3H$.
A small band at about 930 cm^{-1} corresponds to the stretching
vibration of the S-O bond with the single-bond character in the

$-S\overset{\displaystyle \nearrow O}{\underset{\displaystyle \searrow O-H}{\equiv O}}$ group, in analogy to the band observed by Zundel at

907 cm^{-1} for polystyrenesulfonic acid (10). This band is a good
measure of the undissociated sulfonic acid groups in the system.
The extremely intense and broad band centered at 2750 cm^{-1} is due
to the acid protons involved in very strong hydrogen bonds. Such

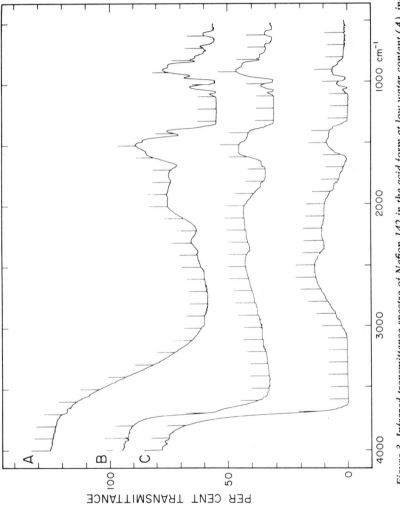

Figure 3. Infrared transmittance spectra of Nafion 142 in the acid form at low water content (A), intermediate (B), and high water content (C). Spectra A and B have been moved up the transmittance scale by 55% and 30%, respectively.

strong bonds are likely to form between $-SO_3H$ groups, with the formation of hydrogen-bonded chains or perhaps dimers of the form $-S\mathbin{\lessgtr}\begin{smallmatrix}O\cdots\cdot H-O\\O-H\cdots\cdot O\end{smallmatrix}\mathbin{\gtrless}S-$. The spectra of hydrated acid membranes show the disappearance of the band at 930 cm^{-1}, which indicates the dissociation of the $-SO_3H$ groups. A complex pattern of intense and broad bands in the OH stretching region is reminiscent of the spectra of aqueous solutions of acids and is characteristic of the strong hydrogen bonding of water molecules to the protonated species H_3O^+ or $H_5O_2^+$. A detailed study of these bands in Nafion has not been made so far.

Information from Absorption Bands of Nafion

The 1060 cm^{-1} Band. This band is due to the symmetric stretching vibration of $-SO_3^-$ groups. Lowry and Mauritz conducted a careful study of the effects of hydration and counterion type on this band using the ATR technique (3). Figure 4 reproduces the spectra of ref. 3 for the fully hydrated and dried samples of Nafion with Li, Na, K, and Rb counterions; Figure 5 shows the spectra of the Li sample at different water contents.

While the spectra of the four hydrated membranes are very similar, the spectra of the corresponding dry samples show significant differences. It is apparent that the degree of hydration, as well as the counterion type, has a significant effect on the 1060-cm^{-1} band. Figure 5 shows that for the Li sample the peak position remains fairly constant until water content falls below about 5 H_2O molecules per ion pair. At this point the peak begins to shift to higher frequencies and broadens substantially. The results of Lowry and Mauritz may be summarized as follows: (i) The magnitude of the shift of the 1060-cm^{-1} peak decreases as the radius of the counterion increases. (ii) The shift begins at lower H_2O/ion ratios for the heavier cations. (iii) No shift occurs in the spectra of the Rb-containing Nafion. These ATR results have been fully confirmed and extended to additional cations by transmission spectra recorded in the author's own laboratory (12). Figure 6 shows our results for the series of alkali-metal cations and the currently available results from both types of measurements are summarized in Table III. The good agreement between the two sets of results is reassuring.

The interpretation of these observations is as follows (3). The changes in the vibrational frequency of the $-SO_3^-$ stretching vibration are due to the polarization of the S-O dipole by the electrostatic field of adjacent counterions. The magnitude of this polarization depends on the radius of the bare cation, being the largest for Li, and practically negligible for Rb and Cs. In the fully hydrated membrane, the peak position is independent of

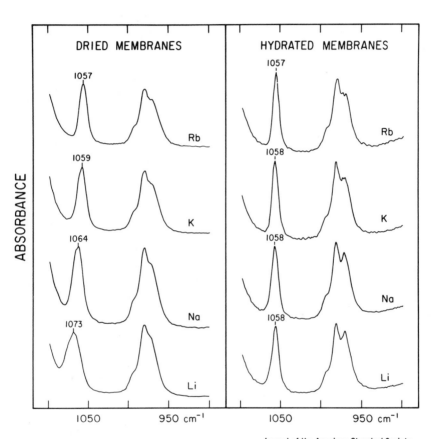

Journal of the American Chemical Society

Figure 4. ATR spectra of Nafion 113 in the range 1100 to 900 cm⁻¹ with different counterions (3).

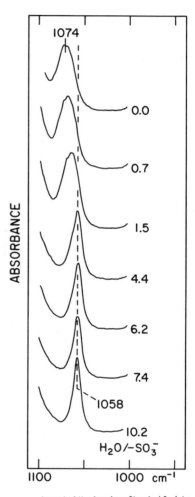

Figure 5. ATR spectra of the Li form of Nafion 113 in the region of 1100 to 1000 cm⁻¹ for different water contents (3).

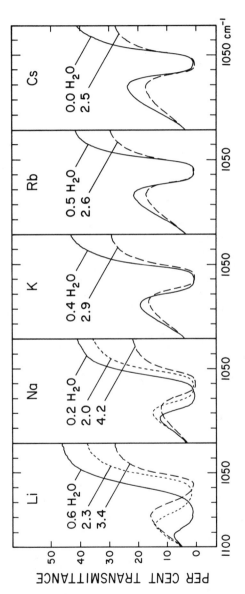

Figure 6. Transmittance spectra of Nafion 142 membranes in the range of 1100 to 1000 cm^{-1} with different counterions and at different water contents, expressed as $H_2O/$—SO_3^- mol ratio.

TABLE III

Shifts of the 1060-cm^{-1} Band due to Hydration

Counterion	Hydrated Membrane		Dry Membrane	
	ATR (ref. 3)	Transmission (This work)	ATR (ref. 3)	Transmission (This work)
Cs	–	1056.0	–	1055.0
Rb	1057	1057.5	1057	1057.0
K	1058	1058.5	1059	1060.0
Na	1058	1059.0	1064	1065.0
Li	1058	1057.5	1073	1071.0

the counterion, as each -SO$_3^-$ group is shielded by several water
molecules from the nearest counterion. As water is being
removed, at some point too few water molecules remain to provide
complete shielding between anions and cations and contact pairs
begin to form. Beyond this point, whose onset depends on the
hydration number and hydration energetics of the cation, a strong
interaction occurs between the sulfonate group and the cation.
For Na and Li counterions the shift of the -SO$_3^-$ stretching
frequency begins when the water content falls below about 5 H$_2$O
per sulfonate group. For K counterions, the water content must
fall below 2 H$_2$O per sulfonate before a barely detectable
frequency shift occurs (1.5 cm^{-1}), while for Rb and Cs no shift
is observed even for completely dried material.

The behavior of the 1060-cm^{-1} band indicates that contact
pairs do occur in Nafion at low degree of hydration and that the
dynamic equilibrium shifts in the direction of greater
dissociation with increasing water content.

The 980-cm^{-1} Band. This band appears to be due to the
symmetric stretching vibration of COC groups in the pendant side-
chains -CF$_2$CF-O-CF$_2$CF- of Nafion (2). In the ATR spectra of
Lowry and Mauritz (3) this band resolves into three components:
main peak near 980 cm^{-1}, a secondary peak near 970 cm^{-1}, and a
shoulder at about 995 cm^{-1}. An inspection of spectra of ref. 3,
reproduced in Figure 4, reveals appreciable changes in the shape
of this band during the hydration–dehydration cycle: the
970-cm^{-1} component becomes more prominent and better resolved in
the spectra of hydrated membranes, this being more noticeable
with Li and Na than with K and Rb. The transmission spectra in
this range suffer from very strong absorption, but nevertheless

in the spectra recorded in the author's laboratory (Figure 7) the occurrence of three components can be inferred. As in the ATR spectra, the low-frequency component gains in relative intensity with respect to the main peak and becomes better resolved at increasing water content. Small shifts of this band, by several wavenumbers, have also been reported when an alkali metal counterion is replaced by a transition-metal ion (2).

The mechanism responsible for these changes is not at present clear, but it does indicate some interaction of the cations with the ether oxygens, particularly for Na, Li, and polyvalent cations. One may conclude that in hydrated Nafion some portions of the side-chains are exposed to the electrostatic field of the cations, and hence that the side-chains penetrate to some extent into the ion clusters.

Information from Absorption Bands of Water

Spectra of water in the sodium salt of Nafion were studied in considerable detail in ref. 4, with special attention to the spectral profiles of the stretching fundamentals of isotopically isolated HDO. Isotopic isolation is achieved experimentally by observing the OD stretching fundamental of a sample with D/H ratio of 1/10 or less, or alternatively the OH stretching fundamental of a sample with H/D ratio of 1/10 or less. These bands are much simpler and thus easier to interpret structurally than the stretching fundamentals of either H_2O or D_2O. The spectra described in ref. 4 have been fully confirmed by further work in the author's laboratory, except that the samples described as "fully hydrated" have now been found to correspond to water content below 3 $H_2O/-SO_3^-$ whereas full hydration for sodium Nafion corresponds to some 14 $H_2O/-SO_3^-$. With this re-interpretation, the results of ref. 4 will be summarized in the next two sections.

HDO Bands at Low Water Contents. Figure 8, reproduced from ref. 4, shows the region of the OH stretching fundamental of Nafion 125 and 142 in the sodium form. The two spectra are very similar, showing that at least at low water contents the physical state of water in Nafion does not change appreciably with the sulfonate content of the polymer.

Unlike the single, broad OH stretching band of HDO in the spectra of liquid water (16) and hydrophilic polymers (17), the spectrum of HDO in Nafion at 1.6 H_2O per sulfonate consists of two bands, marked A and B in Figure 8. The frequencies of these bands, 3660 and 3520 cm^{-1}, are much higher than those of HDO in liquid water. The corresponding OD stretching absorption of HDO in Nafion exhibits an analogous doublet, at 2695 and 2588 cm^{-1} (Figure 9). It can be concluded that OH groups of water in Nafion at low levels of hydration occur in two types of surroundings.

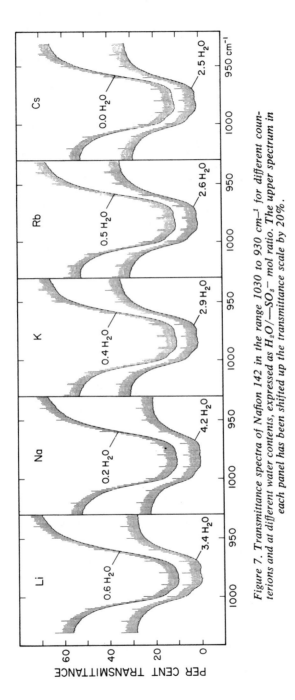

Figure 7. Transmittance spectra of Nafion 142 in the range 1030 to 930 cm^{-1} for different counterions and at different water contents, expressed as $H_2O/{-}SO_3^-$ mol ratio. The upper spectrum in each panel has been shifted up the transmittance scale by 20%.

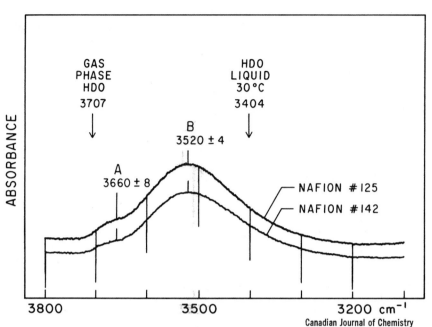

Canadian Journal of Chemistry

Figure 8. The OH stretching region of HDO in the spectrum of deuterated (95% D, 5% H) Nafion 125 and 142 membranes in the sodium form. Water content: approximately 1.6 water molecules per —SO₃⁻ group (4).

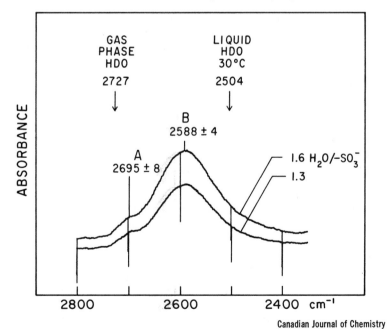

Canadian Journal of Chemistry

Figure 9. The OD stretching region of the spectrum of HDO at low deuteration (10% D, 90% H) in Nafion 125 membranes in the sodium form (4).

The main band, labelled B, is clearly associated with those OH (or OD) groups of water in Nafion which form hydrogen bonds. Two types of hydrogen bonds are possible, water•••water and water•••sulfonate (Figure 10). The occurrence of only one band with no resolved substructure indicates that these two possible types of O–H•••O bonds are of comparable average strength. Similarly, there is no spectroscopic distinction between the stretching frequencies of water molecules which are interacting through their lone electron pairs with the sodium ions (H_2O^1 and H_2O^2 in Figure 10) and those which only interact with other water molecules (H_2O^3 and H_2O^4 in Figure 10). The situation is not unlike that observed in the spectra of HDO in salt solutions (16). The large shift of band B to high frequency with respect to the liquid-phase value (3404 cm^{-1} for OH stretch and 2504 cm^{-1} for OD stretch) shows that the mean strength of hydrogen bonds is substantially lower than in liquid water.

The smaller band, labelled A, is at a frequency sufficiently close to that of the gas-phase HDO (3707 cm^{-1} for OH stretch and 2727 cm^{-1} for OD stretch) that it must be considered due to OH groups not involved in O–H•••O hydrogen bonding. It was shown in ref. 4 that the frequencies are in fact very nearly those that would be expected for water molecules on the periphery of the aqueous medium, with one or both OH or OD groups exposed to the fluorocarbon, but also engaging in some water-water and water-ion interactions.

H_2O and D_2O Bands at Low Water Contents. Figure 11 shows the OH stretching region of the spectrum of H_2O in undeuterated Nafion while Figure 12 shows the OD stretching region of D_2O in highly deuterated Nafion (about 95% D, 5% H). Similarly to Figures 8 and 9, these spectra, reproduced from ref. 4, refer to the sodium salt of Nafion and to water contents of 0.3 to 1.6 $H_2O/-SO_3^-$.

The H_2O spectrum in this region consists of four bands, well resolved at the lowest water contents. For D_2O at low water contents five distinct bands are evident. The assignment of these H_2O and D_2O bands and their relation to the HDO bands A and B have been analyzed as follows (4).

If, as has been concluded from the HDO spectrum, OH groups in Nafion occur in two types of spectroscopically distinct surroundings, A and B, which differ widely as to the strength of hydrogen bonding, then the corresponding H_2O spectrum should consist of a superposition of contributions from two stretching fundamentals, ν_1 and ν_3, of H_2O molecules in each of three distinct types of surroundings: A•••HOH•••A, A•••HOH•••B, and B•••HOH•••B. For brevity, these will be called AA, AB, and BB. In addition, the lower-frequency portion of the OH stretching region of the H_2O spectrum may also contain a contribution from the overtone $2\nu_2$ of the bending vibration, at about $2 \times 1620 = 3240$ cm^{-1}. This overtone may complicate the observed spectrum by

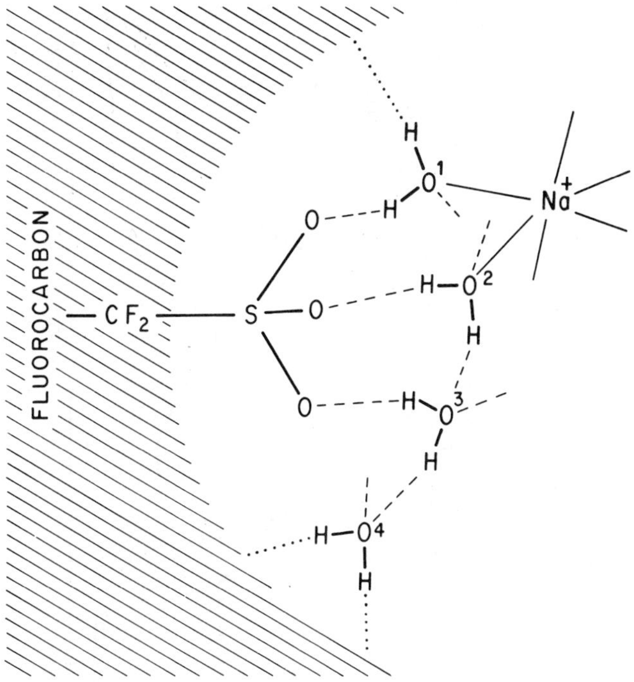

Canadian Journal of Chemistry

Figure 10. Hydrogen-bonded (- - -) and nonhydrogen-bonded (• • •) interactions of water molecules in a portion of hydrated ion cluster in Nafion (4).

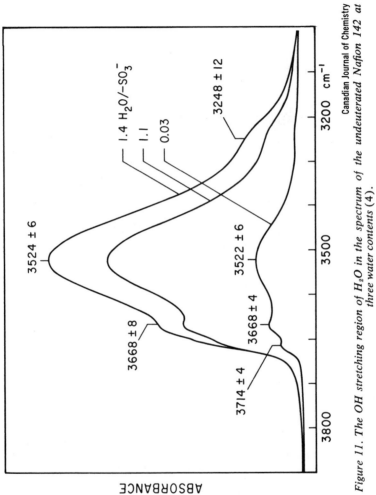

Figure 11. The OH stretching region of H_2O in the spectrum of the undeuterated Nafion 142 at three water contents (4).

Canadian Journal of Chemistry

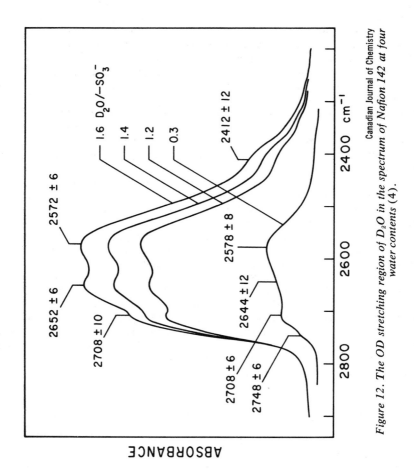

Canadian Journal of Chemistry

Figure 12. The OD stretching region of D_2O in the spectrum of Nafion 142 at four water contents (4).

entering into Fermi resonance with the lower-frequency H_2O
stretching fundamental, v_1, resulting in intensity redistribution
and frequency shifts (18). An analogous situation may be
expected to arise for D_2O.

Using well-known correlations between stretching frequencies
of HDO and those of H_2O and D_2O, derived by Schiffer et al. (19),
the v_1 and v_3 frequencies of H_2O molecules of types AA, AB, and
BB were calculated from the observed A and B frequencies of HDO.
These calculated frequencies, which must be considered
first-order approximations, since they neglect Fermi resonance
with $2v_2$ and coupling of like vibrations on neighbouring
molecules, are shown schematically in Figure 13. The observed
H_2O spectrum in Figure 11 shows a broad band at about 3248 cm^{-1},
which clearly corresponds to the $2v_2$ overtone, allowing for a
small downward shift due to Fermi resonance. The remaining three
observed features, at 3714, 3668, and 3524 cm^{-1}, are then
assigned to the v_1 and v_3 fundamentals of AA, AB, and BB
molecules as shown by the broken lines in the upper part of
Figure 13. Apparently the bands due to $v_3(BB)$, $v_1(AB)$, and
$v_1(BB)$ overlap into a single unresolved band, while $v_1(AA)$ is too
weak to be observed in the midst of other, more intense
absorptions. As may be expected from the HDO spectrum, most of
the intensity arises from the absorptions of molecules of type
BB, with a smaller contribution from AB, and a very small but
observable contribution from AA.

Similarly, the observed D_2O spectrum (Figure 12) shows a
band at about 2412 cm^{-1}, which is due to the $2v_2$ overtone. This
leaves four features which are then assigned to the partly
superposed v_1 and v_3 fundamentals of the three types of D_2O
molecules, as shown by broken lines in the lower part of Figure
13. The reason for one more band being resolved in the D_2O
spectrum than in the H_2O spectrum is the larger separation in D_2O
between the v_1 and v_3 fundamentals in molecules of type BB. As a
result, D_2O and H_2O spectra present a different appearance in the
stretching region. As before, the observation of the D_2O band at
2748 cm^{-1} shows non-negligible occurrence of water molecules of
type AA, at least at lowest hydrations.

HDO Bands at Higher Water Contents. The highest water
contents that have so far been reached with the vapor equilibrium
technique correspond to about 5 $H_2O/-SO_3^-$. Spectra at higher
water contents, up to 14 $H_2O/-SO_3^-$ (which corresponds to membrane
saturation) were only obtained with the sandwiched film technique
and are uncertain because of spectral contribution from the thin
liquid film of squeezed out water. Typical spectra of the OD
stretching fundamental of HDO in the sodium salt of Nafion,
obtained by both techniques, are shown in Figure 14. Increasing
the water content can be seen to result in an increased intensity
of band B and gradual shift of its frequency from 2588 cm^{-1} to
about 2530 cm^{-1} (cf. Figure 15). The frequency of band A remains

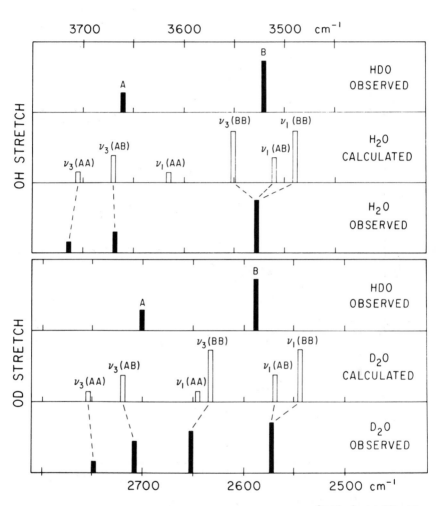

Canadian Journal of Chemistry

Figure 13. Comparison of the observed stretching frequencies of H_2O and D_2O in Nafion with those calculated from the corresponding HDO frequencies (4).

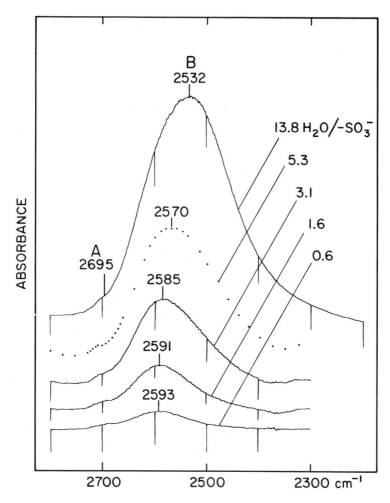

Figure 14. The region of OD stretching fundamental in the spectrum of HDO in Nafion 125 (sodium form, 95% H, 5% D). The top spectrum was obtained by the sandwiched film technique, the others by the vapor equilibrium technique. The point-by-point spectrum was recorded with 3-min time lapses between points to minimize effects of beam heating.

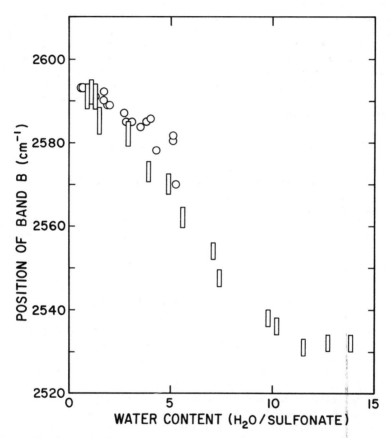

Figure 15. Plot of the frequency of the B component of OD stretching of HDO in Nafion 125 as a function of water content. Key: [], sandwiched film technique; ◯, vapor equilibrium technique.

at 2695 ± 8 cm^{-1} and its intensity relative to that of band B
decreases. These findings must be considered preliminary, but at
least for the range 1 to 5 $H_2O/-SO_3^-$ the same trends have been
observed by both methods, and qualitatively these trends are
expected to continue up to saturation level.

The decrease of the intensity of band A, relative to that of
band B, indicates that at increasing water contents an increasing
fraction of water molecules form O-H•••O hydrogen bonds. The
shift of band B to lower frequencies indicates increasing average
strength of hydrogen bonds with increasing water content. The
constancy of the peak frequency of band A is in line with its
identification with OH groups in O-H•••CF$_2$ environments. The
stretching frequencies of these groups, which do not participate
in hydrogen bonding, are not expected to change much either with
the size of hydrogen-bonded networks or with the average
hydrogen-bond strength.

Fraction of OH Groups Exposed to the Fluorocarbon Environment

In principle, the ratio of the areas of bands A and B of HDO
in Nafion should provide a measure of the relative proportions of
OH groups in O-H•••CF$_2$ and O-H•••O environments, and thus some
insight into the geometry of the aqueous regions in Nafion.
Unfortunately, because of the current experimental uncertainties
discussed in an earlier section, a quantitative estimate does not
yet appear possible, although an order-of-magnitude estimate can
certainly be made at low water contents.

It is difficult to determine the individual areas of bands A
and B accurately because of overlap and uncertain profiles of the
two component bands. The best possible estimate made in ref. 4
was that bands A and B have an area ratio of the order of 1:13.
This of course refers to Nafion containing under 2 H_2O per $-SO_3^-$.
In order to calculate the actual ratio of OH groups in the two
environments, one has to allow for the well-known increase of the
infrared absorptivity of OH (or OD) oscillators with hydrogen
bonding, which has been mentioned above. Using the approximate
linear relation between OD frequency and infrared absorptivity
from ref. 20, one estimates that OD groups giving rise to band B
at 2580 cm^{-1} (3520 cm^{-1} for OH) absorb about four times more
intensely than those which give rise to band A at 2695 cm^{-1}
(3660 cm^{-1} for OH). This means that the ratio of OH groups in
O-H•••CF$_2$ environments to those in O-H•••O environments is of the
order of 4:13, hence a fraction 4/(13+4)=0.24 of all OH groups do
not participate in hydrogen bonding at these low water contents.

At higher water contents the relative areas of bands A and B
are increasingly difficult to estimate from the spectrum because
of the uncertainties already mentioned. It appears that the total
numbers of OH groups exposed to the fluorocarbon increase slowly
if at all with increasing water content, while the fraction of
the OH groups which are so exposed diminishes rapidly.

Geometry of the Hydrated Ion Clusters in Nafion

Studies of low–angle X–ray scattering and of other physical properties of Nafion by Yeo and Eisenberg (21), Gierke (22), and others (23–25), suggest that hydrated Nafion is a two–phase system, consisting of hydrated clusters of ions embedded in the surrounding fluorocarbon medium. From observed Bragg spacings, and assuming that the clusters are spherical in shape, Gierke et al. estimated that in fully hydrated Nafion the clusters are about 4.0 nm in diameter, with the diameter decreasing to 1.9 nm at very low water contents (26).

Some structural deductions about the clusters may be made a priori. In the hydrated sodium form of Nafion at 14 H_2O per $-SO_3-Na^+$ there is sufficient water to complete the coordination of the sodium cation (which requires six H_2O) and of the polymer–bound sulfonate anion (which also requires about six H_2O), especially as water molecules may be shared between the cation and the anion (cf. Figure 10). The ion pairs $-SO_3^-Na^+$ can therefore be expected to be completely surrounded by water, with the individual ions separated most of the time. This is indeed confirmed by the infrared data on the $1060-cm^{-1}$ band, which also show that a minimum of about 5 H_2O molecules are needed to complete the process of ion hydration. The lone pairs of electrons on the water molecules within the clusters may interact either with a sodium ion or with an OH group of another water molecule. Similarly, the OH groups may interact either with an oxygen atom of an anion or with the oxygen of another water molecule (Figure 10). Because of electrostatic repulsion, the periphery of the cluster cannot be covered completely with $-SO_3^-$ ions, so that some water molecules must find themselves in direct contact with the fluorocarbon phase. The occurrence of band A in the stretching region of the spectrum of HDO and of bands AA and AB in the corresponding spectrum of H_2O and D_2O gives direct evidence for the occurrence of such $O-H\cdots CF_2$ contacts. Some $O-H\cdots CF_2$ contacts may also arise through the penetration of parts of the pendant side–chains into the hydrated cluster. The observed changes of band shape of the $980-cm^{-1}$ band with water content may in fact be indicative of such penetrations.

At low water contents (below 2 H_2O per $-SO_3^-$) roughly one quarter of all OH groups of water appear to be exposed to the fluorocarbon. This corresponds to small clusters with an appreciable fraction of water molecules of type 1 and 4 in Figure 10. Simple model building leads to a cluster size of the order of 1.2 nm, assuming a spherical shape and neglecting the volumes occupied by the ions (4). This is not inconsistent with Gierke's estimate of 1.9 nm for very low water contents (26).

At higher water contents, as the clusters grow larger, the fraction of fully hydrogen–bonded water molecules (type 2 and 3 in Figure 10) increases. The fraction of OH groups in $O-H\cdots CF_2$ environments could not be calculated but clearly it decreases

gradually with increasing water content. (This was not realized
in ref. 4, leading to an incorrect conclusion). The spectra can,
in fact, be reconciled with the cluster estimates of Gierke (26),
the increase of band B with respect to A being clearly the result
of an extension of the size of the hydrogen-bonded networks and
the diminution of the relative importance of the non-bonded OH
groups at the surface of the cluster.

It must be stressed that even in fully hydrated Nafion, the
average hydrogen-bond strength does not approach that in liquid
water. This is indicated by the finding that the center of band
B of HDO is at 2530 cm^{-1}, which is still at least 26 cm^{-1} above
its value for liquid water. (2530 cm^{-1} is the lower limit; the
effect of the contribution of the liquid film squeezed out from
the membrane would be to lower the observed value toward that of
liquid water). Weaker hydrogen bonding is related to
incompleteness of the hydrogen-bonded networks, deduced from the
occurrence of band A. Cooperativity of hydrogen bonding between
water molecules is well known (27). Single water molecules are
relatively weak hydrogen-bond donors and acceptors and the
strength of hydrogen bonding diminishes rapidly with decreasing
size of water clusters (28). Even for water clusters as large as
4.0 nm in diameter, the exposure of OH groups on the cluster
surface to the fluorocarbon would be sufficient to diminish the
average strength of the remaining hydrogen bonds throughout the
cluster. In this connection, it should be remembered that the
mean minimum distance of a point inside a sphere from the surface
is only 0.125 of the sphere's diameter. Hence, molecules in a
spherical cluster of 4.0 nm diameter are on the average only
0.5 nm distant from the fluorocarbon phase; this represents less
than two hydrogen-bond lengths.

Far Infrared Spectra

After the present chapter had been completed, the author
learned of the recent and as yet unpublished study of the far
infrared spectra of Nafion by S. L. Peluso, A. T. Tsatsas, and
W. M. Risen (29). These workers examined the spectral region of
300 to 50 cm^{-1} for Nafion membranes with group Ia (except Li^+)
and IIa counterions. They discovered a broad, well-defined band
in all the spectra, though in the case of Na^+ and Ca^{+2} partly
masked by the 205-cm^{-1} band of the polymer main chain. The
band centers were at approximately 150, 111, and 92 cm^{-1} for K^+,
Rb^+, and Cs^+, respectively, and at approximately 240, 155, and
125 cm^{-1} for Mg^{+2}, Sr^{+2}, and Ba^{+2}, respectively. Within each
cation group the frequency was found to vary nearly linearly with
the inverse of cation mass, showing that the vibration is due
largely to cation motion. The frequencies of group IIa
counterions were found to be higher than those of the group Ia
counterions of similar mass, showing that the cation-site forces
are essentially ionic in nature. It is likely that a detailed

report of the results of Peluso et al. and further far-infrared
studies done as a function of the water content will add greatly
to our knowledge of the internal structure of Nafion.

Literature Cited

1. Lopez, M.; Kipling, B.; Yeager, H. L. Anal. Chem. 1976,
    48, 1120.
2. Heitner-Wirguin, C. Polymer 1979, 20, 371.
3. Lowry, S. R.; Mauritz, K. A. J. Am. Chem. Soc. 1980, 102,
    4665.
4. Falk, M. Canad. J. Chem. 1980, 58, 1495.
5. Harrick, N. J. "Internal Reflection Spectroscopy", Wiley-
    Interscience, New York, 1967.
6. Motojima, T.; Ikawa, S.-I.; Kimura, M. J. Quant.
    Spectrosc. Radiat. Transfer 1981, 25, 29.
7. Wilemski, G. J. Quant. Spectrosc. Radiat. Transfer 1978,
    20, 291.
8. Swenson, C. A. Spectrochim. Acta 1968, 24A, 721.
9. Moynihan, R. E. J. Am. Chem. Soc. 1959, 81, 1045.
10. Zundel, G. "Hydradation and Intermolecular Interaction";
    Academic Press: New York, 1969; p. 11.
11. Katiyar, R. S.; Krishnan, R. S. Indian J. Pure Appl.
    Phys. 1968, 6, 686.
12. Falk, M. Unpublished results.
13. Miller, F. A.; Harney, B. M. Spectrochim. Acta, 1972,
    28A, 1059.
14. Liang, C. Y.; Krimm, S. J. Chem. Phys. 1956, 25, 563.
15. Quezado, S.; Kwak, J. C. T.; Falk, M. Work in progress.
16. Wyss, H. R.; Falk, M. Can. J. Chem. 1970, 48, 607.
17. Falk, M.; Poole, A. G.; Goymour, C. G. Can. J. Chem.
    1970, 48, 1536.
18. Scherer, J. R. In Advances in Infrared and Raman
    Spectroscopy. Edited by R. J. H. Clark and R. E. Hester.
    Vol. 5. Heyden and Son, London, 1978. Chapt. 3.
19. Schiffer, J.; Intenzo, M.; Hayward, P.; Calabrese, C.
    J. Chem. Phys. 1976, 64, 3014.
20. Glew, D. N.; Rath, N. S. Can. J. Chem. 1971, 49, 837.
21. Yeo, S. C.; Eisenberg, A. J. Appl. Polymer Sci. 1977, 21,
    875.
22. Gierke, T. D. Paper presented at the 152nd National
    Meeting of the Electrochemical Society, Atlanta, GA,
    October 1977.
23. Hopfinger, A. J.; Mauritz, K. A.; Hora, C. J. Paper
    presented at the 152nd National Meeting of the
    Electrochemical Society, Atlanta, GA, October 1977.
24. Roche, E. J.; Pineri, M.; Duplessix, R.; Levelut, A. M.;
    J. Polym. Sci. Polym. Phys. Ed. 1981, 19, 1.
25. Eisenberg, A. Pure Appl. Chem. 1976, 46, 171.

26. Gierke, T. D.; Munn, G. E.; Wilson, F. C. J. Polym. Sci.
    Polym. Phys. Ed. (in print).
27. Barnes, P.; Finney, J. L.; Nicholas, J. D.; Quinn, J. E.
    Nature 1980, 282, 459.
28. Hankins, D.; Moskowitz, J. W.; Stillinger, F. H. J. Chem.
    Phys. 1970, 53, 4544.
29. Peluso, S. L.; Tsatsas, A. T.; Risen, W. M. personal
    communication. This work is based on the Ph.D. Thesis of
    S. L. Peluso, Brown University, 1980, and is to be published
    shortly.

RECEIVED August 21, 1981.

Mössbauer Spectroscopy of Perfluorosulfonated Polymer Membranes

Structure of the Ionic Phase

B. RODMACQ, J.M.D. COEY,[1] and M. PINERI

Equipe Physico-Chimie Moléculaire, Section de Physique du Solide, Département de Recherche Fondamentale, Centre d'Etudes Nucléaires de Grenoble, 85 X, 38041 Grenoble Cédex, France

Spectroscopic techniques are valuable in studies of the structure of polymers because they give information about the environment of the probe atom on a microscopic scale. Mössbauer spectroscopy is particularly useful because the absorption spectrum is entirely due to one isotope of a single chemical element, most commonly ^{57}Fe. By introducing iron into polymers containing acid groups, it is possible therefore to examine the ionic phase specifically. Its structure can be defined to some extent, and interactions of the cations with their surroundings can be determined.

There have been Mössbauer studies of Nafion membranes by several groups (1-6), besides some work on other ionomers (7,8) and a body of results on polymers where iron or tin is introduced into the polymeric matrix as a probe impurity (8). This chapter will concentrate on the information that has been extracted from the Mössbauer studies of Nafion membranes, and their variations as a function of parameters such as temperature, applied magnetic field, water content and chemical treatment. The results will be discussed in terms of the information they provide about the structure of the ionic phase in perfluorosulfonate materials.

Interpretation of Mössbauer spectra

Mössbauer spectroscopy is now a well-established if somewhat specialized spectroscopic method, explained in numerous text books (9). However, it may be unfamiliar to many workers in the field of polymer science, so we will begin by briefly summarizing the essential features of the method, and some aspects of the interpretation of the spectra. More details are given in a recent paper (5).

[1] Current address: Department of Pure and Applied Physics, Trinity College, Dublin 2, Ireland.

0097-6156/82/0180-0171$06.00/0

The usual experimental method is absorption spectroscopy. It differs from ordinary optical spectroscopy in that the energies of the γ-photons involved are some 10^4 times greater than optical photon energies, and the order of magnitude of the energy-level splittings measured is some 10^8 times smaller. A schematic comparison of Mössbauer and IR spectroscopy is given in Table I.

Table I
Comparison of Infra-red and Mössbauer spectroscopy

	Infra-red	Mössbauer
Energy levels	Vibrational or Electronic	Nuclear
Photon energy	10^{-2} - 1 eV (i.r. light)	10^4 - 10^5 eV (γ-rays)
Source	Broad band, ~ 1 eV	Narrow line, with $\sim 10^{-8}$ eV
Energy selection	Monochromator	Doppler shift
Interactions measured	Interatomic (vibrational) Intraatomic (electronic)	Hyperfine

The radioactive Mössbauer source is monochromatic to a remarkably high degree, better than one part in 10^{12}, but the energy range swept in the experiment is restricted to a very narrow band centred on the γ-energy, E_γ. The sample to be studied is used as an absorber, and it must contain nuclei of the same stable isotope which is emitting in the source. Resonant absorption occurs whenever the difference in energy between the ground and excited states of the nuclei in source and absorber precisely coincide. Each state is split by hyperfine interaction of the nuclear electric and magnetic moments with the electric and magnetic fields created at the nucleus by its surrounding electrons and more distant atoms. The energy of γ-rays emitted from the source is slightly modulated, and the spectrum scanned by varying the Doppler shift obtained by moving the source with a velocity v, of order 10 mm/sec. The small resultant energy shift ΔE is given by the formula $\Delta E/E_\gamma = v/c$, where c is the velocity of light. Structure observed in the Mössbauer spectrum therefore depends on the nature and surroundings of the ion containing the resonant nucleus, and, for this reason, Mössbauer spectroscopy provides information about the structural environment of an ion on a microscopic scale.

Although more than forty elements possess isotopes which have been shown to exhibit the Mössbauer effect, very few of them are ideally suited for Mössbauer spectroscopy. Most work has been done on ^{57}Fe, but ^{151}Eu and ^{119m}Sn have also been used in polymer studies (5,8). We are exclusively concerned with ^{57}Fe here. The source for the resonance is radioactive ^{57}Co in a metallic matrix, which populates the excited state of ^{57}Fe. Absorbers are Nafion membranes exchanged with iron from solutions made from salts of natural iron, containing 2 % of the resonant ^{57}Fe isotope. Alternatively, salts enriched in ^{57}Fe are sometimes needed to improve the signal/noise ratio in the spectrum.

The main information that can be extracted from the absorption spectra is the following :

- Dynamic properties of the lattice in which the resonant atom is bound. The area of the absorption spectrum is governed by the probability of a nucleus absorbing a γ-photon emitted by the source without recoil. The Mössbauer effect is actually the quantum-mechanical result that such a resonant absorption process should have a non-vanishing probability of taking place, given by the recoilless fraction

$$f = e^{-E_\gamma^2 \langle x^2 \rangle / \hbar^2 c^2}$$

where $\langle x^2 \rangle$ is the mean square displacement of the absorbing nucleus, and $\hbar$ is Planck's constant. Observation of the effect is only possible if $\langle x^2 \rangle$ is small, effectively only if the source and absorber nuclei are each bound in a solid matrix, crystalline or amorphous. The absorption area falls off with increasing temperature because of the increasing amplitude of thermal vibrations. Atomic vibrations are often parametrized in terms of a characteristic temperature Θ_D obtained by fitting $f(T)$ to the predictions of the Debye model. Absorption disappears entirely when $\langle x^2 \rangle$ diverges at the melting point, or near the glass transition T_g of a non-crystalline phase. It has been found in many polymers that T_g is closely correlated with the temperature where ln $f(T)$ begins to fall away abruptly (10,11,12).

- Characterization of the electronic configuration of the resonant ions. The oxidation and spin state of the iron ion determine the isomer shift (IS) which depends on the charge density at the nucleus. It is the shift of the centre of gravity of the spectrum, generally quoted with respect to a standard absorber of iron metal. Typical values are 0.2 to 0.5 mm/sec for high spin Fe^{3+}, 0.8 to 1.5 mm/sec for high spin Fe^{2+} and -0.2 to 0.3 mm/sec for low spin Fe^{2+} (9).

- Local site symmetry of the resonant ion. An electric field gradient is produced at the nucleus by electrons of the ion itself and by charges on neighbouring atoms, when the site symmetry is lower than cubic. A common type of iron spectrum is a quadrupole doublet, arising from interaction of the nuclear quadrupole moment with the electric field gradient. The separation of the two peaks is the quadrupole splitting (QS). For instance, QS is 1.74 mm/sec for Fe_2Cl_6 dimers, 0.86 mm/sec for longer chains, and almost zero for crystalline $FeCl_3$ (13). Figure 1 gives an example of the information that can be obtained from quadrupole doublets in

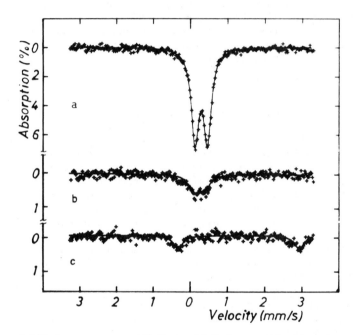

Figure 1. Mössbauer spectra at 100 K of solid (Fe(bipy)₃) (Cl₂) (a), Nafion treated in a 10⁻² M solution of (Fe(bipy)₃) (Cl₂) buffered at pH 7 (b), and the latter in an unbuffered solution (c) (28).

Nafions. Spectra in a) and b) are identical, showing that the low spin $(Fe(bipy)_3)^{2+}$ ion is incorporated into the membrane from a buffered solution, but spectrum (c) is completely different. Its parameters IS = 1.40 (5), QS = 3.25 (10), are typical of the high spin $[Fe(H_2O)_6]^{2+}$ ion. The violet colour of the soaked membrane rapidly bleaches unless the solution or membrane is buffered as the $[Fe(bipy)_3]^{2+}$ is converted by acid to Fe^{2+} and free 2,2'-bipyridyl. The change in coordination of the ferrous ion is obvious from the spectra despite poor statistics caused by very low uptake of the comparatively large $(Fe(bipy)_3)^{2+}$ ions.

-    Magnetic interactions between ions. A wealth of structural information can be derived from the effects of magnetic interactions on Fe^{3+} spectra, particularly when they are recorded at low temperatures with the option of applying an external magnetic field. We consider the evolution of the Mössbauer spectrum of ferric ions as a function of their environments. When distances between ions are large, as in dilute frozen aqueous solutions for example, spin-spin interactions are weak and electronic relaxation times are long for ferric ions. Well defined paramagnetic hyperfine structure can be observed, and the spectrum is the superposition of contributions from the three Kramers doublets $|\pm5/2>$, $|\pm3/2>$, $|\pm1/2>$ for the Fe^{3+} ion (S = 5/2). The spectrum superficially resembles a six-line magnetic pattern with hyperfine field $H_{hf} \cong 580$ kOe, but its shape is extremely sensitive to applied external magnetic fields, even when they are quite weak ($\cong 100$ Oe) (14). When the ions get closer, at distances of about 10 to 15 Å, the relaxation time decreases because of spin-spin interactions and the magnetic hyperfine spectrum progressively disappears, transforming into a central peak or doublet with a small quadrupole splitting (15). Continuing to still smaller distances, there are two possibilities when the ferric ions are on nearest-neighbour sites, at distances of about 3 Å. In the first case, a strong interaction between an isolated pair of ions leads to the formation of dimers and usually gives a doublet with a large quadrupole splitting, because of the anisotropy of the environment (16). In the second case, there is formation of small iron-concentrated clusters which may order magnetically even when they are amorphous (17), leading to superparamagnetism above a certain blocking temperature. Below the blocking temperature a magnetic sextuplet appears with a hyperfine field of about 450 kOe. Above the blocking temperature, the spectrum is a normal quadrupole doublet. The temperature at which blocking occurs allows an estimate to be made of the cluster size.

Thus one can see that two very different cases (well-isolated atoms and concentrated clusters) can apparently give rise to similar Mössbauer patterns. Nevertheless, it is easy to differentiate these two cases by applying an external magnetic field (7, 11,14,18) or by a study as a function of the iron concentration. For isolated ions, increasing the iron concentration should lead to the disappearance of the paramagnetic hyperfine structure as the interionic distances decrease. On the contrary, in the case

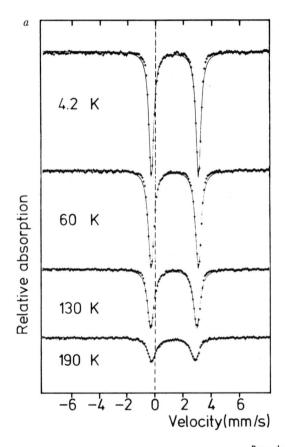

Revue de Physical Appliquée
Figure 2. Mössbauer spectra of a Nafion sample 5% neutralized with Fe²⁺ as a

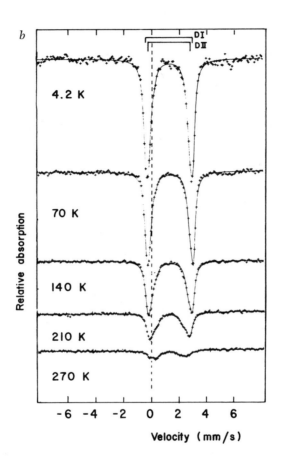

function of temperature (2): a, 28% H₂O by weight; b, 3.6% H₂O by weight.

of iron clusters, increasing the iron concentration should increase the cluster size, increase the blocking temperature and thereby increase the intensity of the magnetic hyperfine component at a given temperature.

With these notions of Mössbauer spectroscopy, we can now go on to consider results obtained on iron Nafions. 125-300 μm films of DuPont Nafion with an equivalent weight of 1100 or 1200 have generally been used. Full details of the experimental procedures may be found in the original articles (2-5).

Fe^{2+} Nafion salts

Influence of water content. Figure 2 shows two series of Mössbauer spectra for a sample neutralized to about 5 % with Fe^{2+} by soaking in an acid solution of $^{57}FeCl_2$. The sample was boiled in water for the spectra in figure 2a) and dried in vacuum at 20°C for the spectra in figure 2b). Table II gives the correlation between weight gain and sample treatment.

Table II
Water content of a Nafion sample 5 %
neutralized with Fe^{2+} after various treatments

Treatment	% H_2O by weight
Boiled H_2O	28
soaked H_2O	17.5
96 % relative humidity	7.8
68 % relative humidity	6.6
dried in vacuum at 20° C	3.6
dried in vacuum at 170° C	0.8

Data on the 28 % water sample were fitted by computer to a single quadrupole doublet whose parameters are plotted as a function of temperature in figure 3. Values of isomer shift and quadrupole splitting of the doublet are quite close to those found for frozen aqueous solutions of $FeCl_2$ and $FeSO_4$ (19). They are typical of the fully hydrated $[Fe(H_2O)_6]^{2+}$ ion. The temperature where the f-factor falls to zero is around 220 K for the water-saturated membrane, but 300 K for the dried one. Asymmetry is noticeable in the spectra of figure 2b) particularly above 200 K. Data on the dried sample at each temperature were fitted to two quadrupole doublets identified with ferrous iron in sites with different degrees of hydration (2).

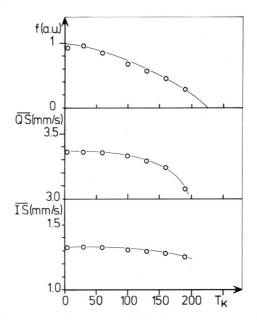

Figure 3. Temperature dependence of Mössbauer parameters of a ferrous Nafion deduced from a one-site fit of spectra in Figure 2a.

Continuing this analysis further, we show in figure 4 the
evolution of the Mössbauer parameters at 4.2 K as a function of
water content. Spectra were all fitted to a single quadrupole
doublet. There is a rapid variation of the isomer shift and quadrupole splitting when the water content falls below 6 %, and a
plateau above this value. There is also a marked increase in
linewidth below this hydration level.

It should be noted that 6 % H_2O by weight corresponds to 7-8
water molecules per ferrous ion, assuming that the absorbed water
is uniformly distributed among the acid and neutralized groups.
This figure is close to the value of 6 water molecules needed to
form the $[Fe(H_2O)_6]^{2+}$ complex. Moreover, eight is likely to be
an overestimate according to data on water sorption in Nafion by
Takamatsu et al. (20) who found that neutralized groups tend to
absorb less water than acid groups.

The variations observed in the spectra with changing water
content (figure 4) imply a modification of the environment of the
ferrous ion below 6 % water. $[Fe(H_2O)_6]^{2+}$ complexes are progressively dehydrated, but it is difficult to establish a precise,
quantitative correlation between the hyperfine parameters and the
number of water molecules in the vicinity of the ion. The influence of the structure of the Nafion itself becomes increasingly felt as the water concentration decreases, and certainly leads
to a very broad distribution of environments for the ferrous ion,
as may be seen from the variation in linewidth as the water content approaches zero.

Influence of temperature. A study of the temperature-varia-
tion of the Mössbauer spectra of samples with different water
contents has been carried out (5). Results are presented in
figure 5, where lnf is plotted as a function of T for water concentrations ranging from 3.6 % to 17.5 %. The temperature T_o where
the f factor falls to zero is plotted as a function of water
content in the insert. The shape of the curve is similar to that
found for all the hyperfine parameters (figure 4).

Below 160 K, the curves in figure 5 corresponding to samples
of different water content superpose fairly well on the lnf(T)
curve calculated from the Debye model with Θ_D = 140 K. Above
160 K, data deviate from the theoretical line, and deviations are
greater as the water content of the sample increases. Addition
of water to a dehydrated sample therefore tends to decrease
the rigidity of the lattice in which the iron ions are bound at
temperature above 160 K. An anomaly, typical of a glass transition has been found in the DTA curves of hydrated Nafion membranes
in the same temperature range (21). We indicated in the introduction that T_g in polymers is generally correlated with the temperature where a break occurs in the lnf(T) curve, and the same is
true of frozen solutions (10,11,12,22), but the glass transition
in Nafion membranes cannot be that of PTFE, which occurs around

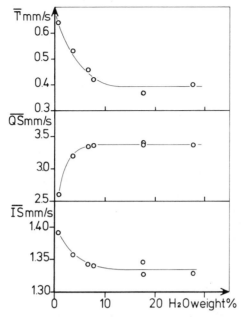

Figure 4. Low temperature (4.2 K) parameters of a ferrous Nafion as a function of water content (one-site fit).

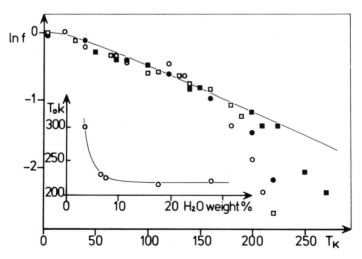

Figure 5. ln f vs. T for a ferrous Nafion as a function of water content. Key: ■, 3.6% H_2O; ●, 6.6% H_2O; □, 7.8% H_2O; ○, 17.5% H_2O. The solid line corresponds to the theoretical ln f curve with $\theta_D = 140$ K. Insert shows the temperature, T_0, at which the f-factor falls to zero.

400 K. It must be a glass transition associated with the ionic
phase.

The most remarkable feature of the data is that the f factor
falls to zero so far below 273 K in fully hydrated samples. This
must rule out any simple picture that places the ions in a normal
aqueous phase. A reduction of 50 K in the freezing point of
water cannot reasonably be provoked by salt in solution nor by
the presence of water in micropores.

Influence of iron concentration. Spectra have also been
recorded for a sample exchanged with Mohr's salt, neutralized to
about 60 % and containing 17 % water. Hyperfine parameters are
identical to those obtained for the less neutralized samples exchanged with ferrous chloride. There is no evidence in spectra
at 4.2 K for magnetic hyperfine structure at this neutralization
level and the lack of any significant magnetic interactions is
also evident from the susceptibility which follows a simple Curie
law down to 4.2 K. It follows that the ferrous ions are physically isolated from each other over a wide range of iron
concentration.

Fe^{3+} Nafion salts

Influence of chemical parameters. Ferric iron appears in a
wider variety of structural forms in Nafion membranes than ferrous
iron. To illustrate the point, we show in figure 6 a representative series of spectra taken at 100 K for Nafion membranes exchanged and treated in various ways. a)-d) are spectra of Nafion 115
soaked in 0.2 M aqueous solutions of different ferric salts. All
are composed of the same two doublets, A with IS = 0.25(1),
QS = 0.44(3) and B with IS = 0.34(1), QS = 1.66(2). Only the relative proportion changes somewhat, from 38 % of doublet B for
the sulphate to 55 % for the nitrate. Quenching in liquid nitrogen increases the proportion of doublet B.

The distinctive quadrupole splitting of doublet B, unusually
large for Fe^{3+}, indicates a highly asymmetric cation environment.
Knudsen et al. (23) have found an identical spectrum in frozen
aqueous solutions of $Fe(ClO_4)_3$, which they attribute to oxygen-
bridged decaqua dimers $[(H_2O)_5Fe-O-Fe(H_2O)_5]^{4+}$. Decisive evidence
that doublet B is due to a dimer with strong antiferromagnetic
coupling is presented in the next section. Figure 6 also shows
the effects of dehydration (g) and concentration of the solution
used for cation exchange (f). Dehydration greatly increases the
proportion of dimers. Figure 7 illustrates the concentration
dependence in more detail for $FeCl_3$ solutions. The relative proportion of dimers, and the total absorption area are plotted as
a function of molarity. Samples exchanged in dilute solutions were equilibrated by repeated immersion in fresh solution
because it was found that the form of the spectrum changes after
each of the first few soakings. These data show that the

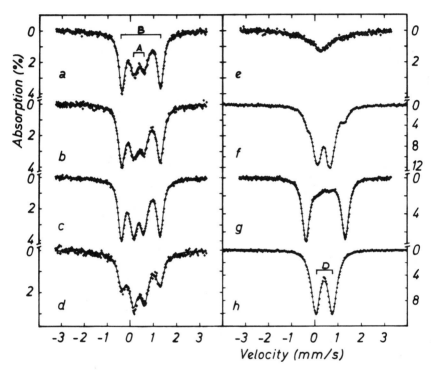

Figure 6. Mössbauer spectra at 100 K of a series of ferric Nafions. Nafions a–e were exchanged in 0.2 M aqueous solutions of ferric nitrate (a), chloride (b), perchlorate (c), sulfate (d), and acidified ferric nitrate (e). For Nafion f the membrane was repeatedly soaked in 0.02 M FeCl₃. Nafion g and h are sample b either dried in vacuum at 20° C (g) or re-exchanged with KCl (h) (6).

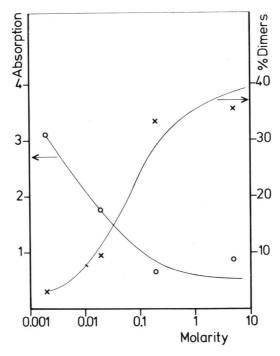

Figure 7. Variation of the total absorption area A (○) and the fraction of dimers (×) present in the Mössbauer spectra of samples equilibrated with FeCl₃ solutions of different concentrations (6).

proportion of dimers is less, the weaker the solution. Uptake of iron is greatest from the weak solutions. The small uptake from very concentrated solutions, an unexpected result at first sight, is probably related to their acidity. Exchanged iron may be completely removed by rinsing the membrane in concentrated HCl. Samples dilute in iron ($\sim$ 2Fe/100 SO_3^-) could be obtained by soaking the Nafion in a solution of ^{57}Fe metal in concentrated HCl.

It should be remarked that samples highly neutralized by use of aqueous $FeCl_3$ salts which have been analysed chemically and by neutron activation, sometimes contain an excess of iron beyond what is needed for complete neutralization of the sulfonate groups. This may be due to the presence of Cl^- ions in the membranes. However, the spectra of figure 6a)-d), which have the same basic structure regardless of the salt used, indicate that anions from the aqueous salt solution are not directly involved in the coordination sphere of the ferric cations.

Also shown in figure 6e) is the spectrum obtained when the Nafion was soaked in a colourless solution of acidified ferric nitrate. The solution contains hexaqua $[Fe(H_2O)_6]^{3+}$ ions, and these ions appear in the Nafion giving a characteristic broad paramagnetic relaxation spectrum at 100 K (24). Finally the spectrum in figure 6h) is typical of a membrane exchanged first with $FeCl_3$, then with a solution of an alkali or alkaline earth chloride. Iron, once introduced, cannot be displaced from the Nafion by subsequent treatment with another salt, but its structural form is completely changed. The dimers of doublet B have disappeared, and a new doublet D with IS = 0.37(2), QS = 0.70(5) mm/s appears in its place.

Bauminger et al. (3,4) find a spectrum at 4 K composed of doublets A and B for Nafion 110 equilibrated with a methanol solution of $FeCl_3$. Parameters were the same as for the aqueous exchanges, but the intensity ratio was somewhat different. These authors have also examined Redcat poly(ethylene sulphonic acid) and sulphonated polysulphone membranes where they again find two doublets, but A has QS = 0.60(3) mm/s.

Figure 8 shows the variation of the spectrum at very low temperatures. Below 4 K, doublet A progressively disappears to the profit of a magnetic pattern C which has a hyperfine field H_{hf} = 560(20) kOe. Doublet B remains unchanged, as expected for a dimeric species. From the relative variations of subspectra A and C, the authors concluded that iron was present in clusters of various sizes. Clusters larger than a certain critical size which depends on temperature give a magnetic spectrum because of superparamagnetic blocking. Using the superparamagnetic model which was previously applied to styrene-butadiene-vinylpyridine terpolymers (7) they estimated the distribution of cluster diameters for various membranes and found good agreement with those evaluated by Gierke on the basis of small angle x-ray scattering experiments (25).

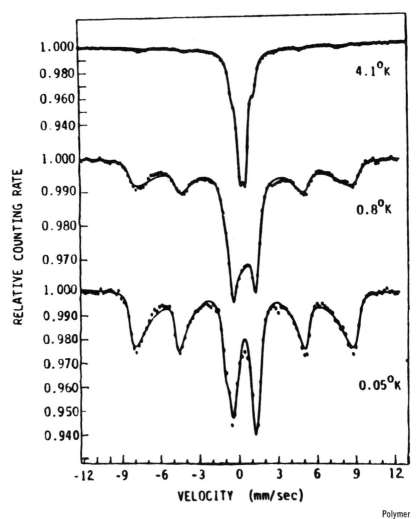

Figure 8. Mössbauer spectra of iron in Nafion equilibrated in methanol solution (4).

This interpretation appears to be at variance with the results on Fe^{2+} Nafions, summarized in the previous section, which
give no evidence for magnetic clustering. However, we have seen
in the introduction that a magnetic hyperfine spectrum at low
temperatures can also arise from isolated paramagnetic ferric
ions. To arrive at a correct interpretation of the magnetic
hyperfine pattern appearing at low temperatures in the Nafions,
an external magnetic field must be applied. We therefore made a
systematic study of ferric Nafion as a function of iron concentration, using an external magnetic field when necessary ($\underline{5}$).

Influence of concentration. The structural form of the iron,
besides depending on chemical treatments, is rather different in
slightly and greatly neutralized membranes. We discuss two extremes before considering samples with intermediate iron concentration and the effect of changing the water content of the Nafion
membranes ($\underline{5}$).

We first consider a sample dilute in iron ($2Fe^{3+}/100SO_3^-$) made
with Fe^{57}. As with the Fe^{2+} Nafion salts, no absorption was apparent at room temperature and it was necessary to cool the sample
below about 200 K in order to obtain a reasonable recoilless
fraction. The spectrum obtained at 4.2 K is shown in figure 9a).
Apart from a small amount of Fe^{2+} contamination giving peaks in
the same position as figure 2a, the spectrum consists of a magnetic hyperfine pattern with a hyperfine field of magnitude 574 kOe.
It is not a simple magnetic sextuplet however. Extra structure
is visible at -7.6 and 4.3 mm/s. The spectrum is very similar
to those reported by Knudsen for dilute frozen aqueous solutions
of Fe $(NO_3)_3$ at this temperature ($\underline{14}$). That author explained his
spectra as being due to slow relaxation of isolated $[Fe(H_2O)_6]^{3+}$
species.

To confirm that this interpretation holds for the Nafion salt
we applied a parallel magnetic field H_{app} = 39 kOe to our sample.
The resulting spectrum is shown in figure 9b). The spectral lines
have become much sharper and the spectrum can be fitted by three
four-line hyperfine patterns having fields of 535, 305 and 76 kOe
and relative intensities 7 : 2 : 1. These fields are represented
by the bar diagram in figure 9b). The spectra consist of only
four lines because two out of the usual six are suppressed by the
application of a parallel magnetic field. Since the hyperfine
field for Fe^{3+} ions is negative (i.e. oppositely directed to the
spin moment), the magnitude of the field measured at the nucleus
decreases in an external field by exactly the right amount
(-574 + 39 = -535). Both the fields and the relative intensities
are as expected for the three pairs of Zeeman levels issuing from
the Kramers doublets $|\pm5/2\rangle$, $|\pm3/2\rangle$ and $|\pm1/2\rangle$ of the Fe^{3+} ion
in the slow relaxation regime. This confirms that the spectrum
of dilute Fe^{3+} Nafion is due to well isolated Fe^{3+} ions and not
to magnetically ordered clusters.

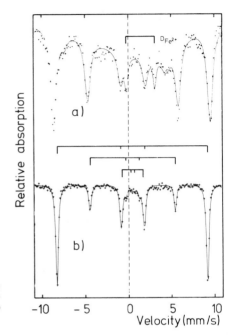

Figure 9. Mössbauer spectra at 4.2 K of a Nafion sample containing 2 Fe^{3+} ions per 100 SO_3^-: a, $H_{ex} = 0$ kOe; b, $H_{ex} = 39$ kOe parallel to the γ-ray direction.

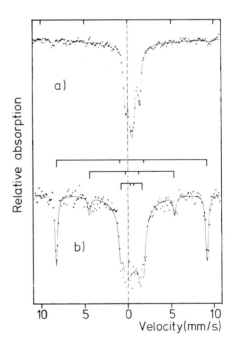

Figure 10. Mössbauer spectra at 4.2 K of a Nafion sample containing 43 Fe^{3+} ions per 100 SO_3^-: a, $H_{ex} = 0$ kOe; b, $H_{ex} = 39$ kOe parallel to the γ-ray direction.

At the other extreme we have a sample neutralized with 0.2 M natural iron chloride solution ($43Fe^{3+}/100$ SO_3^-). The spectrum at 4.2 K is shown in figure 10a). Apart from a small magnetic hyperfine component, the spectrum consists essentially of the usual two ferric quadrupole doublets A and B. The spectrum obtained in an applied parallel magnetic field of 39 kOe (figure 10b) helps to clarify the interpretation. There appears a magnetic hyperfine component identical to that in figure 9b) which is attributed to isolated iron ions. The hyperfine splitting of the two outermost lines again corresponds to a field of 535 kOe. This component comprises 30 % of the total absorption and corresponds to the weak magnetic hyperfine absorption already visible without the field and a broad central relaxation spectrum. Most of the absorption in the applied field is in the central part of the spectrum with a hyperfine field less than 100 kOe.

The moment induced by an applied field H_{app} on any paramagnetic ferric iron experiencing no significant exchange interaction is given by the Brillouin function with $S = 5/2$

$$\frac{<S>}{S} = \mathcal{B}_S \left(\frac{g\mu_B \, S \, H_{app}}{kT}\right)$$

Paramagnetic ferric ion able to relax rapidly among its S_z states should give a four-line magnetic pattern with hyperfine field of 440 kOe at 4.2 K in a 39 kOe parallel applied field since this field is sufficient to produce 85 % saturation of the moment. No fast relaxing isolated iron is present. In fact all the central part of the spectrum must represent iron experiencing strong antiferromagnetic interactions in groups with no net moment. This is certainly true for the dimers of doublet B which contribute to the absorbtion in the central part of the spectrum, but measurement of the areas shows that much of the absorption of doublet A also appears in the central part of the spectrum in the field. Therefore the possibility of attributing doublet A exclusively to almost isolated ions (i.e. paramagnetic ferric ions without strong magnetic interactions) is excluded. Instead it must be mainly attributed to iron in groups containing even numbers of antiferromagnetically coupled ferric ions.

To summarize thus far, we have distinguished four components in the spectra at two extremes of iron concentration.
i) A paramagnetic hyperfine spectrum due to isolated Fe^{3+} ions and not to magnetically ordered clusters.
ii) A ferric doublet B with a large quadrupole splitting associated with dimers.
iii) A second ferric doublet A associated mainly with groups having the particular structure mentioned above. A small contribution from nearly isolated ions is not excluded.
iv) A third, ferrous, doublet due to Fe^{2+} ions.

The variation of the relative proportions of the three ferric components as a function of iron concentration for samples which have been soaked in water is shown in figure 11. The Fe^{2+}

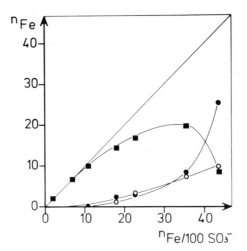

Figure 11. Variation of the proportions of the three ferric components as a function of iron concentration (T = 4.2 K, % H_2O = 17.5). *Key:* ■, *isolated ions;* ○, *dimers;* ●, *larger groups.*

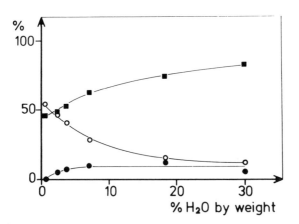

Figure 12. Variation of the proportions of the three ferric components as a function of water content (T = 4.2 K, 23 Fe^{3+}/100 SO_3−). *Key:* ■, *isolated ions;* ○, *dimers;* ●, *larger groups.*

component is neglected as it is small and does not vary in any systematic way. The contribution due to isolated ions is the only one present at low concentrations. It rises to a maximum at 33 Fe/100 SO_3^- and then falls. The second contribution scales exactly with the square of the concentration thus confirming its attribution to dimers. The increase of the third contribution with concentration is more rapid than the dimers, suggesting it is due to larger groups.

We had good evidence from the results on Fe^{2+} as a function of water content that the water is associated with the ionic phase. Given the volume fraction of water in the wet polymer (35 %), the average distance between ions at 33 Fe/100 SO_3^- is calculated as 12.5 A. It is precisely at this distance that paramagnetic hyperfine splitting (slow relaxation) is found to disappear in frozen solutions (15). The idea that the iron is present in an aqueous phase is therefore confirmed.

Influence of water content. Another way of varying the Fe^{3+}-Fe^{3+} distance is to change the water content of the sample. Results for the 23 Fe/100 SO_3^- sample are shown in figure 12. There is the same general tendency for component (i) to decrease and component (ii) to increase as the Fe-Fe distances are reduced. A difference compared with the data as a function of iron content (figure 11) is that the third component seems to decrease when iron is most concentrated in the aqueous phase. The quadrupole splitting of doublet B drops markedly when the water content falls below about 6 %. A systematic study was carried out on samples dried at 150° C as a function of iron content. In this case, the Mössbauer spectra were well fitted by only two subspectra , a magnetic sextuplet corresponding to isolated ions and a doublet with a quadrupole splitting of order 0.8 mm/s.

Discussion

A feature common to Mössbauer spectra of all Nafions is the disappearance of the Mössbauer absorption at a temperature usually well below 0° C. For samples at ambient humidity or above, T_o is around 225 K for Fe^{2+} and around 245 K for Fe^{3+} (5). There is not much difference in T_o for the two doublets A and B, so ferric ion in both environments belongs to the same phase. Results on ferrous iron indicated that the ions are fully hydrated when the water content of the membrane exceeds 6 wt %, and partially hydrated at lower values. We conclude that the ions are entirely located in an aqueous phase in the Nafion membranes, and they are not directly incorporated in the crystalline or amorphous solid phases of the polymer. This is quite consistent with small angle scattering results (26).

The aqueous phase behaves quite differently from a simple ionic solution. DSC traces show a peak near 0° C only in the samples containing the most water. However, there is a broad increase in specific heat in the range 200-250 K in all samples (21). Disappearance of Mössbauer absorption in this range of temperature is understandable if it is associated with a glass transition in the aqueous phase. The unusual properties of the aqueous ionic phase may result from the presence of anions attached to the ends of the side chains. The "stirring" motions of these groups suppress the normal freezing of the aqueous solution, and the agitation is only arrested at a temperature around T_o. Infrared spectra confirm that hydrogen bonding is weaker for water in Nafion than in salt solution (27).

There are some differences in the microstructure of the ionic phase depending on the exchanged ion, the concentration and pH of the solution used for the exchange, and the water content of the membrane.

For Fe^{2+}, there is no evidence for magnetic interactions at low temperatures, whatever the ion concentration. This indicates that the ferrous ions are well-separated from each other and do not tend to form pairs or groups. Changes in all Mössbauer parameters below 6 wt % water reflect a modification of the fully-hydrated ion environment.

For Fe^{3+} the microstructure depends quite critically on the degree of neutralization and the water content, although no qualitative influence of the anion was detected in a comparison of samples neutralized with different ferric salts. At low iron concentrations the spectrum is identified with isolated ferric ions with no other iron neighbours within about 12 Å. The magnetic splitting of the Mössbauer spectrum at low temperatures was unambiguously shown to be paramagnetic hyperfine structure by means of measurements in an applied field. Furthermore the hyperfine spectrum disappears and the paramagnetic central peak increases as the iron content is increased. This is just as expected if hyperfine splitting appears only for isolated single ions.

Only in the case of reexchanged samples (doublet D of figure 6h) have we found a magnetic hyperfine pattern at 4.2 K with $H_{hf} \cong 460$ kOe which could be interpreted in terms of large clusters of ferric ions (17).

As the iron content increases or the water content decreases, the proportion of isolated ions diminishes to the benefit of dimers and larger groups. All the data shows that doublet B must be associated with dimers, identified as oxygen bridged decaqua species. The other doublet A in the centre of the spectrum cannot be associated with a single type of iron in every case. When the iron is dilute, it is reasonable to attribute it mainly to isolated ions with no iron present in nearest-neigbour positions, but some at greater distances. The paramagnetic hyperfine structure collapses into a central doublet as temperature is raised. However, the experiments in a magnetic field for samples more concentrated in iron (figure 10) and for dried samples show that most of doublet A does not arise from quasi-isolated paramagnetic ions, but from groups with no net moment. Groups involving an even number of antiferromagnetically-coupled ferric ions formed by association of the dimers would have the necessary magnetic properties. Such groups may be expected to block magnetically at very low temperatures $\lesssim$ 1 K. Bauminger et al. (figure 7) have observed the transformation of doublet A into a magnetic hyperfine spectrum in this temperature range, although they did not determine how much of the magnetic spectrum was due to the paramagnetic hyperfine splitting of isolated ions present in their sample.

In conclusion, we have demonstrated how it is possible to build up quite a detailed picture of the microstructure of the ionic phase of neutralized Nafion membranes by exploiting the magnetic properties of iron ions, and obtaining Mössbauer spectra at low temperatures in an external magnetic field. Results common to the two Mössbauer cations (Fe^{2+} and Fe^{3+}) concerning the anomalous character of the aqueous phase can reasonably be expected to apply for any other cation. The tendency for Fe^{3+} ions to form pairs and larger groups is not shared by Fe^{2+}, and it may be a special feature of small tripositive ions.

Acknowledgement

Part of this work was supported by the National Board for Science and Technology (Ireland) under research grand URG 18/78. Authors are grateful to Drs. T. Gierke, P. Meakin, C.J. Molnar of Du Pont Company for supplying Nafion samples used in the experiments.

Literature cited

1. Yeo, S.C. PhD Thesis, McGill University, Montreal 1977.
2. Rodmacq, B.; Pineri, M.; Coey, J.M.D. Rev. Phys. Appl. (Paris) 1980, 15, 1179.
3. Bauminger, E.R.; Levy, A.; Labenski de Kanter, F.; Ofer, S.; Heitner-Wirguin, C. J. Physique (Paris) 1980, 41, C1-329.
4. Heitner-Wirguin, C.; Bauminger, E.R.; Levy, A.; Labensky de Kanter, F.; Ofer, S. Polymer 1980, 21, 1327.
5. Rodmacq, B.; Pineri, M.; Coey, J.M.D.; Meagher, A. to be published in J. Polymer Sci., Polymer Phys. Ed.
6. Meagher, A. M. Sc. Thesis, Dublin 1981.
7. Meyer, C.T. J. Physique (Paris) 1976, 37, C6-751.
8. Goldanskii, V.I.; Korytko, L.A. "Applications of Mössbauer Spectroscopy", vol. 1, R.L. Cohen Ed., Academic Press, 1976, p. 287.
9. Greenwood, N.N.; Gibb, T.C. "Mössbauer Spectroscopy", Chapman and Hall, London, 1971.

10. Champeney, D.C.; Sedgwick, D.F.   Chem. Phys. Lett. 1972, 15, 377.
11. Reich, A.; Michaeli, I.   J. Chem. Phys. 1972, 56, 2350.
12. Vasquez, A.; Flinn, P.S.   J. Chem. Phys. 1980, 72, 1958.
13. Litterst, F.J.; Bröll, W.; Kalvius, G.M.   J. Physique (Paris) 1974, 35, C6-415.
14. Knudsen, J.E.   J. Phys. Chem. Solids 1977, 38, 883.
15. Morup, S.; Knudsen, J.E.; Nielsen, M.K.; Trumpy, G,   J. Chem. Phys. 1976, 65, 536.
16. Buckley, A.N.; Herbert, I.R.; Rumbold, B.D.; Wilson, G.V.H.; Murray, K.S.   J. Phys. Chem. Solids 1970, 31, 1423.
17. Coey, J.M.D.; Readman, P.W.   Nature 1973, 246, 476.; Earth Planetary Sci. Lett. 1973, 21, 45.
18. Chappert, J.   J. Physique (Paris) 1974, 35, C6-71.
19. Nozik, A.J.; Kaplan, M.   J. Chem. Phys. 1967, 47, 2690.
20. Takamatsu, T.; Hashiyama, M.; Eisenberg, A.   J. Appl. Polym. Sci. 1979, 24, 2199.
21. Pineri, M. et al., to be published.
22. Simopoulos, A.; Wickman, H.; Kostikas, A.; Petrides, D.   Chem. Phys. Lett. 1970, 7, 615.
23. Knudsen, J.M.; Larsen, E.; Moreira, J.E.; Faurskov Nielsen, O. Acta Chem. Scand. 1975, A29, 833.
24. Morup, S. "Paramagnetic and Superparamagnetic Relaxation Phenomena studied by Mössbauer Spectroscopy", Polyteknisk Forlag, Lyngby 1981.
    Morup, S.; Knudsen, J.E.   Chem. Phys. Lett. 1975, 40, 292.
25. Gierke, T.D.   Electrochemical Society Fall Meeting, Atlanta, Georgia, Oct. 1977.
26. Roche, E.J.; Pineri, M.; Duplessix, R.; Levelut, A.M.   J. Polym. Sci. Polym. Phys. Ed. 1981, 19, 1.
27. Falk, M.   Can. J. Chem. 1980, 58, 1495.
28. Kelly, J.M. ; Murphy, M.J. unpublished results.

RECEIVED September 9, 1981.

Morphology of Perfluorosulfonated Membrane Products

Wide-Angle and Small-Angle X-Ray Studies

T. D. GIERKE,[1] G. E. MUNN, and F. C. WILSON

E. I. du Pont de Nemours & Co., Inc., Plastic Products and Resins Department, Experimental Station, Wilmington, DE 19898

The morphology of the ionomer resin, from which "Nafion" (registered trademark of the E. I. du Pont de Nemours and Co.) perfluorinated membrane products are made, was studied with wide angle and small angle x-ray diffraction techniques. A reflection observed in the small angle x-ray scan from hydrolyzed polymer is attributed to ionic clustering. The effects of equivalent weight, cation form, temperature, water content, and tensile draw on this reflection were studied and are discussed.

"Nafion" perfluorinated membranes are constructed from per-fluoinated resins containing covalently bonded ion exchange sites. A typical perfluorinated resin, used in these membranes, possesses the general chemical structure

$$\left[\begin{array}{c} [\ (CF_2 \ CF_2)_n \ - \ CF_2 \ \underset{|}{CF} \ -] \\ \\ (OCF_2 \ CF-)_m \ OCF_2 \ CF_2 \ SO_2F \\ \\ CF_3 \end{array} \right]_x$$

where the value of m can be as low as 1. The value of x, in the above formula determines the equivalent weight (EW) of the resin. Typical values range between 6 and 14 and correspond to an EW range of from 1000 to 1800 grams/equivalent. The SO_2F group is easily hydrolysed to form the strongly acidic perfluorosulfonic acid exchange site. In this form, the resin is extremely hydrophilic and can absorb as much as 30 water molecules per exchange site and more than double its volume.

As in other ionomers, the ion exchange sites in "Nafion" membranes are observed to aggregate and form clusters. Ionic clustering in "Nafion" membranes has been indicated by a variety of physical studies including dielectric relaxation (1), small angle x-ray scattering (1-4), neutron scattering (4), electron micro-

[1] Current address: Parkersburg, WV 26101.

scopy (2,5), NMR (6), IR (7,8), Mossbauer spectroscopy (9,10) and
several transport studies (2,11,12). In this paper we shall
review the results of small angle x-ray scattering, SAXS, experiments. We shall confine our remarks to a description of the pertinent experimental results. In a later section, a model for
ionic clustering will be proposed and the effects of ion clustering will be proposed and the effects of ion clustering on ion
transport will be discussed.

   Experimental. Our experiments were conducted of films of
known equivalent weight which were 0.1 to 0.3 mm thick. Samples
are easily hydrolysed by heating in any convenient basic medium,
e.g. sodium hydroxide. Samples in various cation forms are prepared using standard ion exchange techniques. The amount of solvent the sample absorbs depends on the thermal history of the
sample (13) and, unless otherwise stated, all samples were conditioned by boiling one hour in water. The amount of solvent
absorbed by the polymer was determined gravimetrically. Dry
polymer densities were determined using standard bouyancy techniques on samples dried for 18 hours in a nitrogen flushed vacuum
oven maintained at 110°C. Our values agree with other reported
values (13,14). The values reported by Roche et al (4), for
similar thermal histories, are about a factor of two larger.
   To obtain x-ray diffusion results from swollen hydrolyzed
samples, methods were developed to inhibit sample dehydration
during the course of the experiment. This was achieved by sealing the samples, thermally, in a bag constructed from oriented
polypropylene film. Oriented polypropylene was selected because
it possesses low permeability to water and is also essentially
"transparent" in the small angle x-ray scans. With such a construction, we determined that the weight of a swollen sample
changed by less than 0.2% during the period required to obtain
the small angle x-ray scans.
   The small-angle data were gathered on a Kratky small-angle
"camera" equipped with a NaI(Tl) scintillation counter and a Ni
filter for CuKα radiation. Pulse-height-analysis was set to
accept 90% of the CuKα radiation symmetrically, and the x-ray
source was a special short (7mm) line focus tube (Siemens) so
that there was no vignetting of the source by the x-ray tube
window. After subtraction of instrumental background, all the
scans were normalized to the intensity expected from a sample of
optimum thickness ($I_{transmitted}/I_o - 1/e$).
   The normalized small-angle intensities (except for highly
oriented samples) were desmeared according to the procedure of
Schmidt and Hight (15,16). The desmeared intensities were
multiplied by the square of the scattering angle -- this can be
considered as the application of the small-angle Lorentz correction or the calculation of the Invariant argument, $h^2I(h)$,
where h is the scattering wave vector. The effect of this data
analysis on the observed variation in amplitude with scattering

angles is shown for a typical scan in Figure 1, where the amplitudes of the three traces in this figure are in arbitrary units. (In this manuscript the scattering angle, 20 is always expressed in degrees.) Even at low water contents, a maximum, in the observed intensity, is normally detected at scattering angles of 1<20<3.

Wide angle data were obtained on a Phillips Electronics wide angle power diffractometer with counting electronics equivalent to that described above. In most SAXS and WAXD scans, we have included, at the top of the figure, a scale of Bragg spacings to aid the reader in comparing these data to other scattering experiments.

Results. The small angle x-ray scans from hydrolysed polymer are fairly complex and the results are consistent with the existence of three distinct phases (3,4). Only part of this information, however, is directly related to ionic clustering and to better distinguish features in the SAXS associated with ionic clustering it is appropriate to first examine some experimental results from a wide angle x-ray diffraction (WAXD) experiments.

WAXD scans from unhydrolysed polymer, $-SO_2F$, reveal that the polymer is semicrystalline. (Figure 2) This result is similar to the behavior observed in copolymers of tetrafluoroethylene, (TFE) with hexafluoropropylene or perfluoropropylvinyl ether. As the EW increases, the molar ratio of TFE to comonomer increases and the amount of crystallinity also increases. The amount of crystallinity, deduced from the relative intensity of the amorphous halo and crystalline peak, ranges between 0 and 40%, however, these values represent qualitative estimates rather than quantitative results.

The degree of crystallinity in the polymer is only moderately altered, compared to the effect of equivalent weight, when the polymer is hydrolyzed, as demonstrated in Figure 3. Although the crystallinity does decrease somewhat upon hydrolysis, it is clear the hydrolyzed polymer is still partly crystalline. Similar results are obtained for polymer in other cation forms. Thus, any ionic clustering which exists in the polymer does not completely disturb the crystalline portion of the matrix.

The effect of temperature on the WAXD scans is shown in Figure 3. (Note in the scans of this figure, the scattering angle increases from right to left.) The intensity of the crystalline reflection decreases gradually with increasing temperature until it has essentially disappeared above 270°C. The melting point of the polymer, determined by differential scanning calorimetry, is about 275°C. This behavior is in marked contrast to that observed in highly crystalline polytetrafluoroethylene where the intensity of the crystalline peak does not change with temperature except over a range of several degrees at the melting point. This result suggests that melting in "Nafion" occurs over a very broad temperature range.

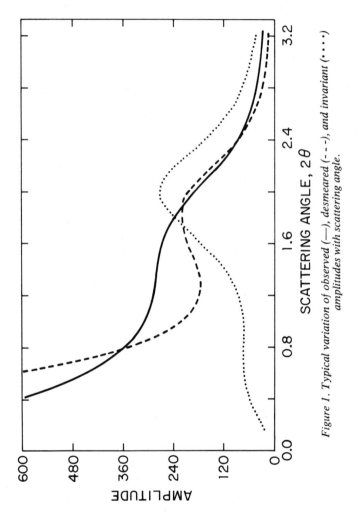

Figure 1. Typical variation of observed (——), desmeared (- - -), and invariant (· · · ·) amplitudes with scattering angle.

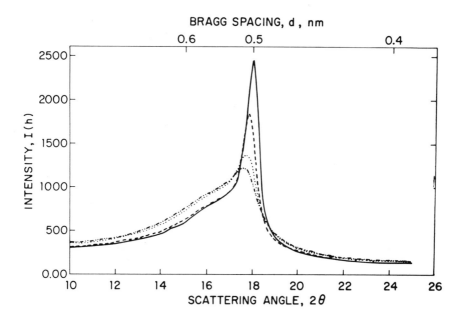

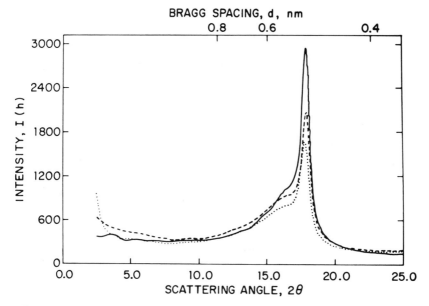

Figure 2. Wide angle x-ray scans from unhydrolyzed polymer. Top, with various equivalent weights. Key: —···—, 1100 EW; ····, 1200 EW; - - -, 1500 EW; —, 1800 EW. Bottom, the effect of hydrolysis on 1800 equivalent weight polymer. Key: —, SO₂F form; - - -, H⁺ form, dry; ····, H⁺ form, wet.

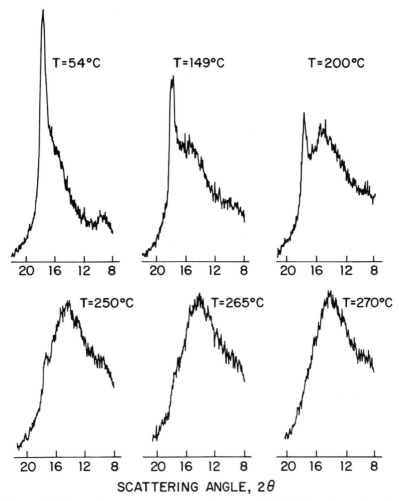

Figure 3. Wide angle x-ray scans from unhydrolyzed 1379 EW polymer showing the effect of temperature. Note scans are from left to right.

The WAXD scan from a fiber which was extruded from the polymer has recently been obtained by Starkweather (17). He observes
that the crystal structure of "Nafion" is very similar to that
observed in p-TFE and the peak observed in Figures 2 & 3 correspond to the 100 reflection. From the width of the 002 reflection, Starkweather concludes that the crystallite size along the
chain axis (4.4 nm) is larger than the average separation of
sidechains (2.0 nm). Starkweather proposed a fringed micellar
structure for the crystalline portion of the polymer consisting
of two rows of a hexagonal lattice with the sidechains extending
outwardly from both sides of the crystalline domain.

In summary, the results of this section indicate that unhydrolysed polymer is semicrystalline and that the crystal structure is very similar to that observed in p-TFE. In addition, the
hydrolysis of the resin and consequent formation of ionic clusters only moderately changes the degree of crystallinity.

Small Angle X-ray Results from Non-Ionized Polymer. SAXS
has also been used to study the morphology of non-ionized polymer.
Figure 4 shows the SAXS scans from several polymers with different EW's. Like the crystalline peak in the WAXD scans, the intensity of this feature decreases with decreasing EW. A similar
reflection is also observed in tetrafluoroethylene/hexafluoro-
propene copolymer and tetrafluoroethylene/perfluoropropylvinyl
ether copolymer. Figure 5 shows the variation of the apparent
Bragg spacing of this reflection with the tetrafluoroethylene/
comonomer molar ration. These results indicate that this feature
is associated with the fluorocarbon matrix and should not be
attributed to ionic clustering.

Some of the properties of this reflection are interesting.
The intensity of this reflection increases as temperature increases, shown in Figure 6, until the polymer melts and the
reflection disappears. On cooling, the reflection is again observed in the SAXS scan. Clearly the intensity of this reflection is related to the electron density difference between crystalline and amorphous portions of the polymer, and is thus not
to be confused with ionic clustering. The effect of tensile draw
on the SAXS scan is demonstrated in Figure 7. After moderate
draw, the Bragg spacing is larger in the machine direction and
smaller in the transverse direction when compared with an undrawn
sample. At higher extension, as observed in a fiber, this reflection is only detected in the meriodional scan. (Note that raw
data is shown for highly extended samples, not invariant scans.)
This implies a periodicity along the fiber axis, which, from the
WAXD scan, corresponds to the direction along which the molecular chains are aligned. These results are similar to those
observed in polyethylene for the reflection attributed to "chain
folding" (18, 19). Our results, however, are not comprehensive
enough to conclude that chain folding also exists in the perfluorosulfonic acid ionomer resin.

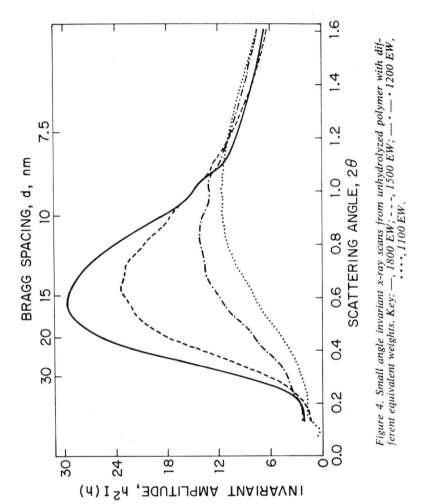

Figure 4. Small angle invariant x-ray scans from unhydrolyzed polymer with different equivalent weights. Key: —, 1800 EW; - - -, 1500 EW; — · 1200 EW,
· · ·, 1100 EW.

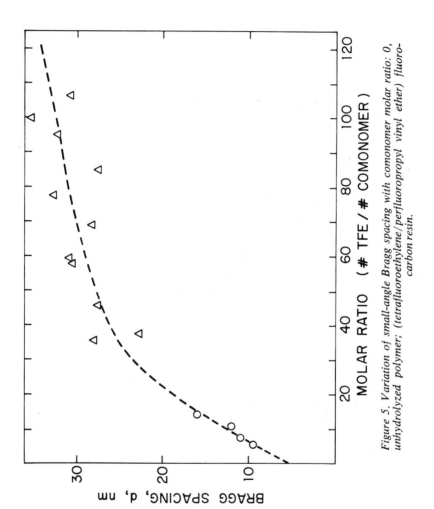

Figure 5. Variation of small-angle Bragg spacing with comonomer molar ratio: 0, unhydrolyzed polymer; (tetrafluoroethylene/perfluoropropyl vinyl ether) fluoro-carbon resin.

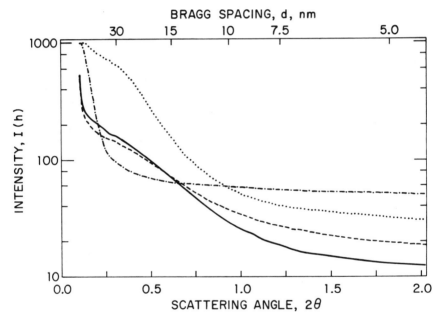

Figure 6. Small angle x-ray scans from 1790 EW, unhydrolyzed polymer showing effect of temperature. Key: ——, as received; - - -, once melted, 20° C; • • • •, 208° C; — • — •, 288° C.

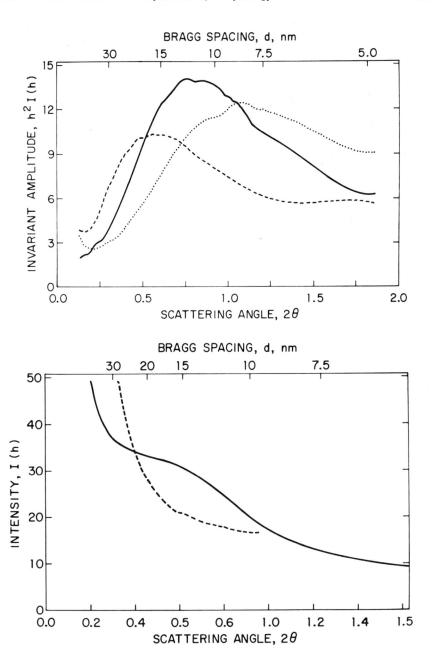

Figure 7. Small angle invariant x-ray scans from unhydrolyzed polymer showing effect of strain. Top, 1200 EW film drawn 2 times. Key: ——, undrawn control; - - -, drawn, machine direction; • • • •, drawn, transverse direction. Bottom, 1179 EW spun fiber. Key: - - - -, equatorial scan; ——, meridional scan.

Roche et al (4) observe a reflection in their small angle
neutron scattering experiments which is essentially identical to
these SAXS results. These authors also attribute this feature to
an interference between crystalline portions of the polymer.

Small Angle X-ray from Hydrolyzed Polymer. When the polymer
is hydrolyzed, a new reflection is observed in the SAXS scan, as
shown in Figure 8, which corresponds to a Bragg spacing of from
3 to 5 nm. It is this reflection which has been attributed to
ionic clustering in "Nafion" (1-4). The origin of this reflection is currently the subject of debate and we will defer our discussion of this topic for a subsequent chapter of this text (20).
We first will present the results of our SAXS experiments.

Figure 9 shows the variation of SAXS scans in hydrolzed polymer with equivalent weight, EW. Unlike the WAXD scans and SAXS
scans from unhydrolyzed polymer, the intensity of the reflection
at 5.0 nm increases with decreasing EW. This variation in intensity with EW reflects the increase in water content of swollen
polymer as EW decreases (1) and is thus consistent with ionic
clustering. The position of this reflection occurs at larger
Bragg spacing, lower scattering angles, as the EW decreases. In
fact, the Bragg spacing is linearly related to the exchange capacity, (1000/EW), of the polymer as shown in Figure 10.

The effect of temperature on a SAXS scan is shown in Figure
11. Note that the reflection at 4 nm persists above the melting
point of the polymer, 265°C. Similar behavior was observed in
ethylene/methacrylic acid ionomer (21, 22) and is strong support
for the existence of ionic clustering in the dry state of the
polymer. Indeed, we will demonstrate shortly that this reflection is observed in dry polymers at room temperature neutralized
with heavier metal ions.

The effect of tensile draw on the "cluster" reflection is
shown in Figure 12 and should be compared with Figure 7 for unhydrolyzed polymer. Under moderate strains, 1.75 times, it is
seen that the "cluster" reflection is observed mostly in a direction normal to the strain. This is also true in the SAXS scan
from fiber where the reflection is principally observed in the
equatorial scan. This implies a periodicity which is normal to
the fiber axis. Thus, the periodicity associated with this
reflection tends to be orthogonal to the periodicity associated
with the reflection observed in the meridional SAXS scan of unhydrolyzed fiber. Similar effects of sample orientation have
recently been reported for ethylene-methacrylic acid ionomer (23).

The effect of different cations on the SAXS scan is shown
in Figure 13. Very little change in the position of the reflection is seen with changing cation. The intensity of the reflection decreases as the cation weight increases. This is a result
of both the change in water absorption and a change in contrast
for x-ray scattering between the ionic clusters and fluorocarbon
matrix, since the electron density of the cluster will increase

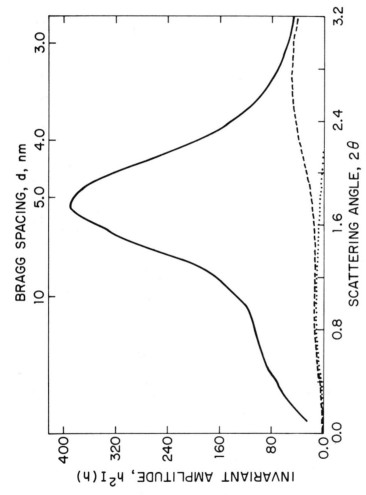

Figure 8. Small angle invariant x-ray scans from 1100 EW polymer showing the effect of hydrolysis. Key: ——, SO_3H, wet; – – –, SO_3H, dry; ···, SO_2F.

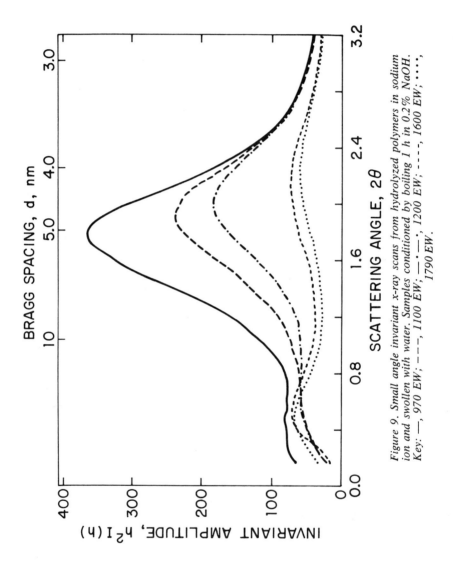

Figure 9. Small angle invariant x-ray scans from hydrolyzed polymers in sodium ion and swollen with water. Samples conditioned by boiling 1 h in 0.2% NaOH. Key: ——, 970 EW; – – –, 1100 EW; —·—, 1200 EW; – - –, 1600 EW; ·····, 1790 EW.

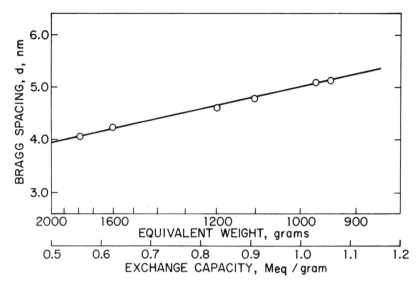

Figure 10. Variation of small angle Bragg spacing with exchange capacity in hydrolyzed polymer swollen with water.

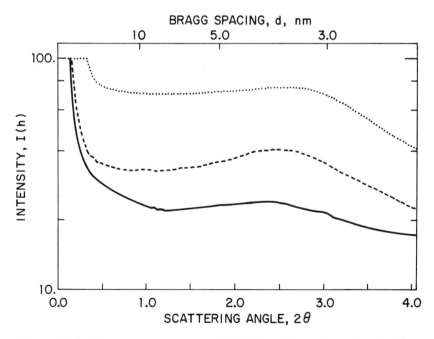

Figure 11. Small angle x-ray scans from 1200 EW, hydrolyzed dry polymer in silver ion form showing effect of temperature. Key: —, 23° C; - - - -, 100° C; ••••, 290° C.

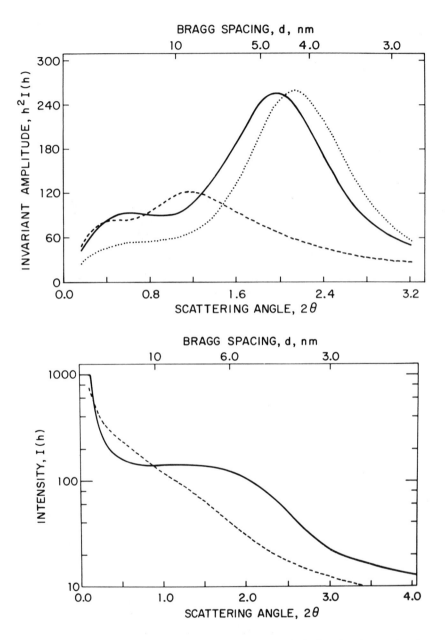

Figure 12. Small angle invariant x-ray scans from hydrolyzed polymer showing effect of strain. Top, 1200 EW polymer in sodium ion form swollen with water and 1.75 draw ratio. Key: —, undrawn control; – – –, drawn, machine direction; ••••, drawn, transverse direction. Bottom, 1179 EW spun fiber in sodium ion form and swollen with water. Key: —, equatorial scan; - - - -, meridional scan.

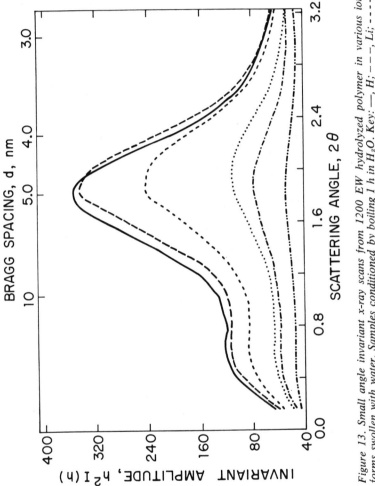

Figure 13. Small angle invariant x-ray scans from 1200 EW hydrolyzed polymer in various ion forms swollen with water. Samples conditioned by boiling 1 h in H_2O. Key: ——, H; — — —, Li; - - - -, Na; • • •, K; — • —, Rb; — • • —, Cs.

for heavier cations which are also less hydrophilic (13). In
fact it is possible, by varying water content and cation form
simultaneously, to make this reflection "disappear" altogether.
A similar result can be obtained by changing the salt concentration of the solution with which the polymer equilibrates. Again,
the intensity of the reflection decreases as the concentration
increases because the electron density of the cluster is increasing. Interestingly, even the proton form of the polymer possesses the reflection attributed to clustering. This observation
is in marked contrast to the results observed in hydrocarbon
analogues (20, 21).

The effect of water content on the SAXS scans from swollen
polymer is illustrated in Figure 14. The decrease in intensity
with decreasing water content again reflects the increasing
relative electron density of the cluster. Note, however, that
the Bragg spacing increases with increasing water content. This
is shown more clearly in Figure 15. Extrapolation to the dry
state yields Bragg spacing of about 3_+ nm. Results have also
been obtained on 1200 EW polymer in H^+ and Ag^+ ion form and a
Bragg spacing in dry polymer of 3.0 nm was also obtained. The
results for the Ag^+ ion form are particularly interesting. As
the water concentration is lowered, the intensity of the reflection first decreases, disappears, and then increases so that the
reflection is observed in the dry state of the polymer. Clearly
this variation of intensity with water content is again the
result of changing the electron density of the cluster with
respect to the fluorocarbon matrix.

Roche et al (4) also report SAXS results as a function of
water content and their results are essentially identical to
those shown in Figure 14.

Discussion. The results in Figures 8-15 provide strong
support for the existence of ionic clustering in "Nafion". However, details of the arrangement of matter in these clusters
cannot be obtained from small angle x-ray results alone. For
hydrocarbon ionomers, several different interpretations have
been advanced as the cause of the SAXS maximum. These include
a model of spherical clusters on a paracrystalline lattice proposed by Cooper et al (22), the shell core model of Macknight
et al (24) and more recently a lamellar model (23). At present,
there is no consensus about which of these models best describes
clustering in hydrocarbon ionomers. The situation for the perfluorinated ionomers is further complicated because they differ
dramatically in several respects from ethylene/methacrylic acid,
(E/MA), ionomers. The SAXS peak is observed in the acid form
of our polymer while it is not observed in E/MA ionomers when in
acid form. This difference is accounted for by recognizing that
the perfluorosulfonic acid exchange site is a very strong acid
and is completely dissociated in water. This also leads to the
remarkable high degree of water uptake observed in the perfluor-

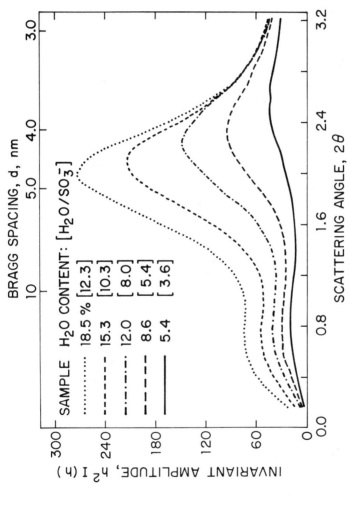

Figure 14. Small angle invariant x-ray scans from 1200 EW hydrolyzed polymer in lithium ion form showing the effect of degree of swelling with water.

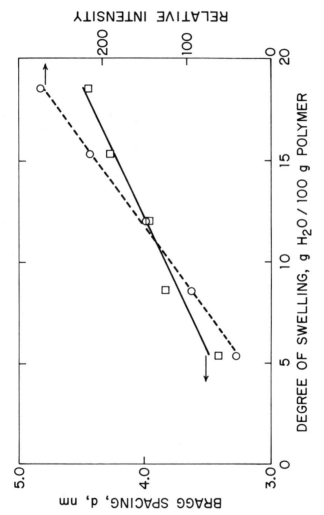

Figure 15. The variation of the small angle Bragg spacing and invariant peak amplitude with water content in 1200 EW hydrolyzed polymer in lithium ion form.

inated ionomer, as much as 50% by volume for 944 EW polymer. This water absorption promotes ionic clustering in the perfluorinated ionomer in contrast to the observed tendency in E/MA ionomer (22). Thus, the details of ionic clustering in the perfluorinated iono- mer may be substantially different from clustering in E/MA iono- mer.

Any model which is proposed for ionic clustering in these perfluorinated ionomers should be consistent with the results of these x-rays studies. It is appropriate, then, to enumerate the pertinent results of these studies.

a.) The polymer is semicrystalline, and the structure of the crystalline phase is similar to the structure observed in p-TFE.

b.) The size of the crystallites are larger than the ave- rage separation between sidechains (16).

c.) Ionic clustering only moderately changes the degree of crystallinity, and ionic clustering exists in the absence of crystallinity.

d.) The effective Bragg spacing associated with ionic clus- tering increases with increasing water content and decreasing equivalent weight but is fairly insensitive to the cation used to neutralize the polymer.

e.) This reflection is observed in the acid form of the resin.

f.) The reflection also is observed in the dry resin.

g.) The cluster separation tends to be normal to the dir- ection of the polymer chains in strained resins.

In a later chapter we will propose a model of ionic clus- tering which we believe is an essential agreement with these results (20). The real strength of the model, however, is that it can be used to understand and describe ion transport through "Nafion" perfluorinated membranes (3, 25, 26).

Literature Cited

1. Yeo, S. C.; Eisenberg, A. J. Appl. Polym Sci. 1977, 21, 875.

2. Gierke, T. D. 152nd Meeting of Electrochemical Society, Atlanta, Georgia, Abstract No. 438, J. Electrochem Soc. 1977, 124, 319C.

3. Gierke, T. D.; Munn, G. E.; Wilson, F. C. J. Polym. Sci, Polym. Phys. Ed, In press.

4. Roche, E. J.; Pineri, M.; Duplessix, R.; Levelut, A. M. J. Polym. Sci. Polym. Phys. Ed. 1981, 19, 1.

5. Ceynowa, J.; Polymer 1978, 19, 73.

6. Komoroski, R. A.; Mauritz, K. A. J. Amer. Chem. Soc. 1978, 100, 7487.

7. Heitner-Wirguin, A. C. Polymer 1979, 20, 371.

8. Falk, M. Am. Chem. Soc, Adv. Chem Series, this volume.

9. Rodmacq, R.; Pineri, M.; Coey, J. M. D. Rev. Phys Appl. 1980, 15, 1179.

10.   Heitner-Wirguin, C.; Bauminger, E. R.; Levy, A.; Labensky de Kanter, F.; Ofer, S. Polymer 1980, 21, 1327.

11.   Will, F. G. J. Electrochem. Soc. 1979, 126, 36.

12.   Lopez, M.; Kipling, B; Yeager, H. L. Anal. Chem 1977, 49, 629.

13.   Grot, W. G. F.; Munn, G. E.; Walmsley, P. N. 141st Meeting of the Electrochemical Society, Houston, Texas, May 1972, Abstract No. 154, J. Electrochem Soc. 1972, 119, 108C.

14.   Takamatsu, T.; Hashiyama, H.; Eisenberg, A. J. Appl. Polym. Sci. 1979, 24, 2199.

15.   Schmidt, P. W.; Height, R. Acta Cript. 1960, 13, 480.

16.   Schmidt, P. W. Acta Cryst 1965, 19, 938.

17.   Starkweather, H. W. to be published.

18.   Geil, P. H. "Polymer Single Crystals Interscience Pub." 1963, pp. 96-98, 313-316, 348-350, 415, and 459.

19.   Wunderlich, B. "Macromolecular Physics", Vol. 1, Academic Press 1973, Ch. 3.

20.   Gierke, T. D., Hsu, W. Y. Am. Chem. Soc. Adv. Chem. Ser. this volume.

21.   Wilson, F. C.; Longworth, R.; Vaughn, D. J. Polym. Prepr., Amer. Chem. Soc., Div. Polym. Chem., 1968, 9, 505.

22.   Marx, C. L.; Caulfield, D. F.; Cooper, S. L., Macromolecules 1973, 6, 344.

23.   Roche, E. J.; Stein, R. S.; Russell, T. P.; MacKnight, W. J. J. Polym. Sci., Polym. Phys. Ed., 1980, 18, 1497.

24.   MacKnight, W. J.; Taggart, W. P.; Stein, R. S. J. Polym Sci., Polym. Symp., 1974, 45, 113.

25.   Gierke, T. D. 153rd Meeting of Electrochemical Society, Seatle, Wash., May 1978) Abstract No. 453, J. Electrochem. Soc. 1978, 125, 163C.

26.   Hsu, W. Y.; Barkley, J. R.; Meakin, P. Macromolecules, 1980 13, 198.

RECEIVED August 24, 1981.

Structure of Sulfonated and Carboxylated Perfluorinated Ionomer Membranes

Small-Angle and Wide-Angle X-Ray Scattering and Light Scattering

TAKEJI HASHIMOTO, MINEO FUJIMURA,[1] and HIROMICHI KAWAI

Kyoto University, Department of Polymer Chemistry, Faculty of Engineering, Kyoto 606, Japan

A series of the Nafion membranes (registered trade mark of E. I. du Pont de Nemours & Co. Inc. for its perfluorinated sulfonic acid products) having different equivalent weights (E.W.), 1100, 1150, 1200, 1400, and 1500 E.W. are chemically modified to prepare a series of sulfonic acid and carboxylic acid form of the perfluorinated ionomer membranes. For a given E.W.(1100), we prepared the membranes in the forms of sulfonic acid, sulfonates (sodium and cesium sulfonates), sulfonyl chloride and sulfoamide as well as carboxylic acid and sodium carboxylates. Small-angle X-ray scattering (SAXS), wide-angle X-ray diffraction (WAXD) and small-angle light scattering (SALS) studies were pursued on the ionomer membranes to clarify their internal supermolecular structures at various levels, ranging from a few nm to a few μm. The SAXS profiles generally exhibit two scattering maxima; one at small scattering angle which is associated with a long identity period D of lamellar crystals and the other at large scattering angle which is associated with the ionic clusters. X-ray crystallinity X_c of the membranes were estimated from WAXD, and spherulitic superstructure, if it exists, was investigated by SALS. The structural parameters D, X_c, size of the ionic cluster, and spherulite size as well as amount of water uptake were investigated as a function of the E.W. and of the functional groups.

The electrochemical (1) and mechanical properties(2,3) of the perfluorinated ionomer membranes as an ion-exchange membrane are obviously influenced by their internal structure of the membranes, especially spatial organization of the ionic sites. In this paper we attempted to carry out very basic studies on the structure of the perfluorinated ionomer membranes in the absence of applied external electric field. Although for practical applications of the membranes it is extremely important to study the structure under

[1] Current address: Japan Synthetic Rubber Co., Inc., Research and Development, 7569 Ikuta, Tama-ku, Kawasaki 214, Japan.

0097-6156/82/0180-0217$08.00/0
© 1982 American Chemical Society

the applied electric feild, this type of the analysis is beyond
the scope of the present investigation.

We shall study wide-angle X-ray diffraction (WAXD) of the
membranes to assess X-ray crystallinity (X_c), small-angle X-ray
scattering (SAXS) to estimate the long identity period (D) of
lamellar crystals from the SAXS maximum at small angles as well as
the size of the ionic cluster from the SAXS maximum at large
angles (defined as "ionic scattering maximum", the spacing as estimated from the scattering angle giving rise to the ionic maximum
being defined as S), small-angle light scattering (SALS) to estimate size (R) of the spherulitic superstructure, if it exists.
The structural parameters X_c, D, S, and R as well as amount of
water uptake by the membranes are measured as a function of the
equivalent weight (the weight of polymer which will neutralize one
equivalent of base(1)) for the membranes having sulfonic acid
groups or carboxylic acid groups. These parameters are measured
also as a function of types of the functional groups (such as carboxylic acid, sodium carboxylate, sulfonic acid, sodium and cesium
sulfonates and sulfonyl chloride) for the membranes having a given
E.W. of 1100.

EXPERIMENTAL METHOD

Test Specimens. A series of the Nafion membranes with
different E.W. (1100, 1150, 1200, 1400, and 1500 E.W.) are chemically modified to prepare a series of sulfonic-acid and carboxylic
-acid forms of the perfluorinated ionomer membranes having the
E.W. corresponding to the original Nafion membranes. (Nafion is a
registered trade mark of E.I. du Pont de Nemours & Co.(Inc.) for
its perfluorinated-sulfonic acid products.) The ionomer membranes
in the forms of sulfonic acid, sodium or cesium sulfonate, sulfonyl chloride and sulfoamide, $-SO_2NHR$ (R; $(CH_2)_3NH_2$) as well as
carboxylic acid and sodium carboxylate were prepared from the
Nafion membranes having 1100 E.W.. The detailed procedure will be
described elsewhere (4). In the carboxylic-acid and carboxylated
ionomer membranes, the residual amount of sulfonate groups were
quantitatively estimated to be minute by analyzing an amount of
sulfur atoms before and after the chemical modification with
fluorescent X-ray spectroscopy.

X-ray and Light Scattering Measurements. X-ray and light
scattering measurements were performed for dry membranes and for
membranes swollen with water. The SAXS and WAXD profiles were
taken with cameras utilizing one-dimensional position sensitive
proportional counter. Cu-Kα radiation (λ = 1.54 A) was used as an
incident beam. A graphite crystal was used for monochromatization
of the incident beam. The pulse-height analyzer in the detector
electronics eliminates the higher order harmonics of white X-ray.
The X-ray beam was generated by a 12 Kw-rotating anode X-ray gen-
erator (RU-z or RU-a, Rigaku, operated at 50 KV and 200 mA). The

SAXS curves were corrected for absorption, air scattering, nonuniformity of the detector sensitivity, slit-length and slit-
width smearings as described in detail in our previous paper(5).
The WAXD curves were corrected for absorption, air scattering and
polarization. The details of the apparatus utilizing the position
sensitive detector (PSD) were also described elsewhere(6).

The depolarized component of the scattered light was also investigated to explore crystalline supermolecular structure of the
membranes. The H_V scattering (taken with vertically polarized
incident beam and horizontally polarized analyzer) depends on
spatial orientation correlation of optically anisotropic scattering elements(7, 8).

Most of the measurements were carried out under so-called
"standard state" to obtain reproducible results, i.e., the
"standard swollen state" is obtained after a half-hour boiling the
membranes in water, and the "standard-dry state" is obtained after
drying the membranes at 107°C for 18 hrs in vacuum oven. Hereafter the measurements under the standard state are simply designated as "dry" or "swollen" state. Some measurements were carried
out under nonstandard states; the membranes dried at room temperature are designated as "room-temperature dry" membranes, and the
specimens immersed in water at room temperature for a few days are
designated as "immersed" membranes.

X-RAY CRYSTALLINITY

Figure 1 shows typical WAXD profiles for a series of the
sulfonic-acid membranes having different E.W. under dry state
where the measured profiles are shown with the data points marked
by dots, while the curves "b" to "d" are decomposed profiles. The
measured profiles were decomposed into the fundamental profiles
"b" to "d" where the decomposition was achieved by least-squares
fit of the measured profile with the reconstructed curve "a" obtained by a superposition of the three fundamental profiles "b"
to "d". In order to attain the least-squares fit, each fundamental profile was assumed to be either Gaussian, Lorentzian, a
linear combination of them or a linear polynomial (for the background scattering profile) with their peak positions, heights, and
widths, a relative abundance of Gaussian and Lorentzian (for the
case of a linear combination of them) and the parameters for the
polynomials as variables. It is seen that the measured profiles
satisfactorily agree with the reconstructed curves "a" for all
the cases.

A measured profile is decomposed into a broad peak at $2\theta \simeq$
16.1° and full-width at half maximum (FWHM) $\simeq$ 4° and a sharp peak
at $2\theta \simeq$ 17.7° and (FWHM) $\simeq$ 1.2°, which are assigned, respectively,
to diffraction maxima associated with noncrystalline and crystalline regions of the membranes in our previous work(4). It is
clearly seen that the crystallinity increases with increasing
E.W.. Figure 2 shows typical change of the profiles with tempera-

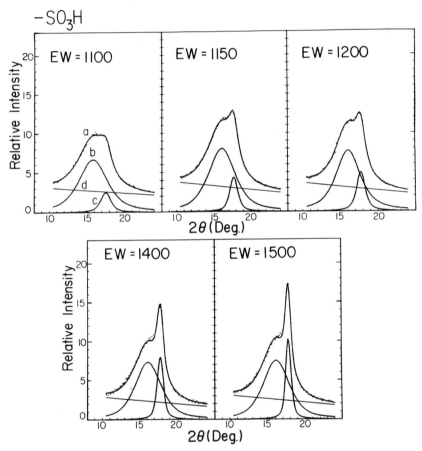

Figure 1. WAXD profiles for a series of the sulfonic acid membranes under dry state. The measured profiles are shown by dots, while Curves b–d are the decomposed profiles by the least squares method, and the Curve a is reconstructed from Curves b–d.

ture for the sulfonic-acid membranes having 1500 E.W.. It clearly shows that the sharp peak looses its intensity while the broad peak is enhanced as a consequence of decreasing crystallinity with temperature. The crystalline peak for the sulfonic-acid membranes having 1100 E.W. completely disappears at 275°C(4). Figure 3 shows an effect of annealing of the membrane having 1500 E.W., showing that an annealing at 275°C increases the crystallinity of the membrane.

We estimated X-ray crystallinity X_c (a weight average crystallinity), the lattice spacing of the crystalline peak, and full-width at half maximum (FWHM) of the crystalline peaks.

$$X_c = \int_0^\infty I_{cr}(s)s^2 ds \; / \int_0^\infty [I_{cr}(s) + I_{am}(s)]s^2 ds \tag{1}$$

where $s = (2 \sin \theta)/\lambda$, 2θ is the scattering angle, and λ is the wavelength of X-ray. I_{cr} and I_{am} are the relative scattered intensity of the sharp and broad peaks, respectively. The results for the series of the sulfonic-acid and the carboxylic-acid membranes are shown in Figure 4 and in Tables I and II. The results for the membranes having 1100 E.W. but different functional groups are listed in Table III.

As seen from Figure 4, (and also from Tables I and II), X-ray crystallinity increases with increasing E.W. of the membrane. Moreover, FWHM decreases, and the lattice spacing also tends to decrease slightly with increasing E.W., indicating that the perfection of the crystals also becomes high with E.W.. For a given E.W. the membranes having the carboxylic-acid groups have a greater crystallinity and a higher perfection on crystals than those having the sulfonic-acid groups. For a given E.W., the crystallinity is a function of cations, as well as anions ($-SO_3^-$ and $-COO^-$), as seen in Table III and Figure 5, i.e., the crystallinity of the membranes decreases in the order of H^+, Na^+, and C_s^+.

The increase of the crystallinity with increasing E.W. for a given functional groups is natural and is a consequence of a decreasing number of noncrystallizable units. On the other hand, the variation of crystallinity with the cations or with the anions for a given E.W. is closely related to cluster formation(9) of the ionic sites which generally perturbs crystallization, giving rise to lower crystallinity. As will be discussed later, ability of forming the ionic clusters through the dipole-dipole interaction between the ion-dipole complexes(10) increases in the order of $-SO_2Cl$, $-COOH$, $-SO_3H$, $-SO_3Na$, and $-SO_3Cs$, resulting in (i) increasing SAXS spacing S (i.e., increasing cluster size) for the "ionic scattering maximum" (the scattering maximum arising from the clusters) in this order, as seen in Table III, and also in (ii) decreasing crystallinity, as seen in Table III and Figure 5.

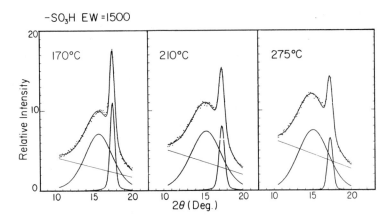

Figure 2. *Temperature dependence of the WAXD profiles for the sulfonic acid membranes having 1500 EW.*

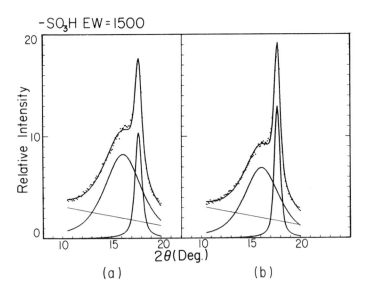

Figure 3. *Effect of annealing of the sulfonic acid membranes having 1500 EW; the membranes as received (a), and the membranes annealed at 275° C (b).*

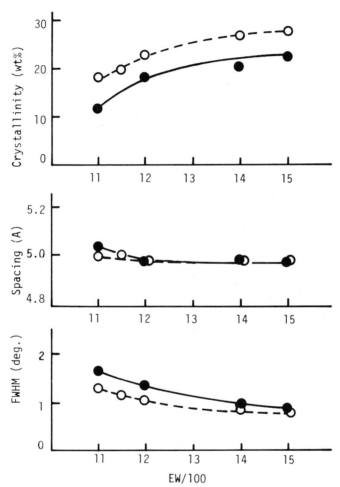

Figure 4. X-ray crystallinity X_c, the lattice spacing, and the FWHM of the crystalline peak as a function of EW for the sulfonic acid (–●–) and carboxylic acid (- -○- -) membranes under dry state.

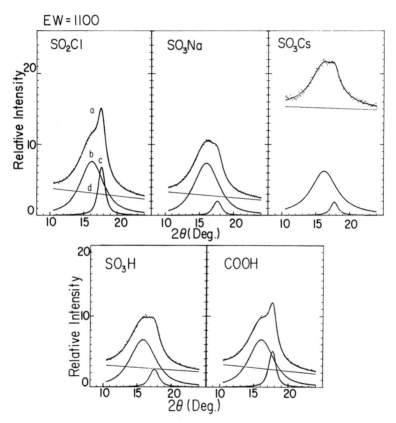

Figure 5. WAXD profiles for various perfluorinated ionomer membranes having 1100 EW under dry state.

Table I. Physical Properties of Sulfonic-Acid Form of
Perfluorinated Ionomer Membranes as a Function of
Equivalent Weight.

| E.W. | Water Uptake | | Crystalline Phase | | SAXS Spacing (nm)[c] | | |
	Overall (wt%)	N[a]	FWHM[b] (deg.)	X_c (wt%)	Lamellar(D) Dry	Ionic(S) Dry	Wet
1100	32	19.6	1.6	12	18.3	2.6	5.5
1150	23	14.7	—	—	—	—	—
1200	20	13.3	1.4	19	18.7	2.7	5.2
1400	10	7.8	1.0	20	22.2	2.8	4.4
1500	8	6.7	0.9	22	24.9	2.8	4.4

a) number of water molecules per one $-SO_3H$ group.
b) FWHM of the decomposed profile due to crystals in the wide-
angle X-ray diffraction profile.
c) spacing estimated from Bragg's equation.

Table II. Properties of Carboxylic-Acid Form of Perfluorinated
Ionomer Membranes as a Function of Equivalent Weight.

| E.W. | Water Uptake | | Crystalline Phase | | SAXS Spacing (nm)[c] | | |
	Overall (wt%)	N[a]	FWHM[b] (deg.)	X_c (wt%)	Lamellar(D) Dry	Ionic(S) Dry	Wet
1100	6	3.6	1.3	18	13.6	—	4.1
1150	5	3.1	1.2	20	15.5	—	4.0
1200	4	2.6	1.1	23	14.6	—	3.8
1400	3	2.3	0.9	27	16.9	—	3.4
1500	4	3.0	0.8	28	20.2	—	3.5

a) number of water molecules per one $-COOH$ group.
b) FWHM of the decomposed profile due to crystals in the wide-
angle X-ray diffraction profile.
c) spacing estimated from Bragg's equation.

LONG-IDENTITY PERIOD

Figure 6 shows typical SAXS profiles for the various ionomer
membranes having 1100 E.W. under dry state. In general the profiles contain two types of scattering maxima, one at small s =
$(2 \sin \theta)/\lambda \simeq 0.07 nm^{-1}$, and the other at large s = $0.3 nm^{-1}$ (θ and
λ being one half of the scattering angle and wavelength of X-ray,
respectively) as previously found by Girke et al[18] and Roche
et al[19]. We proposed in our previous paper[4] that the small-
angle scattering maximum arises from a long identity period of the
lamellar platelets and the large-angle scattering maximum (desig-

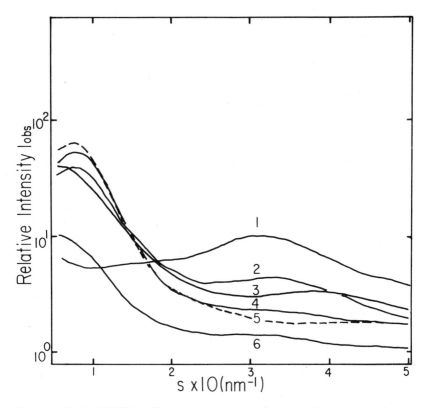

Figure 6. Typical SAXS profiles for the various perfluorinated ionomer membranes having 1100 EW under dry state; S = (2 sin θ)/λ, 2 θ is the scattering angle. Key: 1, —SO₃Cs; 2, —SO₂NHR; 3, —SO₃H; 4, —COOH; 5, —SO₂Cl; 6, —SO₃Na.

Table III. A Comparison of Various Perfluorinated Ionomer
Membranes for a Given Equivalent Weight 1100.

Chemical Modifications	Water Uptake Overall (wt%)	N	Crystalline Phase FWHM (deg.)	X_c (wt%)	SAXS Spacing (nm) Lamellar(D) Dry	Ionic(S) Dry	Wet
$-SO_3H$	32	19.6	1.6	12	14.6	2.6	5.5
$-SO_3Na$	27	16.9	1.5	8	14.6	3.1	5.5
$-SO_3Cs$	13	8.9	1.0	3	—	3.2	5.6
$-SO_2NHR$	3	2.0	—	—	12.6	3.0	3.8
$-COOH$	6	3.6	1.3	18	13.6	—	4.1
$-COONa$	—	—	—	(14)	12.3	(2.8)	4.6
$-SO_2Cl$	1	0.6	1.2	23	13.2	—	—

nated as "ionic-scattering maximum") arises from the ionic clusters. The long identity period D was measured for the series of the ionomer membranes from the position of the small angle scattering maximum $2\theta_{ms}$ by using Bragg's equation,

$$2D \sin\theta_{ms} = \lambda \qquad (2)$$

The results were summarized in Tables I to III and in Figure 7. As seen in Figure 7 and Tables I and II, for a given functional group, the spacing D increases with increasing E.W., which corresponds to increasing crystallinity and perfection of the crystals with E.W. as observed from the WAXD profiles. This tendency is again interpreted to result from decreasing number of the noncrystallizable units with increasing E.W.. For a given E.W., the sulfonic-acid membranes have greater D than the carboxylic acid membranes. This difference may be best interpreted to result primarily from a difference in thickness of the interlamellar amorphous layer. That is, as will be discussed later, in comparison with the carboxylic-acid membranes, the sulfonic-acid membranes have a larger cluster size owing to a greater electrostatic energy released upon cluster collapse. The clusters disturb crystallization to result in smaller crystallinity and thickner interlamellar amorphous layer.

IONIC CLUSTERS

Models For Ionic Clusters. Figure 6 also shows the SAXS profiles for the various ionomer membranes having 1100 E.W. under dry state at large s region where the ionic scattering maximum appears. The membranes having unionized or only weakly ionized groups (e.g., $-SO_2Cl$ and $-COOH$) do not exhibit the ionic scattering maximum, while the membranes having ionized groups (e.g., $-SO_3H$, $-SO_2NHR$ (where R is $-(CH_2)_3NH_2$), and $-SO_3Cs$) exhibit the maximum, indicating clearly that the scattering maximum is associated with the ionic sites, its spatial distribution and organization.

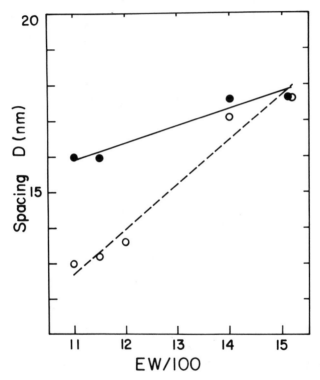

Figure 7. *The long identity periods D for the sulfonic acid (–●–) and carboxylic acid (- -○- -) membranes under dry state as a function of EW.*

In our earlier papers(4, 11) the ionic scattering maximum is proposed to arise from the ionic clusters(9) which are essentially ion-rich regions containing some fluorocarbon chains. The clusters are stabilized by the dipole-dipole interaction between the ion-dipole complexes such as $-SO_3^-\cdots\cdots H^+$(10). The electrostatic energy released upon the cluster collapse tends to be balanced with the eleastic free energy associated with deformation of fluorocarbon chains to form the clusters, giving rise to optimum size of the clusters(12).

We have shown that the two basic models, as shown in Figure 8, can explain the ionic scattering maximum from the perfluorinated membranes(4,11). These two models are nothing other than those proposed to explain the ionic scattering maximum for the hydrocarbon-based carboxylated ionomers. (i) *Two-phase model* (Figure 8(a)), proposed by Cooper et al.(13), in which the ionic clusters are dispersed in the matrix composed of fluorocarbon chains and non-clustered ions (multiplets(12)). It should be noted that in the original two-phase model proposed by Cooper et al. the clusters do not contain polymer chains but are composed only of ionic sites, i.e., multiplets. The scattering maximum is attributed to an ordered spatial organization of the multiplets in a paracrystalline lattice. In this paper, however, we extend the definition of the two-phase model so that the model includes the systems in which the ion-rich regions (defined as clusters in the book by Eisenberg and King (9)) are dispersed in the matrix of intermediate ionic phase composed of fluorocarbon chains and nonclustered ions. The ionic scattering maximum is then attributed to an *inter-cluster interference*, reflecting an average inter-cluster distance. The Debye's hard-sphere type scattering(14) may be the simplest possible model to describe the scattering maximum. (ii) *Core-shell model* (Figure 8(b)), proposed by Macknight, Stein and their coworkers(15,16), in which the ionic cluster (ion-rich core) is surrounded by a shell which is rich in fluorocarbon chains. The core-shell particles are dispersed in the matrix composed of fluorocarbon chains and non-clustered ions. In this model the scattering maximum arises essentially from *intraparticle interference* of the core-shell particle. The scattering angle $2\theta_{m1}$ giving rise to the maximum scattering is related to the short-range order distance. The shape of the core-shell particles has been originally assumed to be spherical(15,16) and modified later to be of lamellar type(17). The core-shell model has been found to be more appropriate than the two-phase model from the studies on variation of the ionic scattering maximum upon swelling the membranes with water(11). In Table I to III are shown the spacing S as estimated from $2\theta_{m1}$ by applying Bragg's equation,

$$2S \sin \theta_{m1} = \lambda \qquad (3)$$

The membranes having sodium sulfonates or sodium carboxylates

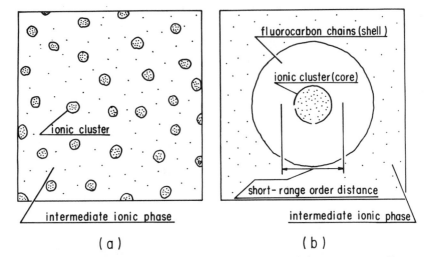

Figure 8. Two models describing the spatial organization of the ionic sites. a: Two-phase model composed of ionic clusters (ion-rich regions) dispersed in a matrix of the intermediate ionic phase, which is composed of fluorocarbon chains and non-clustered ions. The ionic scattering maximum arises from an interparticle interference effect, reflecting an average intercluster distance S. b: Core–shell model in which the ion-rich core is surrounded by an ion-poor shell composed mostly of perfluorocarbon chains. The core-shell particles are dispersed in the intermediate ionic phase. The scattering maximum arises from an interparticle interference effect, reflecting a short-range order distance S of the core-shell particle. Note that the crystalline region was not drawn in the model for the sake of simplification and that the shape of the core–shell particle may not necessarily be spherical.

do not exhibit clearly the ionic scattering maximum, which has
been found not to be due to absence of the ionic clusters but
rather simply due to very small electron density difference between the clusters and the surrounding medium(4).

Effects of Anions and Cations. For a given E.W. and anions,
the spacing S for the sodium-sulfonate membranes is greater than
that for the sulfonic-acid membranes, and for a given E.W. and
cations, the spacing S for the sodium-sulfonate membranes is
greater than that for the sodium-carboxylate membranes. (The
spacing S for the sodium salts was estimated by extrapolating the
spacing measured as a function of degree of swelling with water to
zero degree of swelling). These differences in the spacing may be
interpreted in terms of difference in the dipole-dipole interaction. The larger the interaction, the larger is the electrostatic energy released upon the cluster collapse, and consequently the greater is the cluster size.

It should be noted that the spacing S does not generally have
one-to-one correspondence to the cluster size. The change of S
results from either (i) the change of an average inter-cluster distance (for the two-phase model) or (ii) the change of a short-
range order distance (for the core-shell model). In our systems
the change of the average inter-cluster distance or the short-
range order distance is closely associated with the change of the
cluster size, and, in our discussion above and hereafter, we
assume one-to-one correspondence between the change of spacing and
that of the cluster size for qualitative discussions. One should
note, however, that the short-range order distance in the core-
shell model (which is estimated from the peak position of the
ionic scattering) depends not only on the size of the core and
shell but also the electron densities of the core and shell(11).
The latter effect on S which is small compared with the former is
neglected in our discussion. We are not also concerned with the
estimation of an absolute value of the cluster size. We remain
to state that the cluster size is order of S.

Effect of E.W. The cluster size depends on E.W. as shown in
Figure 9(a) and in Table I, the greater the E.W., the larger the
cluster size. The electrostatic energy released per an ionic site
upon cluster collapse should essentially be constant with E.W..
However the thermodynamic work of the elastic deformation of polymer coils required for the cluster collapse should decrease with
increasing E.W., i.e., with increasing average molecular weight of
molecules between the ionic sites, which leads to increasing cluster size with E.W.. One should note that the free energy for the
elastic deformation of the coils is modified by crystallinity
which is also a function of E.W.. This factor may complicate our
interpretation.

Effect of Temperature. The ionic scattering maximum exists

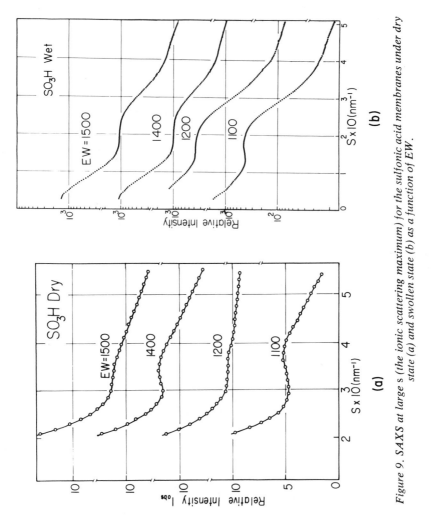

Figure 9. SAXS at large s (the ionic scattering maximum) for the sulfonic acid membranes under dry state (a) and swollen state (b) as a function of EW.

even at temperatures above the melting temperature, as in the
case of ethylene ionomers(15). For example we have shown in our
previous paper(4) that the cesium sulfonate membranes having 1100
E.W. exhibit the ionic maximum at 276°C, slightly above the melting temperature of the membranes. Upon elevating the temperature
the ionic maximum shifts slightly toward larger s, indicating that
the inter-cluster distance or the short-range order distance (and
consequently the cluster size) decreases with temperature. The
decrease in the cluster size with temperature may be rationalized
by increased thermodynamic work for the deformation of polymer
coils to form the cluster (which is proportional to Boltzmann free
energy k_BT). This thermodynamic factor may outweigh the effect of
the thermal expansion which tends to make the clusters bigger.

Effect of Deformation. As we discussed in details elsewhere
(11), the ionic scattering changes with stretching the membranes.
Figure 10 represents typical oscilloscope traces showing variations of the ionic SAXS profiles upon stretching the cesium-
sulfonate membranes having 1100 E.W. under dry state. The scattering profiles were measured by an optical system with which the
effective weighting function becomes isotropic, i.e., with a point
collimating system and with a hight-limiting slit in front of the
PSD. In a direction parallel to stretching direction (a), the
ionic maximum shifts toward smaller angles and its peak intensity
decreases with the draw ratio λ. On the other hand, in a direction perpendicular to stretching direction (b), the maximum
shifts toward larger angles and its intensity increases with λ.

The deformation behaviors have been interpreted in terms of
the two basic models(11), (i) the deformed two-phase model in
which the interparticle distances and the particles, initially
giving rise to Debye's hard-sphere type scattering(14), are affinely deformed under constant volumes (designated as "deformed
hard-particles") and (ii) the deformed core-shell particle model
in which a spherical core-shell particle is affinely deformed
under constant volume into an ellipsoidal core-shell particle.
It has been found(11) that the both models give a similar effect
of deformation and can account for the variations of the SAXS profiles at an early stage of deformation. However at large deformation the simple models cannot account for the observed profiles
and requires some modifications(11). It should be noted that the
variations of the profiles with deformation may also be described
in terms of orientation and deformation of non-spherical clusters
(17).

SWELLING BEHAVIOR

Water Uptake. Amount of water uptaken by the membranes was
measured for various perfluorinated ionomer membranes in the
standard state. The results are summarized in Tables I to III
and also shown in Figure 11 where the percentage of water uptaken

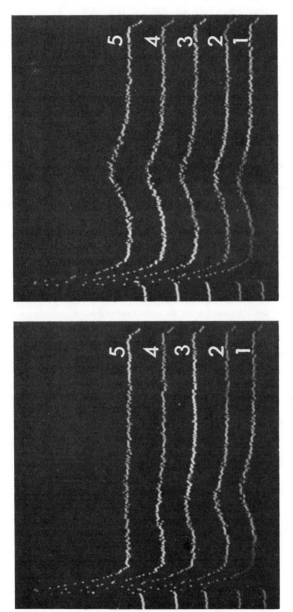

Figure 10. Typical oscilloscope traces showing variations of the ionic SAXS profiles upon stretching the cesium–sulfonate membranes having 1100 EW under dry state in a direction (left) parallel ($\mu = 0°$) and (right) perpendicular ($\mu = 90°$) to stretching direction; λ is draw ratio of the membranes. Key: 1, $\lambda = 1.0$; 2, $\lambda = 1.1$; 3, $\lambda = 1.2$; 4, $\lambda = 1.3$; 5, $\lambda = 1.5$.

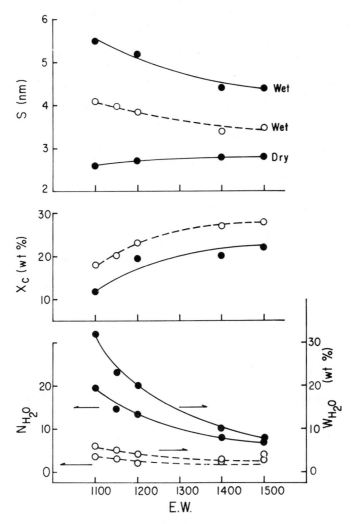

Figure 11. Ionic spacing S, crystallinity X_c, H_2O uptake per a membrane (W) and per an ionic site (N) for the sulfonic acid (−●−), and carboxylic acid (- -○- -) membranes as a function of EW.

by the membranes, W = 100 (weight of water uptaken by the membrane) / (weight of the standard-state dry membrane), and number
of water molecules uptaken per a given functional group, N, are
plotted as a function of E.W., together with other important and
measurable quantities such as percentage crystallinity X_c and the
spacing S measured from the ionic scattering maximum for the sulfonic-acid and carboxylic-acid membranes.

For both the sulfonic-acid and the carboxylic-acid membranes,
W decreases with increasing E.W., due to decreasing number of
ionic sites and also to increasing thermodynamic work of expand-
ing the clusters with water in the medium of a higher rigidity
resulted from a higher crystallinity. The decrease of N with E.W.
which is especially drastic for the sulfonic-acid membranes should
be obviously related to the increased thermodynamic work of the
swelling due to the increased crystallinity of the medium. In
comparison to the carboxylic-acid membranes, the sulfonic-acid
membranes have lower crystallinity and higher affinity to water,
resulting in both N and W for the sulfonic-acid membranes being
higher than those for the carboxylic-acid membranes for a given
E.W..

In Table III it is shown that, for a given anion(e.g.-SO_3^- and
-COO^-) both W and N decrease with the cations in the order of
proton, sodium, and cesium, despite of the fact that the crystallinity decreases with the cations in this order. This result also
may be interpreted in terms of the increasing affinity of the
functional groups to water in the order of cesium, sodium, and
proton for a given anion.

As will be discussed in detail in next section, the water
uptaken by the membranes, expands the ionic clusters into bigger
sizes, resulting in increased ionic spacing S as shown in Figure
9(b). For the sulfonic-acid membranes, although the ionic spacing
in dry state slightly increases with E.W., the spacing in the
swollen state decreases with E.W., which again is related to increased crystallinity with E.W., similarly to the E.W. dependence
of N. The ionic spacings for the wet sulfonic-acid membranes are
greater than those for the wet carboxylic-acid membranes, which
again should be related to smaller crystallinity and higher affinity of the sulfonic acid groups to water in comparison to the
carboxylic-acid groups.

Microscopic Swelling. In this section we shall discuss
change of the ionic scattering maximum with water uptaken by the
membranes. Figure 9(b) shows change of the ionic SAXS maximum
upon swelling the sulfonic-acid membranes having various E.W..
Comparisons with Figure 9(a) clearly indicate that upon swelling
the scattering intensity dramatically increases and scattering
maximum shifts toward smaller s for all the membranes. This
suggests that, upon swelling the membranes, the clusters grow in
size and the density of clusters decreases relative to that of
the surrounding medium as a consequence of preferential water

uptake by the clusters(4,11). It should be noted that the ionic spacings in the wet membranes decrease with increasing E.W. as shown in Figures 9(b) and 11, although they are much larger than those in the dry membranes.

Figure 12 shows the ionic SAXS profiles for various membranes having 1100 E.W. in the standrad swollen state. The neutral membranes (e.g. sulfonyl-chloride membranes) do not naturally exhibit the ionic scattering maximum. In comparison to the ionic maxima in dry state (Figure 6), the ionic scattering maxima for all the membranes (except for sulfonyl-chloride membranes) under swollen state are shifted toward smaller angles, as a consequence of preferential water uptake by the clusters. This preferential water uptake by the clusters generally enhances the electron density difference between the clusters and surrounding medium, resulting in enhanced scattering intensity. For the sulfonated membranes, the wet ionic spacings S or the cluster sizes under swollen state are almost independent of the cations, cesium, sodium, and proton, despite of the facts that the crystallinity or the rigidity of the medium increases in the order of cesium, sodium, and proton and that the cluster size in dry state decreases in this order. This may be interpreted in terms of increasing affinity of the functional group to water in this order and is well correlated with the variations W and N with the cations.

One can control an equilibrium amount of water uptake by the membranes and also the size of the ionic clusters by changing relative humidity of the membranes or by immersing the membranes in aqueous solutions of sodium chloride with different concentrations(11). Figure 13 shows change of SAXS profiles upon immersing the sodium-sulfonated (a) and -carboxylated membranes having 1100 E.W. (b) into the aqueous sodium-chloride solutions with various concentrations.

As clarified in our previous paper(4), the sodium-sulfonated and -carboxylated membranes in the room-temperature dry state do not exhibit the ionic scattering maximum, simply because the electron density of the clusters closely matches to that of the medium. When the clusters preferentially uptake water, their electron densities decrease relative to the medium, giving rise to the enhanced ionic scattering maximum. The amount of water uptake by the membranes and consequently size of the ionic clusters decrease with increasing concentration of sodium chloride, resulting in the ionic scattering maximum looses its intensity and shifts toward larger scattering angle with increasing the concentration. The amount of water uptaken by the sulfonated membranes are much larger than the carboxylated membranes as shown in Table III in terms of W and N. This gives rise to much stronger ionic scattering maximum and less distinct SAXS maximum at small s (associated with interlamellar spacing) for the sulfonated membranes in comparison to the carboxylated membranes. It should be noted that, for the sulfonic-acid membranes, the electron density of the clusters is lower than that of the medium even under dry

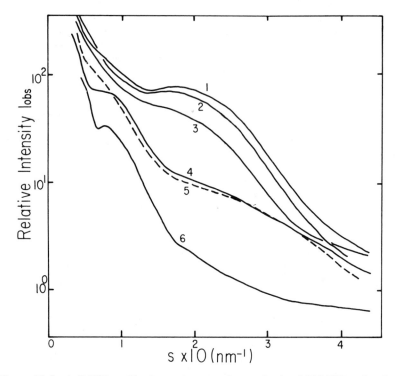

Figure 12. Ionic SAXS profiles for various membranes having 1100 EW under the swollen state. Key: 1, —SO_3H; 2, —SO_3Na; 3, —SO_3Cs; 4, —COOH; 5, —SO_2NHR; 6, —SO_2Cl.

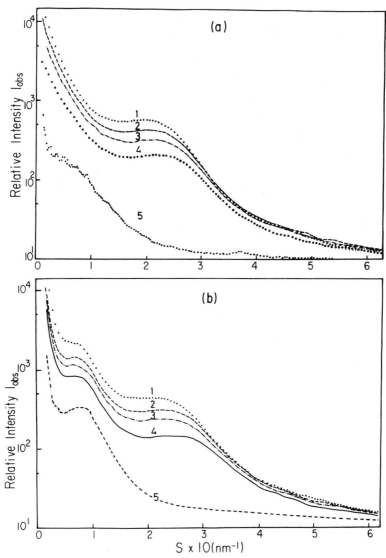

Figure 13. SAXS profiles for the sodium-sulfonated (a), and the sodium-carboxyl-ated (b) membranes having 1100 EW immersed in aqueous NaCl solutions of 0% (1), 5% (2), 10% (3), 20% (4), and dry (5).

state. The density of the clusters is further reduced upon up-taking water, resulting in further increase of the intensity(11). On the other hand in case of the cesium sulfonated membranes, the density of the clusters is higher than the medium under dry state. Therefore the ionic scattering maximum first looses its intensity with water uptake. Upon further increase of amount of water up-take, the ionic intensity becomes zero and then increases(11). The increase of the intensity is due to the fact that the density of the clusters becomes lower than that of the medium.

One can investigate a relationship between the microscopic degree of swelling λ_d as estimated from the ionic scattering and the macroscopic degree of swelling λ_B as estimated from the changes of bulk dimensions or mass of the membranes with water uptake. Figure 14 shows the results obtained for the sulfonic-acid and sodium-sulfonated membranes having 1100 E.W. where λ_d is the ratio of the wet ionic spacing to the dry ionic spacing, and λ_B is the ratio of the bulk dimension in wet to that in dry. The water uptake was controlled either by changing relative humidity (solid circles) of the membranes or by immersing the membranes in the aqueous solution of sodium chloride (open circles).

It is very important to note that λ_d is much larger than λ_B. If the ionic scattering maximum arises from the interparticle interference, the ionic spacing reflects an average interparticle distance. This change of the spacing with water uptake should closely correspond to the change of the bulk dimension, i.e., λ_d should be approximately equal to λ_B. On the other hand, if the ionic scattering amxium arises from a short-range order distance in the core-shell particles, and if water is preferentially up-taken by the clusters, λ_d can be much larger than λ_B. Thus this swelling behavior favors the core-shell model rather than the two-plase model(11).

SMALL-ANGLE LIGHT SCATTERING

Depolarized light scattering studies were pursued in order to investigate crystalline superstructure, a higher-level struc-ture than the lamellar crystallites. Figures 15 and 16 show, respectively, the H_V scattering patterns for the carboxylic-acid and sodium-carboxylated membranes and those for the sulfonic-acid and sodium-sulfonated membranes under room temperature dry, soaked in water and ethanol. All the membranes have 1100 E.W.. The carboxylic-acid and carboxylated membranes which have higher crys-tallinities and well defined interlamellar spacings than the sul-fonic-acid and sulfonated membranes clearly exhibit the spheruli-tic scattering (7), while the sulfonic-acid and sulfonated mem-branes do not exhibit spherulitic scattering but rather an aniso-tropic scattering attributed to local strain existing in the membranes.

Upon swelling the carboxylic-acid and carboxylated membranes, the scattering angle θ_{max} at which the H_V scattered intensity

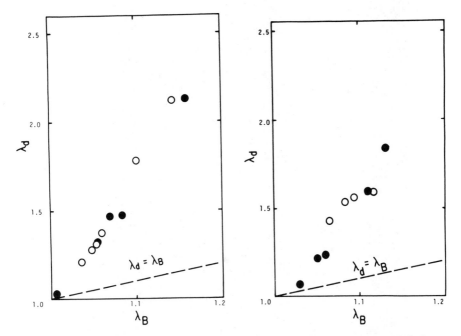

Figure 14. Relationships between the microscopic degree of swelling λ_d and the macroscopic degree of swelling λ_B for the sulfonic acid (left) and sodium-sulfonated (right) membranes having 1100 EW. Amount of H_2O uptake was controlled either by changing relative humidity (●) of the membranes or by immersing the membranes in the aqueous solution of NaCl of various concentration (○).

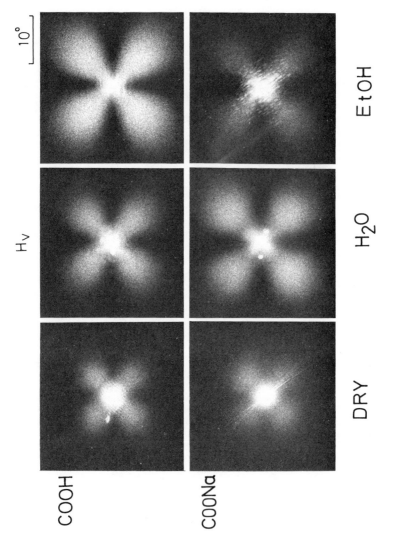

Figure 15. H_V light-scattering patterns for the carboxylic acid and sodium-carboxylated membranes having 1100 EW under room-temperature dry state, and those immersed in H_2O and C_2H_5OH.

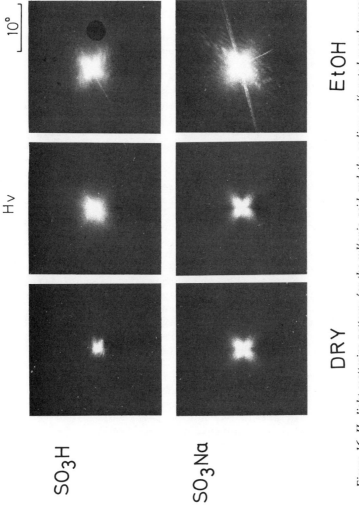

Figure 16. H_V light scattering patterns for the sulfonic acid and the sodium-sulfonated membranes, having 1100 EW under room-temperature dry state and those immersed in H_2O and C_2H_5OH.

reaches to maximum intensity shifts toward larger scattering angles. The greater the degree of swelling, the larger the shift in θ_{max}. The degree of swelling of the membranes for ethanol is much larger than that for water. One can calculate an average spherulite radius R from θ_{max}

$$4.1 = 4\pi(R/\lambda)\sin(\theta_{max}/2) \tag{4}$$

where λ is the wavelength of light in the medium and θ is the scattering angle in the medium.

The shift of θ_{max} toward large angles suggests that the average size of the spherulites R decreases upon swelling the membranes as shown in Figure 17 where L and L_0 are the bulk dimensions of the membranes under wet and dry states, respectively. The decrease of the spherulite size should be a consequence of inhomogeneous swelling of the membranes in the scale comparable to the spherulite size. That is., for the membranes having such a low crystallinity, spherulites may not be volume-filling but rather dispersed in noncrystallized matrix, and it is quite conceivable that the degree of swelling in the interspherulitic amorphous region is greater than that with the spherulite. This inhomogeneous swelling may exert stress to destroy partially the orientation correlation of the optically anisotropic scattering elements of the spherulites, especially in the peripheral regions, resulting in smaller size of the spherulites when observed under the H_V scattering. The inhomogeneous swelling may be envisaged to   occur if number of the ionic clusters in the interspherulitic amorphous region is greater than that within the spherulites.

It should be noted that the change of the spherulite size is reversible. That is, as shown in Figure 18 for the carboxylated membranes, the H_V pattern is expanded upon immersing the membranes in methanol and is contructed to original size upon deswelling the membranes. Upon deswelling, the stress is released to result in reconstitution of the spherulitic orientation correlation in the disordered peripheral regions.

The depolarized pattern due to frozen-in strain in the sulfonic-acid and sulfonated membranes does not vanish but is rather enhanced upon swelling the membranes with water and ethanol (Figure 16). This strain may not be released, unless the crystallinity and the clusters vanish during the swelling process, which does not seem to be quite probable.

CONCLUDING REMARKS

The perfluorinated, carboxylated and sulfonated ionomer membranes form the ionic clusters of a few nm in size, as in the case of the hydrocarbon-based ionomers such as polyethylene,polystyrene and polybutadiene(9). The ionic clusters strongly affect physical properties of the membranes, e.g., the swelling behavior of the membranes (amount of water uptaken by the membranes, W and

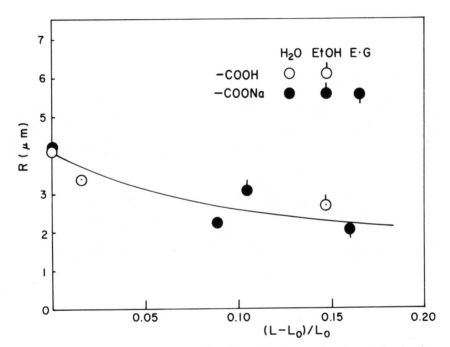

Figure 17. Change of the average spherulite radius R as a function of degree of swelling for the carboxylic acid and sodium carboxylated membranes having 1100 EW. L and L₀ are the bulk linear dimensions under wet and room-temperature dry states, respectively.

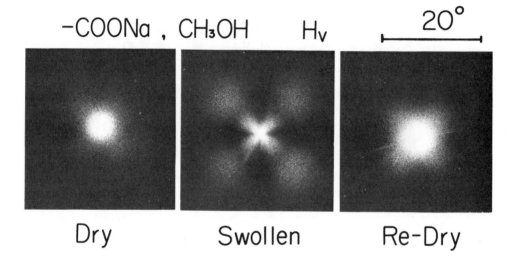

Figure 18. Reversible change of the H_v light-scattering pattern upon swelling and deswelling the sodium-carboxylated membranes having 1100 EW with CH_3OH.

that by a single functional groups, N). They also affect crystallinity X_c, the interlamellar spacing D, and spherulite size R,
and any physical properties associated with these quantities.

The size of the ionic clusters or the ionic spacing S (as
estimated from the scattering angle giving rise to the ionic peak)
are shown to depend on the equivalent weight (E.W.), cations (H^+,
Na^+ and Cs^+), anions ($-SO_3^-$ and $-COO^-$), and temperature. The de-
pendencies of the cluster size on these factors are identical to
those found for the hydrocarbon-based carboxylated ionomers, e.g.,
ethylene ionomers(9,15) and can be basically described by a
theory proposed by Eisenberg(12). That is an optimum size of the
clusters are determined by the two counterbalancing physical factors; (i) the energy released upon the cluster collapse and (ii)
the thermodynamic work of deforming polymer coils which is required for the cluster collapse.

For a given E.W. (1100), in comparison with the membranes
having sulfonic acids, the membranes having carboxylic acids have
smaller ionic clusters due to the weaker electrostatic energy released upon the cluster collapse, leading to greater X_c, better-
defined D and spherulitic structure. The sulfonic-acid and sulfonated membranes do not have spherulitic structure. The greater
X_c of the carboxylic-acid membranes, together with a lower affinity of carboxylic-acid groups to water result in lower W and
smaller N, which in turn results in smaller cluster size in wet
in comparison to the sulfonic-acid membranes.

For a given E.W. and for a given anion, the electrostatic
energy released upon the cluster collapse increases with cations
in the order of proton, sodium, and cesium, resulting in greater
cluster size and lower X_c in this order. Despite of the fact that
X_c increases with the cations in order of Cs^+, Na^+ and H^+, W and
N also increase in this order, which may be best interpreted in
terms of affinity of the functional groups to water being increased in this order. The intrinsic hydration power of the cations,
i.e., the number of water molecules that a given cation can
hydrate, increases in the order of H^+, Cs^+, and Na^+. Therefore
the greater overall affinity of Na^+ to water in comparison to that
of Cs^+ may be explained in terms of this hydration power. The
largest affinity of proton to water, however, cannot be explained
by the hydration power only but should be explained in terms of
detailed morphology of clusters.

For a given functional group, as E.W. increases, the elastic
free energy of polymer coil deformation required for the cluster
collapse decreases, resulting in bigger cluster size under dry
state. Moreover as E.W. increases, fraction of noncrystallizable
ionic groups decreases, resulting in increasing X_c and D. This
increase of X_c results in increased energy required to swell the
ionic clusters, thus giving rise to decreasing N and W with increasing E.W. which, in turn, can account for the decreasing the
ionic spacing S in wet with E.W..

ACKNOWLEDGEMENTS

The authors are indebted to Toyo Soda Manufacturing Co.,
LTD., Yamaguchi-ken, Japan for kindly providing the perfluorinated
ionomer membranes, especially to Drs. T. Hashimoto, N. Kawasaki,
and M. Fukuda. The authors also acknowledge to Mr. T. Takamatsu,
Toyo Soda Manufacturing Co., LTD., for his technical contributions
to this paper.

LITERATURE CITED

1. see for an example, "Perfluorocarbon Ion Exchange Membrane",
   Presented at the 152nd. National Metting, the Electrochemical
   Society, Atlanta, Georgia, October 10-14 (1977).
2. Yeo, S.C.; Eisenberg, A. J.Appl Polym. Sci. 1977,21, 875.
3. Hodge, I.M.; Eisenberg, A. Macromolecules, 1978, 11, 289.
4. Fujimura, M.; Hashimoto, T.; Kawai, H. Macromolecules, 1981,
   14, in press.
5. Fujimura, M.; Hashimoto,T.; Kawai, H. Memoirs Fac. Eng.,
   Kyoto Univ., 1981, 43(2), in press.
6. Hashimoto, T.; Suehiro, S.; Shibayama, M.; Saijo, K.; Kawai,
   H. Polymer J.,1981 13, 501.
7. Stein, R.S.; Rhodes, M.B. J. Appl. Phys., 1960, 31, 1873.
8. Stein,R.S.; Wilson, P.R. J. Appl. Phys., 1962, 33, 1914.
9. Eisenberg, A.; King, M. "Ion-Containing Polymers", Academic
   press, N.Y., 1977.
10. Hopfinger, A.J.; Mauritz, K.; Hora, C.J., in ref.1.
11. Fujimura, M.; Hashimoto, T.; Kawai, H. Macromolecules, 1981,
    1981, 14, in press.
12. Eisenberg, A. Macromolecules, 1970, 3, 147.
13. Marx, C.L.; Caulfield, D.F.; Cooper, S.L. Macromolecules,
    1973, 6, 344.
14. Debye, P. Physik. Z., 1927, 28, 135.
15. Macknight, W.J.; Taggart, W.P.; Stein, R.S. J. Polym. Sci.,
    1974, C45, 113.
16. Kao, J.; Stein, R.S.; Macknight, W.J.; Taggart, W.P.; Cargill
    III G.S. Macromolecules, 1974, 7, 95.
17. Roche, E.J.; Stein, R.S.; Russel, T.P.; Macknight, W.J.
    J. Polym. Sci., Polym. Phys. Ed., 1980, 18, 1497.
18. Girke, T.D. in ref. 1.
19. Roche, E.J.; Pineri, M; Dupplessix, R.; Levelut, A.M.
    J. Polym. Sci., Polym. Phys. Ed., 1981, 19, 1.

RECEIVED September 1, 1981.

Neutron Studies of Perfluorosulfonated Polymer Structures

M. PINERI, R. DUPLESSIX,[1] and F. VOLINO

Equipe Physico-Chimie Moléculaire, Section de Physique du Solide, Département de Recherche Fondamentale, Centre d'Etudes Nucléaires de Grenoble, 85 X, 38041 Grenoble Cedex, France

X-ray and electron microscopy have been widely used to provide direct information on the structure of polymers. The necessary contrast is given by differences in electronical density. X-ray diffraction is the necessary technique for small molecule crystallography and the scattering by an object in solution is also used as a source of structural information. The small angle scattering originates from the existence inside the material of domains whose size is generally the radiation wavelength (1 - 5). A model of structure has been developped from electron microscopy and X-ray studies in Nafions (6). No basic difference exists between X-ray and neutron techniques. Nuclear interactions of neutrons with matter are characterized by the coherent scattering length and the corresponding values for H and D are very different Because of the different origins of the contrast, X-ray and neutron small angle scattering techniques are complementary.

Interactions of neutrons with matter

Properties of the neutron. The free neutron is an elementary particle with zero charge and spin 1/2, liberated for example during the process of fission of a heavy nucleus. In a nuclear reactor, the neutrons are thermalized by the atoms of the moderator yielding a Maxwellian distribution of velocities v peaked at some $\bar{v}$ such that the average (kinetic) energy $\bar{E}$ is

$$E = \frac{1}{2} m_n \bar{v}^2 = \frac{3}{2} k_B T$$

[1] Current address: C.R.M., 6 rue de Boussingault, 67083 Strasbourg, France

0097-6156/82/0180-0249$08.50/0

where m_n is the mass of the neutron.
Neutrons can also be considered as plane waves of wave number $\underset{\sim}{k}$ or
wavelength $\lambda = 2\pi/k$. The relationships between particle and wave
aspects are :

$$E = \frac{\hbar^2 k^2}{2m_n}$$

and

$$v = \frac{\hbar k}{m_n}$$

For thermal neutrons (T = 300K), we have $\bar{E}$ = 26 meV and
$\bar{\lambda}$ = 1.8 Å. It is important to note that these values are of the
same order of magnitude as the intermolecular energies and
molecular dimensions, respectively.

Neutron-nucleus interaction. A neutron interacts with a nucleus
via nuclear and magnetic forces. For the nuclear part, since
nuclear interactions are very short range compared to the(thermal)
neutron wavelength, it can be shown that the interaction
potential between a neutron located at r and a nucleus i located
at $\underset{\sim}{r_i}$ can be written as

$$V(\underset{\sim}{r}) = \frac{2\pi\hbar^2}{m_n} \, b_i \, \delta(\underset{\sim}{r} - \underset{\sim}{r_i})$$

In this expression (the so-called Fermi pseudo-potential), the
scattering length b_i characterizes the interaction and is independent of neutron energy, b_i can be positive or negative
according to the attractive or repulsive nature of the interaction
The theorical calculation of b_i is very difficult and in practice
it is determined experimentally. Concerning the magnetic interaction, the neutron interacts with the spins via the dipole-
dipole coupling. For diamagnetic systems, it is always negligible
compared to the nuclear interaction and will be considered in the
following.

Coherent and incoherent scattering lengths and cross-sections.

Consider an assembly of a given atomic species i, with
many isotopes possessing a nuclear spin. The scattering length b_i
will change from one atom to another, since the interaction
depends on the nature of the nucleus and on the total spin state
of the nucleus neutron system.
The average $< b_i >$ of b_i over all the isotopes and spin
states is called the coherent scattering length. We thus have

$$b_i^{coh} = < b_i >$$

$$b_i^{incoh} = [< b_i^{\,2} > - < b_i >^2]^{1/2}$$

From these definitions, it is clear that b^{coh} and b^{incoh} can be changed merely by modifying the relative concentration of the various isotopes. This fact is of great practical importance in neutron experiments (isotopic substitution). The coherent and incoherent scattering cross-section are defined by

$$\sigma_i^{\,coh} = 4\pi (b_i^{\,coh})^2$$

$$\sigma_i^{\,incoh} = 4\pi (b_i^{\,incoh})^2$$

In the following table are listed the values of these quantities in barns (1 barn = 10^{-24} cm^2) for a few common atoms.

Table I

atoms	spin	σ^{coh}	σ^{incoh}
^{16}O	0	4.2	-
^{12}C	0	5.5	-
^{14}N	1	11.6	$\sim$ 0.3
D	1	5.6	2.0
H	$\frac{1}{2}$	1.76	79.7

Small angle neutron scattering experiment.

Information about the structure of a material is obtained from small angle scattering by analyzing the coherent scattered intensity which is given by

$$I(\underset{\sim}{Q}) = <\sum_i \sum_j d_i \, d_j \, \exp[-i\underset{\sim}{Q}.(\underset{\sim}{r}_i - \underset{\sim}{r}_j)]>$$

where Q is the scattering vector whose magnitude is given by $4\pi\sin\theta/\lambda$, where θ is half the scattering angle and λ is the wavelength of the incident radiation. The vector $\underset{\sim}{r}_i$ is drawn from an arbitrary origin to a point which possesses a coherent scattering factor d_i. For a system with two scattering components the previous equation can be simplified to give

$$I(\underset{\sim}{Q}) = C(d_1 - d_2)^2 \, S(\underset{\sim}{Q})$$

where

$$S(\underset{\sim}{Q}) = <\sum_i \sum_j \exp[i\underset{\sim}{Q}.(\underset{\sim}{r}_i - \underset{\sim}{r}_j)]>$$

where d_1 and d_2 denote the scattering factors of region 1 and 2, C is a constant, and $<\,>$ denotes an equilibrium average. For small-angle neutron scattering (SANS), the scattering factor is proportional to the coherent neutron scattering-length density β^{coh}, giving the result that

TABLE II

Scattering parameters of Nafion-Water system

Group	b $(10^{-14} \, cm \, \overset{\circ}{A}{}^{-3})$	e $(mole \, el/cm^3)$	d (g/cm^3)
CF_2 amorphous	3.84	0.91	1.9
CF_2 crystalline	4.45	1.05	2.2
RSO_3H	2.5-3.7	0.5-0.75	1-1.5
RSO_3Na	2.7-3.9	0.53-0.8	1.1-1.6
H_2O	-0.56	0.55	1
D_2O	6.46	0.55	1.1

$$(d_1 - d_2)^2 = (\beta_{1_{coh}} - \beta_{2_{coh}})^2$$

The coherent scattering length density is defined by

$$\bar{\beta}_{coh} = \rho \sum_K n_K b_K / \sum_K n_K m_K$$

where ρ is the mass density, and n is the number of atoms with coherent scattering length b and mass m. For SAXS the contrast is determined by the difference in electron density

$$(d_1 - d_2)^2 = (\sigma_1^e - \sigma_2^e)^2$$

Values of coherent neutron scattering length densities and electron densities relevant to the Nafion-water system are given in Table II. Values for CF_2 are given corresponding to mass densities of 2.2 and 1.9 g/cm^3 for crystalline and amorphous materials, respectively. An excellent SANS contrast exists between either amorphous or crystalline CF_2 and H_2O. A difference of 4 10^{-14}cm.$\text{\AA}^{-3}$ is generally considered good enough to give a reasonable signal. The difference between the side groups and the CF_2 backbone is less well defined owing to uncertainties in the mass densities but is certainly much small. Thus the SANS technique is much less sensitive to aggregation of the ionic groups than it is to aggregation of water molecules. A similar situation exists for SAXS with relatively large electron density difference between CF_2 and water and a smaller contrast between CF_2 and the ionic side groups.

Information about both the size and the geometry of the scattering particles can be obtained from different scattering domains. In the Guinier range with QR < < 1 the scattering function can be written

$$S(Q) \sim 1 - \frac{Q^2 R^2}{3}$$

where R is the radius of gyration of the diffusing object. In the intermediate range the scattering function can be written

$$S(Q) \sim \frac{1}{Q^n}$$

Table III

Sample   Treatment

Code	Form	Thermal treatment	Hydration level	$\Delta W/W$	N_{H_2O}/SO_3R
A	acid	as received	dried 23°C	0.02	1.7
B	acid	as received	15 % R.H.	0.04	3.4
C	acid	as received	50 % R.H.	0.07	5.8
D	acid	as received	90 % R.H.	0.15	12.5
E	acid	as received	soaked (1hr)	0.20	16.7
F	acid	as received	boiled (1hr)	0.45	37.5
G	Na^+	as received	dried 23°C	0.02	1.7
H	Na^+	as received	15 % R.H.	0.04	3.4
I	Na^+	as received	50 % R.H.	0.06	5.0
J	Na^+	as received	90 % R.H.	0.11	9.2
K	Na^+	as received	soaked (1hr)	0.20	16.7
L	Na^+	as received	boiled (1hr)	0.60	50.0
M	Na^+	quenched	dried 200°C	0	0
N	Na^+	quenched	dried 23°C	0.02	1.7
O	Na^+	quenched	50 % R.H.	0.05	4.2
P	Na^+	quenched	83 % R.H.	0.09	7.5
Q	Na^+	quenched	soaked (15min)	0.14	11.7
R	Na^+	quenched	soaked (24hr)	0.16	14
S	Na^+	quenched	boiled (1hr)	0.36	30

with n = 1 for rods, n = 2 for a gaussian coil or flat disk and
n = 3 for a two density system.

The specificity of neutrons in structure determination
of hydrophilic materials has its origin in this large difference
between the H and D scattering lengths. It is therefore possible
to only change the contrast between phases without changing the
structure of the studied material. Detailed analysis has been
done concerning this contrast variation method which is mainly
used in biological systems. In a simple two phase system with
hydrophilic diffusing particles in an hydrophobic matrix it will
be shown that a simple superposition of the diffusion curves
can be obtained with different H_2O/D_2O relative concentrations.

Small angle neutron experiments

Nafion 120 acid samples were obtained from the Dupont
Company. Neutralization was carried out by soaking for several
days in a 2 M aqueous solution of NaOH. The neutralization was
measured to be of the order of 77 % by flame photometry. A list
of samples and corresponding thermal and hydration treatments is
given in Table III. The samples as received were given no additional thermal treatment and are semicrystalline. Quenched
samples were made by heating samples to 330°C for one hour and
then passing hydrogen gas over the film for quenching. Quenching
speeds obtained by such a procedure are one or two orders of
magnitude larger that usual quenching in liquids. Before hydration
samples were dried at 23°C under flowing nitrogen gas and then
were hydrated at different humidity levels in a hydration cell.
Different relative humidities (R.H.) were achieved by controlling
the temperature of a water bath through which air is passed
before entering the cell. Higher absorption was achieved by
soaking or boiling samples for various times. Table III lists the
weight of water absorbed relative to the dry weight $\Delta W/W_D$ along
with the corresponding number of water molecules per sulfonate
group NH_2O/SO_3R. The dry weight corresponds to that of a sample
dried at 200°C.

The influence of water content on the scattering curves
is shown in fig. 1 for the acid Nafion [6]. A maximum corresponding to a Q value of $3.5\ 10^{-2}\ \text{Å}^{-1}$ (Bragg spacing of 180 Å) is
apparent for the low content samples and this maximum becomes a
shoulder when increasing the water content. A similar maximum
has been reported by Gierke [7] in a small angle X ray experiment.
Essentially analogous results in this Q range were found for the
Na^+ samples as shown in fig. 2, indicating that the neutralization
has little effect on the structure giving rise to the scattering.

From SAXS studies it was found that the samples possess
a degree of crystallinity of the order of 15-20 % based on an
analysis of the crystalline diffraction ring at 5.0 Å and the
amorphous halo at 4.9 Å. These correspond to rings observed for
polytetrafloroethylene. Dark field electron microscope studies of

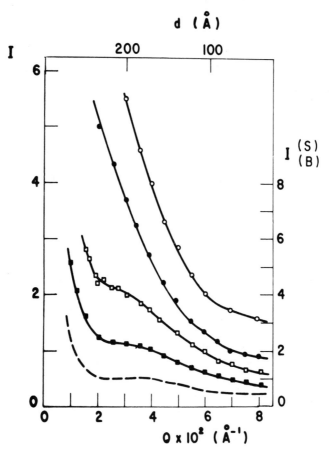

Figure 1. SANS curves in region of the first scattering maximum as a function of humidity for semicrystalline Nafion acid samples. Key: – – –, A, dry; ■, C, 50% R.H.; □, D, 90% R.H.; ●, E, soaked; ○, F, boiled.

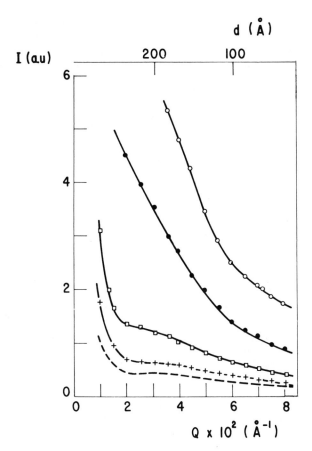

Figure 2. SANS curves in region of the first scattering maximum for Nafion Na+ (not quenched) samples. Key: — — —, G, dry; +, I, 50% R.H.; □, 90% R.H.; ●, K, soaked; ○, L, boiled.

these Nafion polymers also show regions with dimensions of
several hundred Ångströms that give rise to crystalline diffraction patterns. In amorphous films prepared by the quenching
technique the 180 Å Bragg maximum is eliminated (Fig. 3).

From these results the 180 Å peak can clearly be assigned
to interference between crystalline structures. Lamellar structures with a periodicity of ∿ 1000 Å are known to exist in polytetrafloroethylene [8]. Copolymerization would be expected to
reduce the structural repeat unit if one considers that the large
side groups would be excluded from the crystalline regions. For
a random copolymer an average distance of 15 Å would exist between
side groups which is much smaller than the observed periodicity.
Therefore it must be concluded that either a very non-random
distribution of side groups is present at least in some portion of
the material or that some side groups are somehow included in the
crystalline region.

A second SANS maximum is observed at larger Q for all
samples. The intensity of scattering increases and the position
of the scattering maximun shifts to lower angles when increasing
water content. Such a behavior is shown in fig. 6.

From these initial results we can point out the following
conclusions :

- The perfluorinated bakbone of Nafion polymer forms
superstructures which give rise to a scattering maximun corresponding to a Bragg distance of 180 Å.
- The microstructure of the acid and salt forms of the
membranes is very similar. This point makes a drastic difference
between these membranes and the other ion containing polymers
which exhibit some phase separation only after neutralization.
- The scattering signal arising from the hydrophilic
ionic regions exhibits a maximum over a wide range of water
concentrations. The interpretation of this maximum is still under
discussion. Models involving an internal ionic aggregate structure of a paracrystalline ordering of the aggregates are possible
but difficult to understand because of the absence of any known
range ordering potential.

The technique of isotopic replacement (H_2O/D_2O) has then
been used to define if there are more than two phases in the
quenched material and also to define the relative amount of water
in the ionic phase. If the quenched material can be described as
a two phase system, for a well defined amount of water absorbed
we can change the relative contrast between phases by changing
the relative percentage H_2O/D_2O of the absorbed water. The intensity ratio corresponding to two different H_2O/D_2O values is
therefore independent of the scattering vector and the scattering

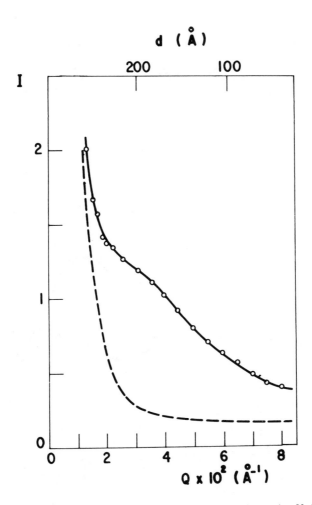

Figure 3. SANS curves in region of the first scattering maximum for Nafion Na⁺ (90% R.H.) samples. Key: ◯, J, not quenched; – – –, P, quenched.

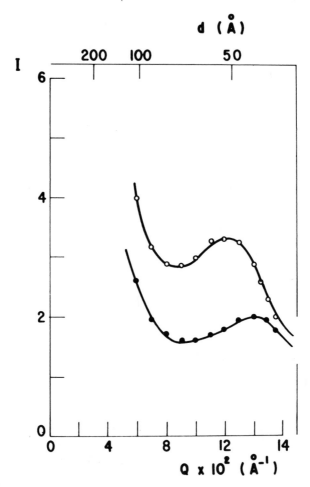

Figure 4. SANS curves in region of the second scattering maximum for Nafion acid samples. Key: ●, E, soaked 1 h; ◯, F, boiled 1 h.

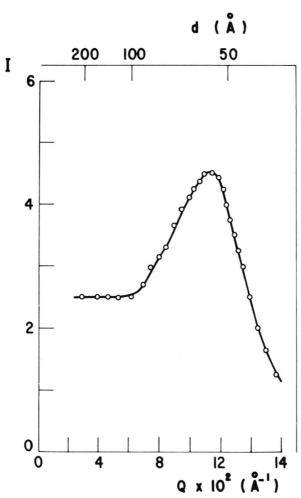

Figure 5. SANS curve in region of second scattering maximum for Nafion Na+ quenched sample S boiled 1 h.

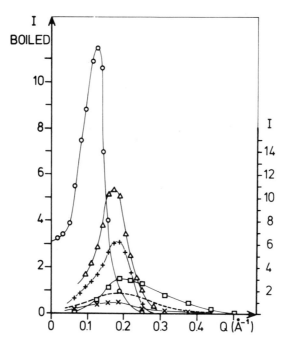

Figure 6. SANS curves in region of the second scattering maximum for Nafion Na+ samples. Key: ×*, N, dried 23° C;* —*, 50% R.H.;* □*, P, 83% R.H.;* +*, Q, soaked 15 min;* △*, R, soaked 1 h;* ○*, S, boiled.*

curves differ only by a constant multiple. For systems involving
more than two phases this will not generally be the case. No
structural changes are expected to occur on isotopic replacement.

A superposition of SANS curves for samples hydrated at
100 % R.H. with different mixtures of H_2O/D_2O is shown in fig. 7.
These samples contain about 9 % by weight water. The Q range
studied corresponds to the zero order scattering region. Before
superposition a constant value corresponding to the incoherent
scattering intensity has been subtracted from each curve. A
correction has also been made for differences in the sample
transmission factors. Subsequently a constant of multiplication has
been chosen to give the best possible overall superposition. A
good superposition is achieved particularly for Q > 0.01. An
analogous superposition is shown in fig. 8 for samples soaked
24 hours in a mixture of H_2O and D_2O and which contain 16 % by
weight water. For a sample soaked in a mixture of 12.5 % $H_2O/87.5$
D_2O essentially no scattering was observed above a constant
background. This result in itself indicates that the scattering
system contains essentially two phases or contrast regions. The
matching concentration is known as the isopicnic point.

A superposition of data is shown in fig. 9 for samples
boiled 1 hour which contain 36 % by weight water. Here the region
of the scattering maximun has been explored. In this case the
isopycnic point was determined to correspond to a concentration of
25 % $H_2O/75$ % D_2O.

In fig. 10 are plotted the ratio $\dfrac{\beta_1 - \beta_2}{\Delta\beta_{H_2O}} = \alpha$ versus $D_2O/$
H_2O. $\beta_1 - \beta_2$ represents the contrast difference bet-
ween two phases and $\Delta\beta_{H_2O}$ corresponds to this value for the polymer containing only H_2O. These α values have been obtained from
the superposition factors used in fig. 7, 8 and 9. A sign reversal
occurs in α values for D_2O/H_2O ratios beyond the isopicnic point.
For a sample hydrated with H_2O a large difference in the scattering
lenght density exists between the phase containing most of the water and the low water content matrix which contains mainly CF_2
units. As the ratio $[D_2O]/[H_2O]$ changes, this contrast decreases
resulting in a decreased scattered intensity. Finally beyond the
isopycnic point the scattering length density of the hydrophobic
phase exceeds that of the matrix and the intensity increases again.

The theoretical linear dependence of α versus D_2O/H_2O for
a two phases system, is only observed for the high water content
samples. A large deviation drom the linearity occurs in the 9 %
water content specimen. The two phase approximation is therefore
less valid in this sample and the superposition which was observed
may result from a fortuitous choice of multiplication constants.

These results indicate that both the zero order scattering component and the scattering maximum arise from the same
scattering length density differences. The zero order scattering

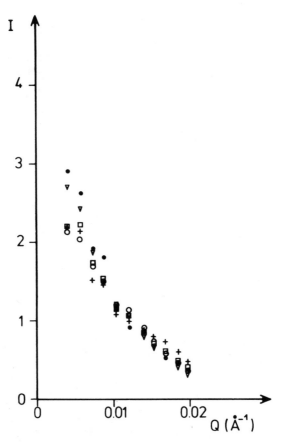

Figure 7. Superposition of SANS curves (in low Q region) for samples P with mixtures of H_2O/D_2O of 0.0 (○), 0.5 (□), 0.75 (+), 0.875 (▽), and 1 (●).

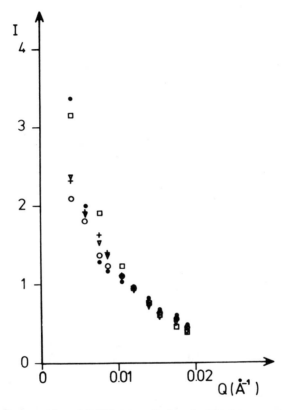

Figure 8. Superposition of SANS curves (in low Q region) for samples R soaked in mixtures of H_2O/D_2O of 0.0 ($\bigcirc$), 0.25 ($\triangledown$), 0.5 ($\square$), 0.75 (+), and 1 ($\bullet$).

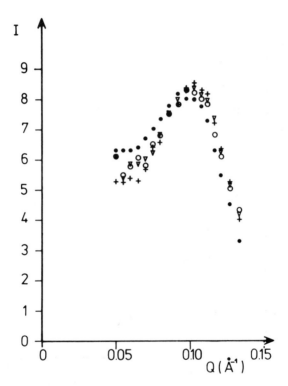

Figure 9. Superposition of SANS curve (in high Q region) for samples S boiled in mixtures of H_2O/D_2O of 0.0 ($\bigcirc$), 0.25 (+), 0.875 ($\triangledown$), and 1 ($\bullet$).

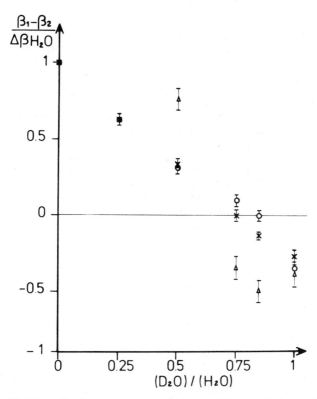

Figure 10. *Multiplication used in superposition of Figures 2–4 as a function of* $[H_2O]/[D_2O]$.

does not therefore arise from impurities. The existence of a
scattering maximum indicates that a spatial ordering exists
between different water rich regions. The zero order scattering
presumably indicates that there is an inhomogenous distribution
of these regions of clusters. A discussion of this component for
other ionomer system has been recently given [9]. Different
models have been tested to obtain a more quantitative idea of the
chemical composition of the two phases [10]. A comparison has
also been made with the known values of the SAXS mean squared
electron density fluctuation.

The model fits show that the Nafion-water system
can be considered as essentially two phases at water content of
15 % by weight or greater with a majority of the water molecules
clustered in one phase and the vast majority of the $-CF_2-CF_2-$
type units in the other phase. The current study cannot rule out
the existence of some non-clustered water. In addition little can
be said about the location of the ionic groups. It would be
expected that a majority of such groups would be found in a region
between phases. The existence of such an interfacial region can be
detected from the behavior at large Q. Our SANS measurements do
not extend to large enough θ to analyze such behavior but appropriate plots from the SAXS data are shown in fig. 11. For a two
phase system with an infinitely sharp boundary and no density
fluctuations within phases the scattered intensity follows
Porod's law which, for the case of slit smeared intensities,
yields :

$$\lim_{Q \to \infty} Q^3 I(Q) = K$$

If a transition zone exists negative deviations from Porod's law
occur while density fluctuations within phases will lead to positive deviations. A detailed discussion of Porod law behavior has
been recently given [11]. Here we observe good agreement with
Porod's law indicating no apparent transition zone for the soaked
and boiled samples. It is possible that the transition zone
thickness is very small compared to the dimension of the phases
themselves and that its effect is not observable. Cutler [12] has
reported transition zone thicknesses of 5 Å for Nafion samples
containing water but gave no plots. Such a value is of the order
of the minimum observable by SAXS. A sample hydrated at 83 % R.H.
shows positive deviations from Porod's law indicating that the
phases are less pure at low water contents. Such positive deviations have been previously observed for other ionomer systems at
low water contents [13]. This result is consistent with the previously stated conclusion that the two-phase model is less valid
at lower water contents. From theses isotopic replacement experiments it is therefore shown that at high water content (15 % by
weight) Nafion membranes are essentially two phases systems. If a

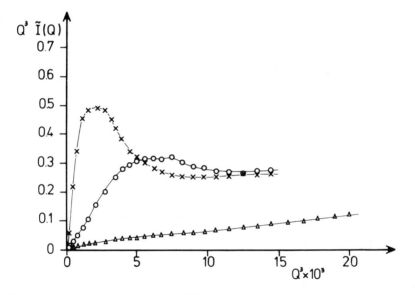

Figure 11. Porod law plots from SAXS soaked P and boiled S samples. Key: ×, *boiled 1 h;* ○, *soaked 1 h;* △, *83% R.H.*

transition zone exists between phases its thickness must be very
limited as indicated by SAXS behavior at large Q. At lower water
contents deviations from the two phase model are observed by both
the SANS isotopic replacement experiments and the SAXS behavior
at large Q.

The neutron quasi-elastic scattering (NQES) method

In an inelastic scattering experiment, one analyzes the
energy of the scattered neutrons, for several scattering angles.
The term quasi-elastic is used when the analysis is restricted
to small energy changes. If the molecular motions are of diffusive
nature, and are sufficiently fast, the NQES spectra generally
appear as a broad curve centered around ω = 0 where ω is the
neutron energy transfer. Various spectra components may often
be separated. The time scale of the molecular motions is given by
the width of the spectra, and the nature of these motions is
contained in the angular dependence of the shape, and of the
relative intensities of the various components. If the scattering
is mainly incoherent, as is generally the case with hydrogen
containing systems, then the data should be analyzed in terms of
self motion of the moving protons. This can be done by direct
comparison of the theorical incoherent scattering law corresponding to a given model, with the spectra. The NQES method yields
in fact a rather severe test of the model since it should fit not
only one, but many spectra.

With one component systems like plastic crystals [14] or
liquid crystals [14, 15] the experimental situation is rather
simple since only the nuclei of interest contribute to the spectra
With a composite system like water absorbed in a membrane, it is
important to be able to separate the contribution to the scattered intensity from the water, and from the polymer. A useful indication on the relative intensities of these contributions may be
obtained by comparison between the neutron diffraction patterns
of the wet and dry membranes. This is because the intensity of
the diffraction curve at a given Q value is approximately the
integral over the energy of the NQES spectrum at the same Q. This
is why it is often useful to make a neutron diffraction experiment
before performing the quasielastic study.

Neutron diffraction results

Diffraction experiments were performed on the D1B diffractometer of the ILL. Fig. 12 shows the results for the wet and
dry samples. The spectra are very similar to those obtained by
X-rays [7]. Their main feature is the existence of a broad peak
between 0.9 and 1.4 $Å^{-1}$ which is due to the lateral packing of the
polymer chain. This broad peak is usually decomposed into a broad
and a sharp component, as schematically indicated by a dashed
line on the spectrum of the dry sample. The components are

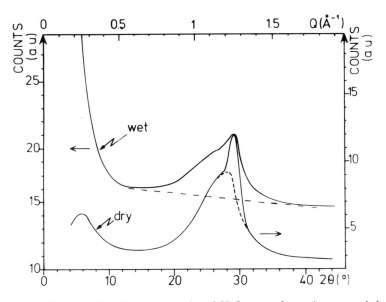

Figure 12. Neutron diffraction spectra of acid Nafion membrane for wet and dry samples. Instrument is D 1B. Incident wavelength is $\lambda_0 = 2.56$ Å. The horizontal axis is expressed either in scattering angles (2 θ) units, or in neutron momentum transfer ($Q = 4\pi/\lambda_0 \sin\theta$) units. Note the different vertical scales for the two samples. The separation between the amorphous and crystalline contributions to the large angle peak is shown for the dry sample. Temperature ∼ 25° C.

attributed to the amorphous and crystalline parts of the membrane, respectively. Despite the similarity between these two
diffraction spectra, there are three important differences :

      i) the broad peak is significantly broader for the wet
sample than for the dry sample. This means that the wet material
is more disordered,
      ii) the wet sample exhibits very intense small angle
scattering below 0.4 $\AA^{-1}$ while the dry sample does not. This
means that this scattering should be essentially attributed to
the water, and this simple result means that the water is not
uniformly distributed in the sample. The extensive study of this
scattering [6,7,10] was mainly aimed at the determination of the
structure and the distribution of the corresponding "clusters",
      iii) the background is much higher for the wet than for
the dry sample (note the different vertical scales in fig. 12).
This extra background should be attributed to the incoherent
scattering of the protons of the water molecules. This background
level is in theory independent of Q. However due to inelastic
effects, a small decrease is observed, as indicated on the figure.

      These results show that the main contribution to the
scattered intensity comes from the water molecules. For
$Q > 0.4$ $\AA^{-1}$, it is reasonable to assume that the corresponding
scattering is incoherent since the broad peak aroud 1.2 $\AA^{-1}$ comes
mainly from the polymer. Consequently, the analysis of the quasi-
elastic spectra may be carried out in terms of self motion of the
protons. For $Q < 0.4$ $\AA^{-1}$ the existence of intense small angle
scattering due to water means that the coherent scattering is no
more negligeable and that it may be important to include collective effects in the analysis. In the present work, we have
restricted the NQES study to Q values between 0.4 and 1.2 $\AA^{-1}$ so
that the latter effects can be neglected.

Neutron quasi-elastic scattering results

      The quasi-elastic experiments were performed on the
multichopper time-of-flight spectrometer IN5 at the ILL. Three
incident wavelenghts were used : 10, 11 and 13 $\AA$ corresponding
to (elastic) energy resolutions $\Delta\omega$ of 18.5, 14 and 9μeV
(1μeV = 1.52 x 10^9rd/s) full width at half maxumum (FWHM), respectively. For each wavelenght, eight spectra were recorded
simultaneously at different scattering angles corresponding to
the Q range 0.4 to 1.2 $\AA^{-1}$.

      Two typical spectra obtained with wet and dry samples
are shown in fig. 13. For the wet sample, the NQES spectra are
composed of a sharp peak reproducing the resolution function of
the instrument superimposed on a broad component having about
100μeV FWHM. The dry sample on the contrary, exhibits only the
sharp peak. This observation immediately tells us that the broad

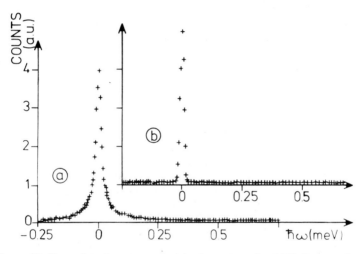

Figure 13. Examples of neutron quasielastic spectra of acid Nafion membrane for Q = 0.59 Å⁻¹. a: Membrane containing 15% H_2O by weight, (13n) dried membrane. b: Reproduces exactly the experimental resolution function obtained with a V sample. Instrument IN5, incident wavelength 10 Å, energy resolution 18.5 µeV FWHM. Temperature ∼ 25° C.

component of the wet sample spectrum is due to the water, and
consequently that the water molecules move on the time scale of
ca. 10^{-11} sec, i.e. practically as fast as in bulk water. In
order to get information on the nature of these motions, it is
necessary to analyze in detail all the NQES spectra.

We first studied the shape and the width of the broad
component. For this purpose, we excluded the central part of the
spectra and fitted a single lorentzian line, convoluted with the
resolution function, to the remaining points of the spectra. The
result is that the experimental curve systematically deviates
from the lorentzian shape, the deviation being more pronounced as
Q increases, in our experimental range. This is illustrated in
Fig. 14a and 14b. This procedure was used to define the experimental width as the width of the lorentzian which yields the best
fit as given by a standard least square fit routine. In fig. 15,
we have shown the experimental width obtained in this way versus
Q^2, for all the spectra recorded. It is seen that the width is
practically constant at low Q and then increases at larger Q. The
meaning of such results can be understood by comparison with the
predictions of the following classical models :

- long range sel-diffusion characterized by a diffusion
coefficient D_t. In this case, the NQES spectra are single
lorentzian lines whose HWHM is equal to $D_t Q^2$. In the log-log plot
of fig. 15, this means that the points are on a straight line
whose slope is 1. Since the experimental line shapes are not
lorentzian and the experimental widths are clearly not proportional to Q^2, this model must be rejected for water in nafion.
At this point it is interesting to compare the present NQES data
with data obtained with bulk water. Such an experiment was
performed on IN5 at 10 Å [16]. In this case (i) the experimental
spectrum is lorentzian within experimental accuracy (cf fig. 14c)
and (ii) the corresponding HWHM is given by $D_t Q^2$ (cf fig. 15),
with $D_t = 2.5 \times 10^{-5} cm^2/s$, which is an accepted value at 28°C for
water [20, 21]. This last result gives confidence, in the ability
of the high resolution NQES method to test quantitatively models
for molecular motions.

- Rotation of the water molecules. Since this rotational
motion is probably nearly isotropic, the incoherent scattering
law is given to a good approximation by the Sears formula [19].
In this model the protons are assumed to diffuse on a sphere of
radius ρ. The diffusion is characterized by a rotational diffusion coefficient D_r. This model predicts that the broad component of the spectra is no longer loretzian : its width is
constant at low Q and then increases at higher Q, as observed
in our experiment. We have thus tried to fit this model to the
spectra. Taking $\rho = 0.95$ Å which is the distance between the
center of mass and the protons [20], we cannot fit the spectra

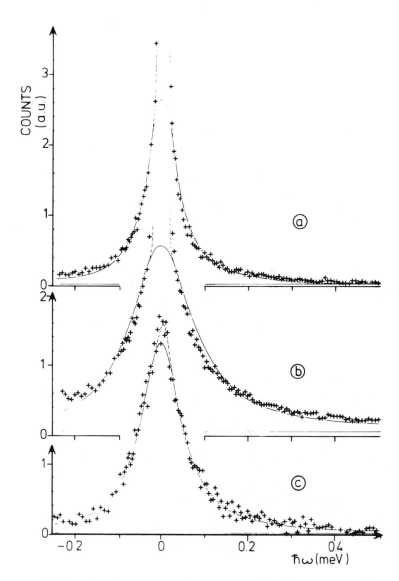

Figure 14. Broadened component of neutron quasielastic spectra, experimental points and best fit Lorentzian line. a: Nafion membrane containing 15% H_2O, Q = 0.59 $\overset{\bullet}{A}^{-1}$. b: Idem, but Q = 1.05 $\overset{\bullet}{A}^{-1}$. c: Bulk pure H_2O at 28° C, Q = 0.59 A^{-1}. The small sharp component in Figure 14c comes from the quartz sample holder. Instrument IN5, incident wavelength 10 $\overset{\bullet}{A}$. Temperature ~ 25° C. Note the systematic deviation from the Lorentzian shape in Figures 14a and 14b.

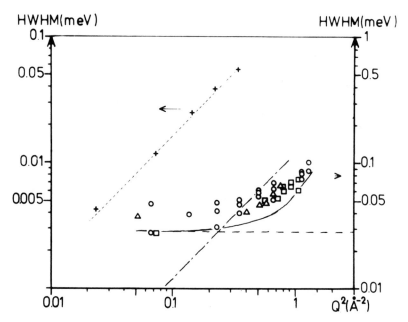

Figure 15. Half width at half maximum of the broadened component of the neutron quasielastic spectra obtained with an acid Nafion membrane containing 15% H_2O by weight. The points are the widths of the best fit Lorentzian lines from spectra obtained with an incident wavelength of 10 Å (○), 11 Å (□), and 13 Å (△). The full line is the theoretical width predicted by the model with diffusion in a sphere (with D = 1.8 × 10⁻⁵ cm²/s and a = 4.25 A). The two theoretical asymptotes for Q → O and Q → ∞ are also shown (compare with Figure 2 of Reference 2 for more details). Half width at half maximum of the best fit Lorentzian lines to the spectra obtained with bulk water at 28° C (incident wavelength 10 Å) is denoted by +. The straight line passing through the points (+) is the theoretical width predicted by the simple self diffusion model with D_t = 2.5 × 10⁻⁵ cm²/s. Note the different vertical scales for the Nafion sample and the bulk water sample.

for any D_r. Taking ρ as an adjustable parameter, it seems that some fit may be found for $\rho \sim 3$ Å. Since this value does not correspond to any "radius" in the water molecule, we have pushed the analysis further, but this result has some interesting meaning as we shall see below. In any case, this shows that the NQES data cannot be explained in terms of the rotation of the water molecules only.

- Translational self-diffusion and rotation of the water molecules. Since the above two models fail to explain the data, one may think of a model which combines both, and which is certainly more realistic. We have tried such a possibility with $\rho = 0.95$ and $D_t = 1.6 \times 10^{-6}$ cm^2/s. This last value is the long range self-diffusion coefficient of water in this membrane, measured by radioactive tracers. We found that no fit is possible with these values whatever D_r is chosen. As for the preceeding section, we find that the fit improves considerably if we take either ρ or D_t as parameters. With D_t fixed, we should increase ρ to ~ 3 Å, as above, and with ρ fixed, we should increase D_r to $\sim 10^{-5}$ cm^2/s. These results suggest that one should think of a model which contains these two features. The simplest one is a model where the water molecules, more precisely the protons, are restricted to diffuse (diffusion coefficient D) in a sphere of radius a, where we expect $D \sim 10^{-5}$ cm^2 and $a \sim 3$ Å.

This model has been developped [21] involving the diffusion of an incoherent particle inside a sphere with an impermeable surface. The main features of this model are (cf ref. 21 for details)

   i) the spectra are composed of a sharp $\delta(\omega)$ peak, superimposed on broad component which is a sum of an infinite number of Lorentzian lines,
   ii) the HWHM of this broad component is 4.33 D/a^2 = constant for Qa < < π and tends to DQ^2 for Qa >>π. Moreover, in these limiting cases, the shape is nearly Lorentzian,
   iii) in the intermediate case Qa $\sim \pi$, the shape is no more lorentzian but is more peaked around ω = 0 than Lorentzian.

A detailled comparison of this model with the experimental data has been made in reference [22]. In fig. 16 are represented the experimental spectra together with the theorical spectra (solid line) calculated using the best fitted values :

D (diffusion coefficient inside the sphere)= 1.8 10^{-5} cm^2/s
a (radius of the sphere) = 4.25 Å
D_t(long range diffusion) = 1.6 10^{-6} cm^2/s.

This last value has been obtained from radioactive tracer measurements.

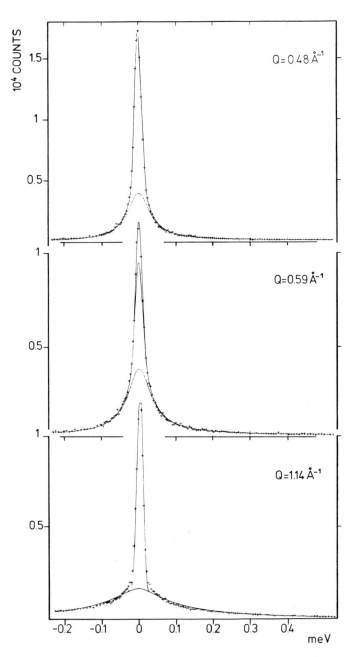

Figure 16. a: Series of experimental spectra together with the best fit theoretical curves calculated. Parameters: D = 1.8 × 10⁻⁵ cm²/s, a = 4.25 Å, Dₜ = 1.6 × 10⁻⁶ cm²/s, λ = 10 Å. The separation between the various components of the spectra are indicated.

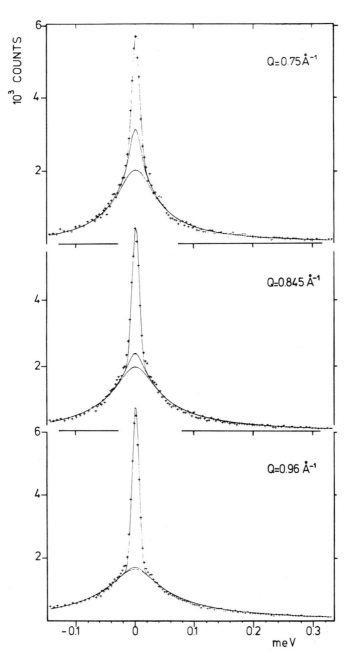

Figure 16. b: Series of experimental spectra together with the best fit theoretical curves calculated. Parameters: $D = 1.8 \times 10^{-5}$ *cm²/s,* a $= 4.25$ *Å,* $D_t = 1.6 \times 10^{-6}$ *cm²/s,* $\lambda = 11$ *Å. The separation between the various components of the spectra are indicated.*

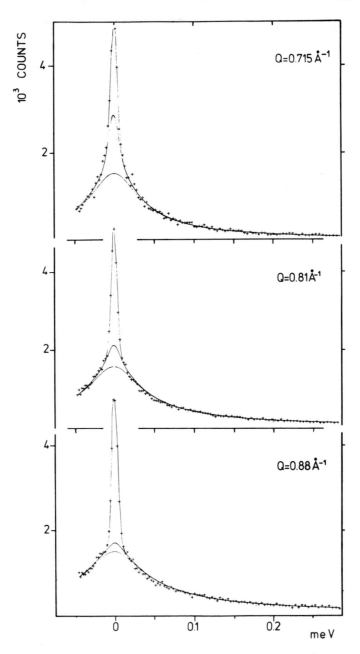

Figure 16. c: Series of experimental spectra together with the best fit theoretical curves calculated. Parameters: $D = 1.8 \times 10^{-5}$ *cm²/s,* $a = 4.25$ *Å,* $D_t = 1.6 \times 10^{-6}$ *cm²/s,* $\lambda = 13$ *Å. The separation between the various components of the spectra are indicated.*

This model therefore accounts well for the experimental results except at large Q where the theorical shape of the broad component deviates from the experimental one. This deviation mainly comes from the limitation of the diffusive description at large Q (finite jump distance between two successive proton positions). We have also examined the possibility of an aniso - tropic shape of the restricted volume. We have found that the possible anisotropy is in any case rather small (at most 1.5 to 2)

These results show the power of the high resolution neutron quasi-elastic scattering technique to study the water mobility in these membranes. The main conclusion is that on a space scale of 8 Å, the water protons move practically as freely as in bulk water, but the longer distance motion is much more difficult. Since the Vander Waals radius of the water protons is about 1 Å, the space should be renormalized to 10 Å for the water molecules.

From these different neutron studies it has been shown that Nafion membranes are composed of small crystalline regions which may act as physical crosslinking points and a hydrophilic phase dispersed in an hydrophobic matrix. The size of the ionic clusters is a few hundred angströms and some inter or intra particle correlation give a maximum in a small angle neutron scattering experiment. Neutron quasi-elastic scattering results have evidenced a local diffusion of water molecules. These small geometries may have two origins : substructure inside the ionic phase or intermediate structure between the hydrophilic ionic clusters. The small geometries in which this local diffusion is observed may correspond to substructure existing inside the hydro- philic ionic clusters.

Acknowledgements

The authors want to thank the Du Pont Company for pro- viding samples and for discussions.

References

1.   Guinier A., Fournier G., Small angle scattering of X-rays
      (1955) New York, Wiley
2.   Luzzati V., Acta Crystallog. 13, 939 (1960)
3.   Kratky O., Pilz I., Rev. Biophys. 5, 481 (1972)
4.   Jacrot B., Rep. Prog. Phys. 39, 911-953 (1976)
5.   Higgins J.S., Stein R.S., J. Appl. Cryst. 11, 346-375
      (1978)
6.   Roche E.J., Pineri M., Duplessix R., Levelut A.M., J. Polym
      Sci. Polymer Physics, 19, 1-11 (1981)
7.   Gierke T. Proceedings of the Electrochemical Society Meet.
      Atlanta, 1977

8.   Bassett D.C., Davitt R., Polymer, 16 (1975)
9.   Roche E.J., Stein R.S., Russell T.P. and MacKnight W.J.
      J. Polym. Sci. Pol. Phys.
10.   Roche E.J., Pineri M., Duplessix R., J. Polym. Sci. Polymer
      Physics to be published
11.   Koberstein J.T. Morra B., Stein R.S., J. Appl. Cryst. 13,
      34-45 (1980)
12.   Cutler S.G., Polymer Preprints 19, 2 (1978)
13.   MacKnight W.J., Taggart W.P., Stein R.S., J. Polym. Sci.
      Polym. Symp. 45, 113 (1974)
14.   Leadbetter A.J. and Lechner R.E. "Neutron Scattering
      Studies" in "The Plastically Crystalline State" Edited
      by J.N. Sherwood J. Wiley and Sons, 285 (1979)
15.   Dianoux A.J., Hervet H. and Volino F., J. Physique, 38,
      809 (1977)
16.   Lechner R.E., unpublished results
17.   Glaser J.A. in "Water, a comprehensive treatise" Edited
      by F. Franks, Plenum Press 1, 215 (1972)
18.   Pruppacher H.R., J. Chem. Phys. 56, 101 (1972)
19.   Sears V.F., Can. J. Phys., 45,237 (1967)
20.   Kern C.W. and Karplus M., in "Water a comprehensive
      treatise" Edited by F.Franks, Plenum Press 1, 21 (1972)
21.   Volino F., Pineri M., Dianoux A.J., de Geyer A., J. Poly.
      Sci., in press.

RECEIVED September 9, 1981.

The Cluster–Network Model of Ion Clustering in Perfluorosulfonated Membranes

T. D. GIERKE[1] and W. Y. HSU

E. I du Pont de Nemours & Co., Inc., Central Research and Development Department, Experimental Station, Wilmington, DE 19898

A model for ionic clustering in "Nafion" (registered trade-mark of E. I. du Pont de Nemours and Co.) perfluorinated membranes is proposed. This "cluster-network" model suggests that the solvent and ion exchange sites phase separate from the fluorocarbon matrix into inverted micellar structures which are connected by short narrow channels. This model is used to describe ion transport and hydroxyl rejection in "Nafion" membrane products. We also demonstrate that transport processes occurring in "Nafion" are well described by percolation theory.

The solvent and ion exchange sites in "Nafion" perfluorinated membranes phase separate from the fluorocarbon matrix to form clusters (1-5). This ionic clustering will not only affect the mechanical properties of the polymer (1), but should also have a direct effect on the transport properties across these membranes (2). In addition the exchange sites in the resin are strongly acidic and the polymer is extremely hydrophilic. Combined with the polymer's exceptional thermal and chemical stability, these properties make "Nafion" membranes particularly suitable for a variety of applications. These include applications as membrane separators in several electrochemical processes (6-9), as a superacid catalyst in organic syntheses (10-12), and as a membrane electrode (13).

The principal application of "Nafion" currently is as a membrane separator in chlor-alkali cells, shown schematically in Figure 1. In this process water is decomposed in the cathode compartment to produce caustic and hydrogen, while saturated brine is fed to the anode compartment where the chloride ion is reduced to chlorine gas. The role of the membrane is to separate the two compartments, allow the facile transport of sodium ions from the anode to cathode compartments, and to restrict the flux of hydroxyl ions across the membrane. In the classical picture of ion exchange membranes (14) where the ion exchange sites are

---

[1] Current address: Parkersburg, WV 26101.

homogeneously distributed, the membrane will fulfill this role
only if the internal molal concentration of ion exchange sites
greatly exceeds the concentration of hydroxide ion in the cathode
compartment. This condition is not satisfied. The internal concentration of exchange sites is generally between 5 and 15 molal,
while the concentration of the caustic produced in the cathode
compartment typically exceeds 15 molal. Indeed we have observed
that traces of calcium in the brine tend to precipitate at the
anode surface, which suggests that the membrane is very basic
throughout its thickness. However, in these same membranes, over
90% of the current may be carried by the sodium ions. It appears
that the hydroxide ion can get into the membrane, but its motion
is highly restricted in relation to the flux of sodium ions. It
seems likely that ionic clustering plays some role in relation
to this behavior.

In this work we propose a model for ionic clustering, which
we have called the cluster-network model (2), to account for
hydroxyl rejection in "Nafion" perfluorinated membranes. In developing this model we have been guided by two requirements: 1.
the model should be consistent with the available data on the
microscopic structure of the polymer (1-5) 2. the model should
be cast in a form which can be used to describe ion transport.
In pursuing the second condition, we make several approximations
in our model which are clearly idealizations. We feel that these
assumptions are reasonable and are justified, in part, by the
utility and success of the model which results.

In the next section we will present the data and arguments
on which the cluster-network model is based. We will also discuss the effects of equivalent weight, ion form, and water content on the dimensions and composition of the clusters. In the
third section we will present a formalism, which follows from the
cluster-network model, based on absolute reaction rate theory (2)
and hydroxyl rejection in "Nafion" perfluorinated membranes.
Finally we will outline the concepts of percolation theory and
demonstrate that ion transport trough "Nafion" is well described
by percolation.

The Cluster on Network Model. Previous small angle x-ray
(1-5) and neutron (4) scattering experiments clearly indicate
that ionic clustering is present in "Nafion". However, details
of the arrangement of matter in these clusters cannot be obtained
from these results alone. For hydrocarbon ionomers several different interpretations have been advanced as the cause of the
SAXS maximum. These include a model of spherical clusters on a
paracrystalline lattice proposed by Cooper et al. (16), the shell
core model of Macknight et al. (17) and more recently a lamellar
model (18). At present there is no consensus about which of
these models best describes clustering in hydrocarbon ionomers.
In considering these various models, we concluded that the shell
core model was highly unlikely because of the unfavorable electro-

static energetics of having isolated ion dipole pairs imbedded in
a fluorocarbon medium which possesses a very low dielectric constant (19).

Of the remaining geometries which could lead to the observed
SAXS reflection, one could include clustering into spheres, cyl-
inders, or sheets. Transmission electron microscopic studies
indicate that the spherical cluster morphology is the most likely
geometry of the three geometries just mentioned (2,20,21). Micrographs from stained, ultramicrotomed sections always reveal metallic clusters approximately circular in shape with diameters of
3-10 nm. An example is shown in Figure 2. No evidence to support
an extensive cylindrical or sheet-like morphology, which would
give noncircular projections in the micrographs, has been observed
by us (21). These results do not exclude the possibility that
the clusters are aspherical. In this respect our observations
are not inconsistent with the model of Roche et al. (18), provided
the aspect ratio of the lamellar structures is not too large.

The model of ionic clustering we believe to be most likely at
present is that of an approximately spherical, inverted micellar
structure. In this model the absorbed water phase separates into
approximately spherical domains, and the ion exchange sites are
found near the interface, probably imbedded into the water phase.
Such a structure satisfies the strong tendency for the sulfonic
acid sites to be hydrated, and at the same time this structure
will minimize unfavorable interactions between water and the
fluorocarbon matrix.

The effect of deformation and orientation on this clustering
can be understood by recalling that the SAXS peak, associated
with clustering, is normal to the molecular chain axis (3,5). In
a WAXD study of this fiber Starkweather has suggested the crystalline portion of the polymer exists as a fringed micelle with a
pseudo-hexagonal lattice (22). Even in the non-crystalline portion of the polymer one might expect short range order of this
nature. We believe ion clusters form on either side of this
fringed micellar structure, and thus the average distance between
clusters is heavily weighted towards a value equal to the sum of
the thickness of the fluorocarbon fringed micelle plus the cluster diameter. This could potentially account for our ability to
detect the SAXS reflection at low water contents. Naturally any
deformation which orients the polymer chains will result in this
most probable cluster separation being observed normal to the
direction of deformation, which is in agreement with observation
(3, 5). We believe this model is in substantial agreement with
experimental observation; however, further experimentation will
be required before the complete details of ionic clustering in
perfluorinated ionomers are defined.

If the ionic clusters are approximately spherical in shape,
then their average size may be estimated from solvent absorption
studies (6) using straight-forward geometric arguments. Suppose,
for the sake of calculation only, that the clusters were distri-

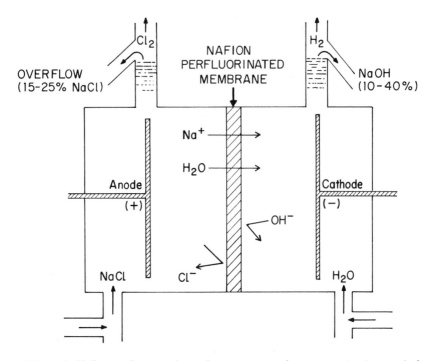

Figure 1. Nafion perfluorinated membrane as a membrane separator in a typical chlor-alkali cell.

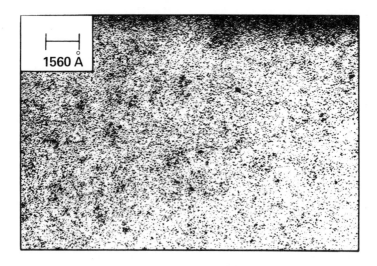

Figure 2. Transmission electron micrograph from 1200 EW cross section, stained by counter-diffusion of Ag^+ and S_n^{+2} ions across film.

buted on a simple cubic lattice with an average lattice constant, d, given by the observed Bragg reflection. We also know that the change in volume, ΔV, which occurs during the swelling of polymer with water is described by the empirical relation (6)

$$\Delta V = \rho_p \Delta m / \rho_w \tag{1}$$

where ΔV is the fractional volume change in (cm^3/cm^3 of dry polymer), m is the fractional weight gain in (g/g dry polymer), ρ_p is the density of dry polymer and ρ_w is the density of water. The values of Δm are either taken from reference 6 or are measured. For a simple cubic lattice the volume of a cluster, V_c, will be given by

$$V_c = \frac{\Delta V}{1 + \Delta V} d^3 + N_p V_p \tag{2}$$

where N_p is the number of ion exchange sites in a cluster and V_p the volume of an exchange site. N_p can be obtained directly from the density and equivalent weight, EW, of the polymer

$$N_p = [N_A \rho_p / EW(1 + \Delta V)] d^3 \tag{3}$$

N_A is Avogadro's number, and the term inside the brackets, [-] represents the number of exchange sites in a cubic centimeter of swollen polymer. The cluster diameter, d_c, is, of course, given by

$$d_c = [6V_c / \pi]^{1/3} \tag{4}$$

Finally, the number of water molecules in our average cluster is given by

$$N_w = [N_A \rho_p \Delta m / 18(1 + \Delta V)] d^3 \tag{5}$$

We emphasize that only estimates of cluster dimension are obtained by using Equations (1-5). Yet, recognizing this limitation one may be able to obtain additional inside for the structure of ionic clusters and how this structure varies with equivalent weight, cation form, water content, and other factors.

Table 1 shows the results of such cluster morphology calculations for polymers with different equivalent weights. The value for V was taken as 68×10^{-24} cm^3 and corresponds to an effective radius for the ion exchange sites of 0.25 nm. This is the same order of magnitude assumed by Hora et al. (23) . As the equivalent weight increases, the cluster diameter, ion exchange sites per cluster, and water per exchange site decrease. Qualitatively, these trends may be understood by recognizing that the crystallinity and polymer stiffness increase with increasing EW. At higher EW's, it will thus require more energy to hydrate each exchange site and to have the exchange sites aggregate.

Table 2 shows the results for cluster morphology calculations for 1200 EW polymer neutralized with various cations. As

Table I. Results of Cluster Morphology Calculations According to Equations 1–5 for Polymers with Difference Equivalent Weights in Na Ion Form. Samples Conditioned by Boiling 1 h in 0.2% NaOH.

EW	Polymer Density ρ, g/cc	% Mass Gain 100 Δm	% Volume Gain 100 Δv	H_2O/Cluster	Fixed Charge/ Cluster	Bragg Spacing d, nm	Cluster Diameter d_c, nm
944	2.088	42.0	87.3	2102	95	5.12	5.09
971	2.093	37.5	78.2	1944	96	5.09	4.97
1100	2.103	23.8	49.8	1200	84	4.78	4.31
1200	2.113	17.8	37.5	863	73	4.55	3.88
1600	2.135	8.1	17.2	370	52	4.22	3.03
1790	2.144	6.3	13.3	266	43	4.07	2.74

Table II. Results of Cluster Morphology Calculations According to Equations 1–5 for 1200 EW Polymers in Different Cation forms. Samples Conditioned by Boiling 1 h in H_2O.

Cation Form	Polymer Density ρ, g/cc	% Mass Gain[a] 100Δm	% Volume Gain 100ΔV	H_2O/Cluster	Fixed Charge/ Cluster	Bragg Spacing d, nm	Cluster Diameter dc, nm
H^+	2.075	33.6	69.7	1690	76	4.98	4.74
Li^+	2.078	29.7	61.7	1430	72	4.82	4.49
Na^+	2.113	21.0	44.3	1120	80	4.78	4.21
K^+	2.141	8.7	18.7	520	89	4.61	3.45
Rb^+	2.221	8.1	17.9	560	103	4.78	3.56
Cs^+	2.304	5.9	13.6	470	120	4.90	3.50

a   Taken from Reference 1.

the cation weight increases, the cluster diameter and water per exchange site decrease, but the number of exchange sites per cluster increases. Clearly, the hydrophilicity of the exchange site is lower with the heavier cations which is consistent with the observation that the heavier cations are more tightly bound to the exchange site (24, 25). One explanation for the increase in number of exchange sites per cluster for the heavier cations might be related to the balance of energy of elastic deformation on one hand and hydration and ion aggregation on the other (26). As the hydration of the individual exchange sites decreases with heavier cations, the elastic strain of the fluorocarbon matrix associated with hydration will also decrease. This will make it possible for additional clustering to occur, with an associated increase in elastic strain, until thermodynamic equilibrium is achieved with the external solvent.

Table 3 shows cluster morphology calculations for 1200 EW polymer with different internal water content. The results in this table provide some insight into the growth of clusters. As the polymer absorbs more water, the cluster diameter, exchange sites per cluster, and waters per exchange site increase. Figure 3 shows more clearly the variation of cluster diameter and exchange site per cluster with water content. As noted earlier, clusters do exist in the dry polymer, and in this sample they are about ~1.9 nm in diameter and contain ~26 ion exchange sites.

The increase in the number of exchange sites per cluster with increasing water content is noteworthy because it suggests that cluster growth does not merely occur by an expansion of the dehydrated cluster. Rather the growth of cluster appears to occur by combination of this expansion and a continuous reorganization of exchange sites so there are actually fewer clusters in the fully hydrated sample. The type of reorganization visualized is shown schematically in Figure 4. This figures illustrates on dehydration how the exchange sites from two clusters (#6-10) could be redistributed to form a third new cluster without a significant translation of polymer chains.

Of course the incentive for obtaining a better understanding of ionic clustering in "Nafion" is to determine the relationship between ion clustering and mass transport. With this in mind we have measured the hydraulic permeability and diffusion coefficient of water through membranes of different equivalent weights. These data are listed in Table 4. These transport measurements were used to estimate the average size of the structural feature controlling transport, or the effective pore diameter, D_p. The diffusion data were analyzed according to equation (6) or (7) (27).

$$(D_w/D^\circ_w) = (A/Ao) = (\pi/4d^2) (D_p - \sigma_w)^2 \qquad (6)$$

or

Table III. Results of Cluster Morphology Calculations According to Equations 1–5 for 1200 EW Polymers with Different H_2O Content.

Cation Form	Polymer Density ρ, g/cc	% Mass Gain $100 \Delta m$	% Volume Gain $100 \Delta v$	H_2O/Cluster	Fixed Charge/ Cluster	Bragg Spacing d, nm	Cluster Diameter dc, nm
H^+	2.075	19.4	41.0	984	76	4.66	4.08
Li^+	2.078	18.5	39.1	886	72	4.55	3.94
Li^+	2.078	15.3	32.3	687	67	4.38	3.70
H^+	2.075	12.9	27.2	505	59	4.13	3.38
Li^+	2.078	12.0	25.3	456	57	4.07	3.27
Li^+	2.078	8.6	18.2	288	53	3.90	2.97
H^+	2.075	6.3	13.3	233	39	3.46	2.52
Li^+	2.078	5.4	11.4	149	41	3.52	2.52

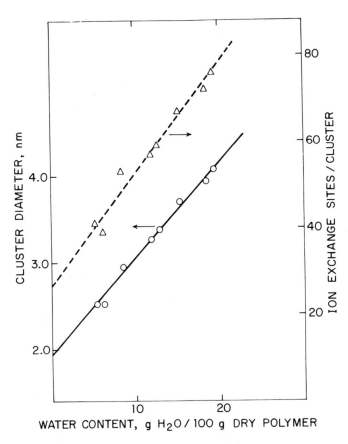

Figure 3. The variation of cluster diameter (○) and ion exchange sites (△) per cluster with water content in 1200 EW polymer.

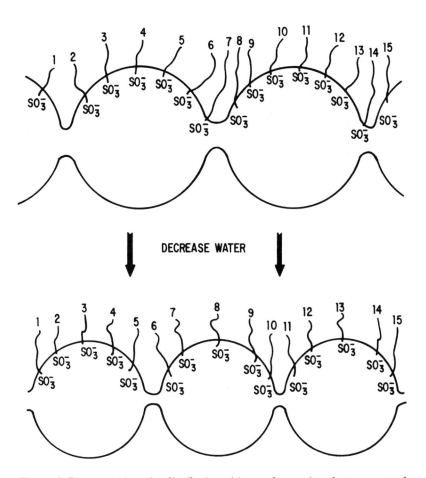

Figure 4. Representation of redistribution of ion exchange sites that occurs on dehydration of polymer.

Table IV. Effective Pore Diameters Derived from H_2O Transport Measurements.

EW	Diff. Coef. [a] $D_w, 10^{-6} \frac{cm^2}{sec}$	Pore Dia. [b] nm	Hyd. Perm. $10^{-13} \frac{cm^3}{dyne-sec}$	Pore Dia. nm
1100	1.28	1.6	--	--
1200	0.69	1.2	2.1	1.3
1400	0.42	1.0	1.2	1.1
1500	0.34	0.9	--	--
1600	0.16	0.7	0.2	0.7

a.) From Reference (28)

b.) $D_w^o = 2.4 \times 10^{-5} \ cm^2/sec$

$$D_p = \sigma_w + [4d^2(D_w/D^\circ_w) /\pi]^{1/2} \qquad (7)$$

where D_w is the measured diffusion coefficient (28), D°_w is the diffusion coefficient of water in water, σ_w is the diameter of a water molecule (taken as 0.35 nm) and d is the distance between clusters which is obtained from the SAXS data (3,5). The hydraulic permeability were analyzed according to the expression (27).

$$D_p = (128 \ t\eta L_p d^2/\pi)^{1/4} \qquad (8)$$

where t is the thickness of the membrane, η is the viscosity of water, and L_p is the hydraulic permeability coefficient. The results for the effective pore diameter are also listed in Table 4. Note that the two experiments result in values which are self-consistent.

Combining the results of these water transport experiments with the inverted micellar structure proposed for the clusters, we arrive at the cluster-network model shown in Figure 5. In this model the clusters are connected by short narrow channels whose dimensions are derived from the water transport measurements. The cluster separation (5.0 nm) is consistent with the SAXS experiments, and the cluster diameters (4.0 nm) are consistent with the results given in Tables 1-3. The significance of the cross-hatched area will be explained shortly. As we will demonstrate, this model of ionic clustering is very useful in describing ion transport in "Nafion".

Absolute Reaction Rate Formalism. Given the structure shown in Figure 5, we can explain the observation described above of high current efficiency in membranes which are basic throughout their thickness. Because of repulsive electrostatic interactions, hydroxide ions will be excluded from the surface of the clusters and connecting channels by the polymeric fixed charges which are assumed, in the model, to be located in these regions. From the theory of the electric double layer (29,30), we know that the effective range of these interactions will be about 0.5 nm at the concentrations that exist inside the membrane. This region is represented by the cross-hatched area in Figure 5. In a large portion of the cluster, the hydroxide ion will be effectively shielded from these interactions by sodium ions, and by Boltzmann statistics the hydroxide concentration in the interior of the cluster will be similar to the external concentration. This would explain why the membrane is basic throughout its thickness when in a chlor-alkali cell. However, for a hydroxyl ion to migrate from one cluster to the next, it would have to overcome a fairly large electrostatic barrier in the channel, which a positive ion like Na^+ will not experience. It is this barrier which would account for the high current efficiency.

These qualitative concepts may be cast into a quantitative formalism using absolute reaction rate theory (31). As the hydro-

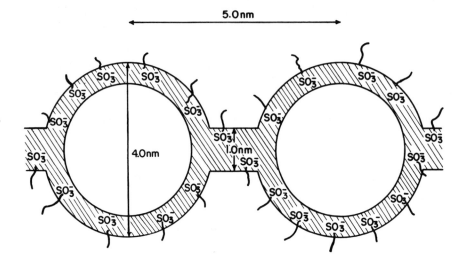

Figure 5. Cluster–network model for Nafion perfluorinated membranes. The polymeric ions and absorbed electrolyte phase separate from the fluorocarbon backbone into approximately spherical clusters connected by short, narrow channels. The polymeric charges are most likely embedded in the solution near the interface between the electrolyte and fluorocarbon backbone. This configuration minimizes both the hydrophobic interaction of water with the backbone and the electrostatic repulsion of proximate sulfonate groups. The dimensions shown were deduced from experiments. The shaded areas around the interface and inside a channel are the double layer regions from which the hydroxyl ions are excluded electrostatically.

xyl ion migrates through the membrane, in a chlor-alkali cell it will encounter an oscillating potential which is low in the cluster and high in the channel. This situation is shown schmatically in Figure 6. The overall potential gradient, $\Delta\phi$, is provided by the voltage drop across the membrane, and the barrier height, α, contains both a term due to the geometric restriction, β, and a term due to the electrostatic repulsion, $<\phi>$. For ionic species M with charge q (M), we may write

$$\alpha(M) = \beta(M) - q(M) <\phi> \tag{9}$$

$\Delta\phi$, α, β, and $<\phi>$ are all expressed in reduced units (units of $k_B T$). In Figure 6, d corresponds to the effective Bragg spacing deduced from the SAXS experiments.

Using absolute reaction rate theory (31) an expression for the ratio of the hydroxide ion flux to sodium ion flux may be derived:

$$-[J(OH)/J(Na)] = \frac{\mu(OH)}{\mu(Na)} \frac{C(OH,n)}{C(Na,0)} \exp [-2.0 <\phi>] \tag{10}$$

In equation (10), J(M) is the flux of species M, C(OH,n) is the concentration of hydroxide ion in the cathode compartment, C(Na,0) is the concentration of sodium ions in the anode compartment, and μ(M) is defined

$$\mu(M) = k_o(M) \exp [-\beta(M)] \tag{11}$$

where k_o(M) is the intrinsic rate of transport of species M. The experimental quanity is the current efficiency, CE, which is defined by,

$$CE = 1.0/(1.0-J(OH)/J(Na)) \tag{12}$$

In equation (10) there are two unknown quantities: the ratio of mobilities, μ(OH)/μ(Na), and the electrostatic contribution to the barrier, $<\phi>$. A value for $<\phi>$ can be estimated by assuming that the channel is a cylindrical capillary of diameter, D_p, with a uniform charge density at the surface, σ_c, giving.

$$\sigma_c = [d\rho_p/(2EW(1+\Delta V))] [\frac{2}{9\pi} (\frac{1+\Delta V}{\Delta V})^2]^{1/3} \tag{13}$$

The radial potential distribution inside the capillary, $\phi(r)$, is then obtained by solving the Poisson-Boltzmann equation for cylindrical symmetry (30). The resulting potential depends on a single adjustable constant which is fixed by the boundary condition on the potential which relates the potential gradient at $r=1/2D_p$ to the surface charge density, σ_c. Then we define

$$<\phi> = \int_0^{1/2D_p} \phi(r)r\,dr \Big/ \int_0^{1/2D_p} r\,dr \tag{14}$$

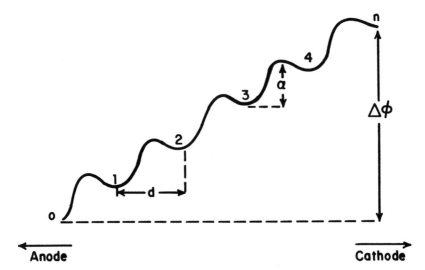

Figure 6. Schematic potential seen by a hydroxyl ion as it moves across a Nafion perfluorinated membrane in a chlor-alkali cell. This potential consists of two parts: a constant sloping portion that arises from the voltage drop across the membrane; and an oscillating part that arises from electrostatic restriction of the hydroxyl ions. Physically, the hills and troughs correspond to the channel and cluster regions, respectively. For simplicity, a one-dimensional, periodic, model potential is used to evaluate the membrane current efficiency although the real potential is three-dimensional and aperiodic.

It should be noted that $<\phi>$ is calculated without any ad-justable parameters, since D_p, and the parameters in equation (13) are experimentally accessible. Thus only the mobility ratio is required for a calculation of current efficiency. In Figure 7 we show the experimental variation of current efficiency with polymer equivalent weight (solid line). The solid circles are the results of the cluster-network-absolute-reaction-rate calcu--lation. The mobility ratio has been fixed by requiring agree-ment between the calculation and experiment at 1400 equivalent weight. The remaining four points were calculated without <u>any</u> adjustable parameters. The agreement is evident.

The dashed line and squares represent the results of simi-lar calculations using the same mobility ratio and neglecting the effect of $<\phi>$. This corresponds to the classical picture of an ion exchange membrane with homogeneously distributed fixed charges. Adjustment of the mobility ratio cannot reproduce <u>simultaneously</u> the observed magnitude and slope in current effi-ciencies. In further support of the cluster-network model we note that we were able to calculate the apparent diffusion co-efficient of NaOH as defined by Berzins (<u>32</u>), using the same para-meters derived above. The results are shown in Table 5. The agreement between the calculated and experimental values is apparent.

<u>Percolation Theory</u>. The possibility that ion transport in "Nafion" membranes might be described by percolation theory was first proposed by Hsu, Barkley, and Meakin (<u>15</u>). The concept behind percolation theory can be conveniently illustrated in a two-dimensional grid with some of its sites randomly occupied (Figure 8). For our case the empty and occupied sites would represent the fluorocarbon phase and the ion cluster phase, res-pectively. At low concentration of coverage the occupied sites, which are represented by the crosses, are well separated into islands; consequently, macroscopic ion flow is impossible. At a higher concentration of occupancy, such as in (b) where the shaded areas are sites previously occupied and the crosses are for sites newly occupied, two developments occur: (1) the con-ductive "islands" grow in size, and (2) they connect to form extended pathways. However, crucial links, such as those lab-elled L in (b), are still missing such that conduction is impos-sible. Finally, above a certain threshold as shown in (c) some of the key missing links have been filled to form continuous channels that pervade the grid; macroscopic conduction is now allowed. Therefore, an ionic insulator to conductor transition occurs at the threshold electrolyte loading, at which the aver-age size of connected ion-clusters (which is the correlation length in our phase transition problem) also becomes macroscopic. At yet higher levels of occupancy such as shown in (d), percola-tion channels criss-cross the board engulfing deadends and fil-ling in missing links leading to progressively higher conductiv-

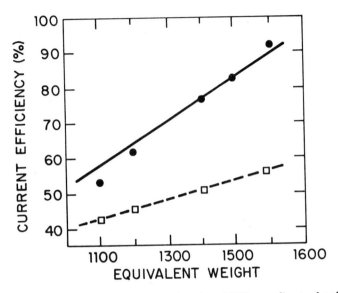

Figure 7. Computed current efficiency as a function of EW according to the absolute reaction rate theory. The closed circles and open squares are results of the Cluster-Network (CN) model and classical Donnan equilibrium (DE), respectively. In comparison the experimental trend (————) is also shown. The adjustable parameter in the CN model is fixed at 1400 EW, whereas no choice of the parameter on DE can simultaneously reproduce both the empirical slope and magnitude.

Table V. Cluster–Network Model Calculations of NaOH Apparent Diffusion Coefficients.

Sample	D (cm^2/sec) Calculated	D (cm^2/sec) Experimental[32]
1150 Sulfonate	1.4×10^{-6}	1.3×10^{-6}
1200 Sulfonate	8.8×10^{-7}	1.1×10^{-6}
1300 Sulfonate	4.0×10^{-7}	6.4×10^{-7}
1500 Sulfonate	1.0×10^{-7}	2.2×10^{-7}
1150 Sulfonamide	2.9×10^{-8}	2.5×10^{-8}

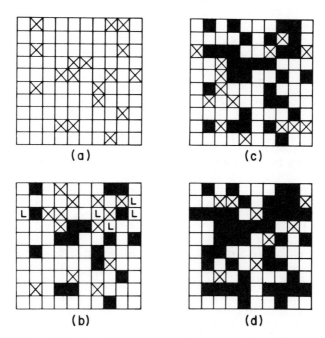

Figure 8. A two-dimensional illustration for the concept of percolation. The shaded and crossed areas correspond, respectively, to sites that were previously occupied and sites that have just been occupied. Those marked L in (b) are empty sites that must be occupied before the onset of ion transport. The percentage of occupancy of the grid are 18, 31, 45, and 53% for Cases a to d, respectively. In this context, the empty and occupied sites would represent the fluorocarbon backbone and the electrolyte phase, respectively.

ity become finite? (2) how does it depend on the electrolyte uptake? Insight can be obtained theoretically by using one of the
techniques listed in Table 6. The mean field theory (33) provides an analytical and exact solution in a six-dimensional (6D)
hyperspace. In 3D the renormalization group (34) and the effective medium (35) techniques complement each other nicely. Last
but not least, computer simulations (36) have been used to provide quantitative information. After lengthy analyses the main
conclusion of the formal percolation theory is similar to the
qualitative description just given, i.e., there is a threshold
value for the volume loading of the aqueous phase below which ion
transport is extremely difficult and improbable because of a lack
of extended pathway. Above and near the threshold the conductivity obeys a simple power law:

$$\sigma = \sigma_o \ (c-c_o)^n \tag{1}$$

where c is the volume fraction of the aqueous phase. The critical exponent n is a universal constant that depends on the spatial
dimensionality only (37). It is equally applicable to any per-
colative system regardless of its chemical, mechanical, structural, and morphological properties. In 3D, n is about 1.5, and
appropriate values for other dimensions are listed in Table 7.
The threshold volume fraction, c_o, depends not only on the spatial dimensionality but also on extrinsic factors. For a 3D, continuous, random mixture, d_o is ideally 15%; values for other
dimensions are listed in Table 7. However, c_o may be easily
affected by the state of dispersion, size, shape, orientation,
and distribution of the conductive phase. Therefore, it is best
determined empirically. Finally, the prefactor σ_o depends on
details of ionic and molecular interactions and can only be computed from specific microscopic models. The important feature to
be appreciated here is that the topological and geometrical information of the cluster connectivity is lumped together in the
$(c-c_o)^n$ factor whereas the microscopic diffusion information and
various interactions are contained in the prefactor σ_o.
    To test this theory, the room temperature conductivity of
"Nafion" perfluorinated resins was measured as a function of
electrolyte uptake by a standard a.c. technique for liquid electrolytes (15). The data obey the percolation prediction very
well. Figure 9 is a log-log plot of the measured conductivity
against the excell volume fraction of electrolyte $(c-c_o)$. The
principal experimental uncertainty was in the determination of
c as shown by the horizontal error bars. The dashed line is a
non-linear least square law to the data points. The best fit
value for the threshold c_o is 10% which is less than the ideal
value of 15% for a completely random system. This observation
is consistent with a bimodal cluster distribution required by
the cluster-network model. In accord with the theoretical prediction, the critical exponent n as determined from the slope of

Table VI. Theoretical Tools Used in the Formal Analysis of Percolation and Their Range of Applicability (d, Dimension; c, Electrolyte Volume Loading).

Technique	Range of Applicability
Mean Field Theory	Exact for $d \geq 6$
Renormalization Group Theory	$c \gtrsim c_o$
Effective Medium Theory	$3c_o \lesssim c \leq 1$
Numerical Simulation	Any c and d

Table VII. Theoretical Values for n and C_o for an Ideal Continuous Mixture (38).

Dimension d	Critical Exponent n	Threshold Loading c_o
1	—	1.00
2	1.2	0.45
3	1.5	0.15
6	2.0	<0.02

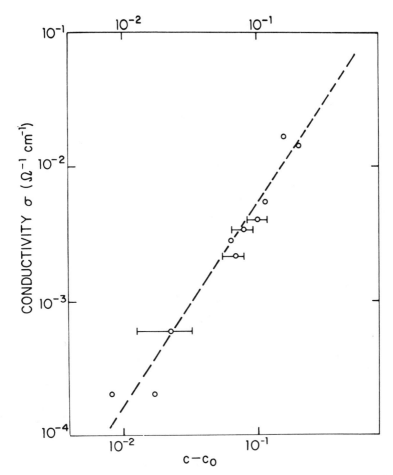

Figure 9. Log-log plot of conductivity vs. excess volume fraction (c-c_O) of the aqueous phase. Typical error bars for the determination of volume fraction are shown. The corresponding errors for conductivity are much smaller and omitted for clarity. The straight line is a fit of the percolation prediction Equation 1 to the data with $n = 1.5$, $c_O = 0.10$, and $\sigma_O = 0.16 \Omega^{-1} cm^{-1}$.

the dashed line is 1.5.   Finally, the prefactor σ_o is about
$0.2\Omega^{-1} \text{ cm}^{-1}$.

Summary.   We have shown that ion transport in "Nafion" perfluorinated membrane is controlled by percolation, which means
that the connectivity of ion clusters is critical. This basically reflects the heterogeneous nature of a wet membrane. Although
transport across a membrane is usually perceived as a one-dimensional process, our analysis suggests that it is distinctly
three-dimensional in "Nafion". (Compare the experimental values
of c_o and n with those listed in Table 7.) This is not totally
unexpected since ion clusters are typically 5.0 nm, whereas a
membrane is normally several mils thick. We have also uncovered
an ionic insulator-to-conductor transition at 10 volume % of
electrolyte uptake. Similar transitions are expected in other
ion-containing polymers, and the Cluster-Network model may find
useful application to ion transport in other ion containing
polymers. Finally, our transport and current efficiency data
are consistent with the Cluster-Network model, but not the conventional Donnan equilibrium.
        Discussions with P. Meakin and contributions by G. Munn,
J. Barkley, and F. Wilson are gratefully acknowledged.

Literature Cited

1.   Yeo, S. C. and Eisenberg, A.   J. Appl. Polym. Sci., 1977,
     21, 875.
2.   Gierke, T. D., 152nd Meeting of Electrochemical Soc.,
     Atlanta, GA, Oct. 1977. Abstract No. 438, J. Electrochem
     Soc. 1977, 124, 319C.
3.   Gierke, T. D.; Munn, G. E.; and Wilson, F. C., J. Polym.
     Sci., Polym. Phys. Ed., In press.
4.   Roche, E. J.; Pineri, M.; Duplessix, R.; and Levelut, A. M.
     J. Polym. Sci., Polym. Phys. Ed. 1981, 19, 1.
5.   Gierke, T. D.; Munn, G. E.; Wilson, F. C. Adv. Chem. Ser.
     "Perfluorinated Ionomes" ----
6.   Grot, W. G. F.; Munn, G. E.; Walmsley, P. N.; 141st Meeting
     of the Electrochemical Society, Houston, Texas, May 1972.
     Abstract No. 154, J. Electrochem. Soc., 1972, 119, 108C.
7.   Leitz, F. B.; Accomazzo, M. A.; Michalek, S. A.; 141st
     Meeting of the Electrochemical Society, Houston, Texas,
     May 1972. Abstract No. 150, J. Electrochem. Soc., 1972,
     119, 108C.
8.   Nuttal, L. J. and Titterington, W. A., Conf. Eelectrolytic
     Production Hydrogen, London 1975.
9.   Will, F. G. J. Electrochem. Soc., 1979, 126, 36.
10. Olah, G. A., Kaspi, J.; Bukala, J. J. Org. Chem. 1977, 42,
     4187.
11. Olah, G. A.; Kaspi, J. Nouv J. Chim., 1978, 2, 581,585.
12. Olah, G. A.; Keumi, T.; Meidar, D. Synthesis, 1978, 929.

13.   Rubinstein, I. Bard; Bard, A. J.   J. Am. Chem. Soc., 1980,
      102, 6641.
14.   Hellferich, F. "Ion Exchange" McGraw-Hill 1962.
15.   Hsu, W. Y.; Barkley, J. R.; Meakin, P. Macromolecules, 1980,
      13, 198.
16.   Marx, C. L., Caulfield, D. F., Cooper, S. L. Macromolecules,
      1973, 6, 344.
17.   MacKnight, W. J.; Taggart, W. P.; Stein, R. S.   J. Polym.
      Sci., Polym. Symp., 1974, 45, 113.
18.   Roche, E. J.; Stein, R. S.; Russell, T. P.; MacKnight, W. J.
      J. Polym. Sci., Polym. Phys. Ed. 1980, 18, 1497.
19.   Eisenberg, A. Macromolecules, 1970, 3, 147.
20.   Ceynowa, J. Polymer 1978, 19, 73.
21.   Gierke, T. D. and Hebert, R. R. to be published.
22.   Starkweather, H. W. to be published.
23.   Hora, C. J.; Mauritz, K. J.; Hopfinger, A. J., 152nd Meeting
      of Electrochemical Society, Atlanta, GA, Oct. 1977. Ab-
      stract No. 439, J. Electrochem. Soc., 1977, 124, 319C.
24.   Lopez, M.; Kipling, B.; Yeager, H. L. Anal. Chem 1977,
      49, 629.
25.   Mauritz, K.; Lowry, S. R. Polym Prepr., Am. Chem. Soc.,
      Div. Polym Chem. 1978, 19, 336.
26.   Hsu, W. H.; Gierke, T. D. to be published.

27.   Bean, C. P. Physics of Porous Membranes - Neutral Pores
      Membranes, a Series of Advances ed. G. Eisenman, VI pgs.
      1-54.
28.   Ferguson, R. private communication.
29.   Adamson, A. W.   Physical Chemistry of Surfaces, Interscience,
      1967, 2nd Ed.
30.   Oldam, J. B.; Young, F. J.; Osterle, J. F.   J. Colloid Sci.,
      1963, 18, 328.
31.   Parlin, R. B.; Eyring, H. Ionic Transport Across Membranes
      Clarke, H. T.; Nachmansohn, D., eds., Academic Press 1954,
      pg. 103.
32.   Berzins, 153rd Meeting of Electrochemical Society, Seattle,
      Wash., May 1978 Abstract No. 452, J. Electrochem Soc.,
      1978, 125, 163C.
33.   Toulouse, G., Nuvo Cimento 1974, 23, 234; Harris, A. B.;
      Fisch, R. Phys. Rev. Lett. 1977, 38, 796.
34.   Kogut, P. M.; Straley, J. P., AIP Conf. Proc 1978, 40, 382
      and references therin.
35.   See for example, Elliott, R. J.; Krumhansl, J. A.; Leath,
      P. L. Rev. Mod. Phys., 1974, 46, 465.
36.   Straley, J. P. Phys. Rev. B 1977, 15, 5733, Cohen, M. H.;
      Jortner, J.; Webman, I. Phys. Rev. B, 1978, 17, 4555 and
      references therein.
37.   For a general discussion of critical exponents and their
      properties, see for example Nelson, D. R., Nature (London)
      1979, 269, 379.
38.   Zallen, R. Proc. 13th IUPAD Conf. of Stat. Phys., Haifa,
      Israel 1977, pp. 310-321.

RECEIVED October 29, 1981.

APPLICATIONS

Electrosynthesis with Perfluorinated Ionomer Membranes in Chlor-Alkali Cells

RONALD L. DOTSON and KENNETH E. WOODARD

Olin Corporation, P. O. Box 248, Charleston, TN 37310

The field of electrochemical science has been quietly revolutionized during this past decade by development and application of a new family of perfluorinated ionomer ion-exchange membrane separators in concert with new cell designs and stable electrode systems for electrosynthesis (1) (2).

These new membranes are much more than structural supports. The perfluorocarbon structures impart oxidative and hydrolytic resistance to the membrane materials while their cationic strength rejects anions. This combination of unusual ionic character and exceptional chemical resistance makes these materials prime candidates for use as electrolytic separators for electrosynthesis (3).

In recent years, a number of electrolytic processes have utilized membranes in producing both anodic and cathodic products. By far, however, the most important application of this technology has been in the chlor-alkali industry. Intense commercial and academic interest has been focused into this field during the past decade so that ion exchange theory as applied to membranes is in a more advanced state than any of the other ion exchange systems. The primary examples of industrial chlor-alkali electrochemistry are found in the production of chlorine, caustic soda and potash, hydrogen and hypochlorite (1) (4).

The three general types of chlor-alkali electrolyzers in use today are mercury, diaphragm and membrane cells. Each one offers certain advantages, and the first two have undergone many changes since their appearance in industry over 80 years ago. During the past decade there has been a resurgence of interest in the design and operation of chlor-alkali cells, and an entirely new type of cell was invented. This new cell, the membrane cell, was developed in response to the new pollution requirements and higher capital, energy and operating costs required by the older types of cells (5) (6) (7).

0097-6156/82/0180-0311$13.50/0

Membrane cells use metal anodes and fluorinated ion-exchange membranes as separators to produce high quality caustic and chlorine in a two chambered cell where chlorine is discharged from the anode in a relatively pure form, having low oxygen content and where the hydrogen is evolved from the cathode in pure form and separated from the chlorine by the membrane. The catholyte in these cells contains 20 to 38 weight percent caustic with 16 to 22 weight percent salt in the anolyte. The successful development of the membrane cell has depended on the structural integrity, dimensional stability, low electrical resistance and high current efficiency over a long operating life of these new membranes. In order to bring this technology into full commercialization, it has also been important to develop a fundamental understanding of the physical and chemical properties of these membrane systems interacting with simple electrodes in the solution environments in which they are used (8).

Membrane Chlor-Alkali Cell Characteristics

The three types of electrochemical cells used for the production of chlorine gas, Cl_2, hydrogen, H_2, and caustic, NaOH, mentioned are unique in different ways. Both flowing mercury cathode and diaphragm type cells have undergone many developments during the past four score years, and each offers certain advantages. During the past decade, however, there has been a resurgence of interest in cells because of increased power, capital and labor costs and questions about pollution and product quality. Considerable changes have been made in cell sizes and materials of construction; and an entirely new type of cell, the membrane cell, was invented and commercialized (9) (10).

Membrane cells are electrochemical cells with metal electrodes set into a two chambered container separated by a wettable perfluorinated ion exchange membrane. The successful development of the membrane cell has resulted from the development of new and improved membranes having long life and a unique property of the membrane called cation permselectivity. This selectivity permits the passage of positively charged sodium ions and water, while it essentially rejects the passage in both directions of negatively charged ions, such as chloride (Cl^-), and hydroxyl ions, (OH^-). The membrane cell's present success has depended on the membrane's ability to operate efficiently over an extended time period.

The overall electrode reactions for the diaphragm and membrane cells are both the same, and given as:

$$
\begin{array}{ll}
\text{Na}^+ + \text{Cl}^- \xrightarrow{\text{Electrical Power}} 1/2 \ \text{Cl}_2 + e + \text{Na}^+ & \text{(anode)} \\
\text{OH}^- + \text{H}^+ + e \longrightarrow 1/2 \ \text{H}_2 + \text{OH}^- & \text{(cathode)} \\
\hline
\text{NaCl} + \text{H}_2\text{O} \longrightarrow 1/2 \ \text{Cl}_2 + 1/2 \ \text{H}_2 + \text{NaOH} &
\end{array}
$$

as with the diaphragm cell, the cathode current efficiency depends on two factors: the amount of hydroxyl back-migration across the separator from the catholyte to the anolyte, and the extent of water splitting in the anolyte or membrane polarization at the brine - membrane phase boundaries depending on brine concentration (6) (7).

Basic Membrane Cell Operation

A membrane cell is similar to a diaphragm cell except that the porous diaphragm is replaced by a non-porous ion exchange membrane, as shown in Figure 1.

Saturated brine is fed to the anode chamber, chlorine gas produced by oxidation of the chloride ion, (Cl^-) at the anode leaves the anode chamber as Cl_2 gas. The weak brine leaves the anode chamber for resaturation. Sodium ions, (Na^+), and water molecules are driven through the membrane as a flow of current by an imposed electrical pressure and ions flow through the permselective cation exchange membrane separator and then into the cathode chamber. The ion exchange membrane prevents passage of chloride ions to the cathode chamber and hydroxyl ions to the anode chamber. Some of the water added to the cathode chamber is electolyzed at the cathode forming hydrogen gas and hydroxyl ions, and these hydroxyl ions combine with the sodium ions to form sodium hydroxide.

Performance of the conventional membrane cell depends on several operating variables such as: caustic strength, brine concentration, cell voltage, cell temperature, current density, brine purity and pH.

The current efficiency of these cells shows a dependence on anolyte and catholyte concentrations. Most cells operate from $1-3\text{KA/m}^2$ with anolyte strengths of 3-3.5N, (176-205 GPL) at 80-90°C and caustic products of 20-40 weight percent. The cell voltage increases dramatically with caustic strength. The ohmic drop of the membrane or its electrical resistance increases with increasing caustic concentration and also with brine concentration but to a lesser extent with brine than caustic strength.

The water transfer coefficient of a membrane depends directly on anolyte and catholyte concentrations. The amount of water transferred across the membrane per mole of sodium transferred, called the water transfer coefficient, decreases with increasing

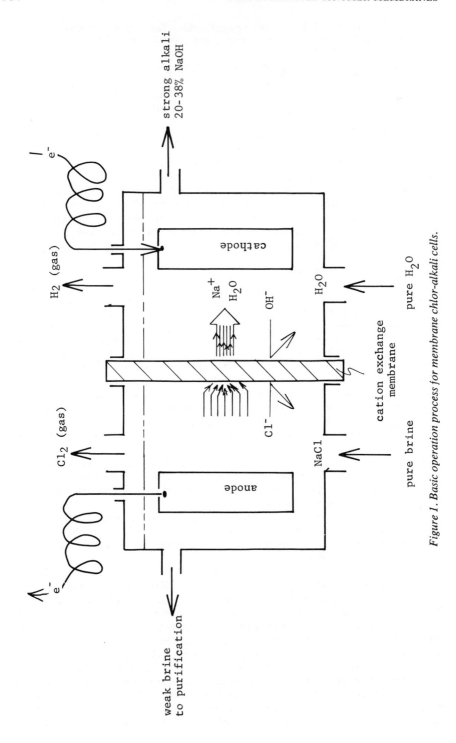

Figure 1. Basic operation process for membrane chlor-alkali cells.

anolyte concentration; however, it changes very little with the catholyte concentration in the range of 30–40% caustic. In order to keep the catholyte concentration constant, the rate of external water addition to the cell must be balanced with the rate of water transfer across the membrane. The reason that the water balance is important is that the composition of the desired corresponding crystalline caustic hydrate, $(NaOH \cdot nH_2O)$, formed in the catholyte must be maintained during cell operation at 25,35, and 40 wt% as seen in the freezing point solubility given in Figure 2. In every case, the known structures of the corresponding crystalline hydrates are retained in concentrated solutions (1).

The pH dependence of membrane conductivity is important. The pH dependence of the membrane conductivity shows that for weak acid membranes below a certain critical pH, the membrane conductivity drops dramatically. This occurs because in the acid form, the membrane is in an undissociated form and the polyelectrolyte becomes much less dissociated than when in the sodium salt form(19).

Membrane Cell Components

Dimensionally stable electrodes in this system serve as conductive, rigid, corrosion resistant electrocatalysts.

Anode

The dimensionally stable anode in this system is composed of an electrically conductive substrate of titanium, having a coating of a defect solid solution containing mixed crystals of precious metal oxides. These substitutional solid solutions are both electrically conductive, electrocatalytic, and dimensionally stable. Within the aforementioned solid-solution host structures the valve metals include: titanium, tantalum, niobium, and molybdenum; while the implanted conductive precious metal guest elements include: platinum, ruthenium, palladium, indium, rhodium, and osmium. There is a close connection between the nature of the defect solid and its catalytic properties in the coatings. At present, the titanium-dioxide ruthenium-dioxide solid solution coatings are preferred.

Cathode

The cathode material may be made of any conductive metal having a surface that is capable of withstanding the corrosive conditions in the cathode chamber of the cell. Useful materials may be selected from a group consisting of stainless steel, nickel, steel, or platinum metals with sintered or otherwise porous coated surfaces that provide catalytic sites showing low overvoltage characteristics for hydrogen evolution. The cathode may be made from foraminiferous expanded metal mesh or screen. A high surface area material is desired with correct geometrical

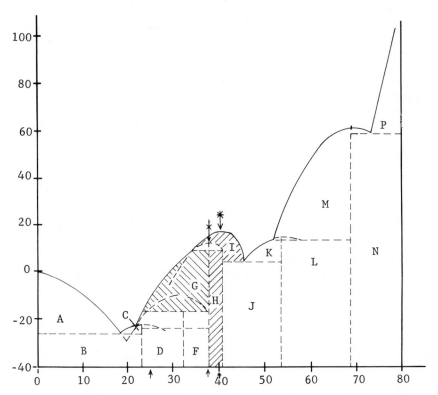

Figure 2. Freezing point solubilities for ice (A), ice + NaOH · 7H₂O (B), NaOH · 7H₂O (C), NaOH · 7H₂O + NaOH · 5H₂O (D), NaOH · 5H₂O (E), NaOH · 5H₂O + NaOH · 4H₂O (F), NaOH · 4H₂O (G, X), NaOH · 4H₂O + NaOH · 3.5H₂O (H, *), NaOH · 3.5H₂O (product) (I), NaOH · 3.5H₂O + NaOH · 2H₂O (J), NaOH · 2H₂O (K), NaOH · H₂O (L, M), NaOH · H₂O + NaOH (N), and NaOH (P).

design so as to provide good gas release hydrodynamics and minimize electrolytic resistance. Facile release of gas is important in the catholyte where membrane-cathode bubble masking within the highly viscous caustic solution is found to dramatically increase voltages as claimed by U.S. Patent 4,105,514.

Membrane

The separator in the chlor-alkali cell is by far the most important component. It allows the free passage of electrical current and keeps reactants and products apart by maintaining sufficient gradients between its phase boundaries.
In the absence of an electrical field and in dilute solutions, the degree of ionic selectivity depends solely on the physico-chemical properties of the membrane, but in the presence of a high intensity electrical field and the resultant large field gradients in concentrated solutions, the dynamic properties of both the membrane and solution interact with the imposed electrical field to provide the anomalous permselectivity observed.

Membrane Properties and Processes

Membranes are characterized by structure and function; that is, what they are and how they perform. The most significant primary structural properties of a membrane are its chemical nature; including the presence of charged species at the molecular level, its microcrystalline structure at the microcrystalline level, and on the collodial level its pore statistics such as pore size distribution and density, and degree of asymmetry (11) (12).
Cation exchange membranes are used in the membrane chlor-alkali cell process and must have good chemical stability. This requirement is satisfied by the perfluoropolymers. The types of membranes that are available for industrial chlor-alkali production are classified as: 1) perfluorosulfonic acid; 2) perfluorosulfonamide: and 3) perfluorocarboxylic acid types.
Large differences in permeabilities of membranes can be attributed to differences in interchain displacement and flexibility related to polar and steric effects. The polar molecules such as polytetrafluoroethylene have a stronger tendency to form rigid associations leading to crystal formation than nonpolar molecules. Polytetrafluoroethylene polymers are highly crystalline products with sharply definable melting points. Oriented specimens of high strength may be obtained, exactly as in the crystalline condensation polymers (13).
For every amorphous polymer there exists a narrow temperature region in which it changes from a viscous or rubbery condition at temperatures above this region, and changes to a hard and relatively brittle one below it. This transformation is equivalent to the solidification of a liquid to a glass; it is not necessarily a phase transition (13)(14).

Not only do hardness and brittleness undergo rapid changes in the vicinity of the glass transition temperature, Tg, but other properties such as the thermal expansion coefficient, heat capacity, and in the case of a polar polymer the dielectric constant also changes markedly over the interval of a few hundred degrees. Tg is regarded as the brittle temperature, or the critical temperature for the glassy state, or second order transition temperature, although no phase transition is involved (14) (15).

Considering phase equilibria in these liquid systems, when a solvent is chosen for a given polymer that becomes progressively poorer or the concentrations or temperature is lowered, eventually a point is reached below which solvent and polymer are no longer miscible in all proportions. At each lower concentration or temperature, mixtures of polymer and solvent over a certain composition range will separate into two phases leading to a partitioning of species between the two phases.

Membranes produced for chlor-alkali applications are generally composites of two equivalent weights of polymer or of one equivalent weight where one surface has been chemically modified to change the nature of the ion exchange grouping. These modifications alter the dynamic properties of the membrane. It has been found from experiments in which NaCl and NaOH solutions are separated by such membranes that the side of the membrane in contact with the NaOH solution dominates in the control of membrane performance as found in its resistance and selectivity. The apparent membrane diffusion coefficients for NaOH have been measured for such systems. Diffusion of sodium ions increases with the lower equivalent weights and decreased NaOH concentrations. Surface modifications can be used to produce films with lower diffusion rates by increasing the activation energy for diffusion (15).

The measurement and control of transport properties for ion exchange membranes is the key element in optimizing the operating conditions for modern chlor-alkali membrane cells. Ideally, a membrane should allow a large anolyte-catholyte sodium ion flux under load, while at the same time the hydroxide ion and water fluxes are kept minimal. Under these conditions, high current efficiency and low membrane resistance can be realized simultaneously in a cell producing concentrated caustic and chlorine gas.

Water, sodium ion, and hydroxide ion concentrations have been measured within the membrane phase as a function of bulk caustic solution concentration and temperature. These internal membrane concentrations are important because of their influence on the membrane polymer morphology, structural memory, plasticity and the resultant effects on its internal resistance, viscoelasticity and material transport. In addition, the self-diffusion coefficient of the sodium ions in various Nafion membranes has been measured as a function of temperature and external caustic concentration using the $^{22}Na^+$ radiotracer isotope. In this way, a true self-diffusion coefficient in the membranes can now be determined without the complicating problems of osmotic flow of water and the resultant gradients in ionic activity ever present in dialysis membranes (15).

Diffusion the Fundamental Process

Diffusion coefficients provide two kinds of information. First, their absolute magnitudes, combined with membrane sodium ion concentrations, are useful indicators of the temperature dependence of ionic self-diffusion and thereby they yield the activation energy for diffusion. They thereby provide insight into the nature of the diffusion mechanism (16). When activation energies are measured for various types of related membranes, the influence of different membrane structural design features can thus be separated and determined directly.

Measurements of the 120, 214 and 295 duPont Nafion films and also fully converted ethylene diamine films are considered to be typical. The 120 polymer is a homogeneous film 10 mils thick of 1200 equivalent weight, (ew), perfluorosulfonic acid resin. The 214 and 295 films are each of 7 mils thick resin of 1150 ew having one surface facing the cathode that has been chemically modified to 1.5 mils depth, and having T-24 backing. The 295 films are the same as the 214 except that they are modified to a 1.5 mil depth and have T-900 backing (17) (18).

Results show that the water uptake decreases and caustic concentration is relatively constant for these materials as the caustic concentration of the solution increases. The temperature dependence of these properties is not pronounced. The self-diffusion coefficient of Na+ in these membranes is strongly dependent on both temperature and caustic concentration. Below certain temperatures, dependent on caustic concentrations, EDA treated Nafion becomes impermeable to sodium ion diffusion. At higher temperatures, diffusion proceeds by a different process with activation energies of 7 to 12 kcal/mol depending on the separator material. The activation energies are insensitive to caustic concentration, but the absolute magnitudes of sodium ion diffusion coefficients are very concentration dependent. Also, differences in the activation energy for 214 and 295 Nafion can be correlated with differences in membrane voltage drops found in operating cells. An overall conclusion from this work is that the fabric backing in these materials is an important factor in increasing the electrical membrane resistance (19) (20).

Several processes occur simultaneously within the membrane phase of an operating cell. Sodium, chloride and hydroxide ions all migrate under the combined effects of concentration and electrical potential gradients with sodium ions as the major current carrier. The flow of sodium ions in a field is accompanied by a net electroosmotic flow of water in the same direction. Chloride ion flux is much smaller than that of sodium

and hydroxide ions, since the membrane presents an effective
barrier to it, and the electrical potential across the separator
opposes the transport of the chloride. Interactions occur among
cations, anions, water and the membrane matrix. The magnitude of
these interactions depends on the membrane properties as well as
the water and electrolyte sorption, combined with capillary
transport through the thin-film quasi-lattice of imbibed
solution. These polymer-solvent interactions define the overall
operational properties of the membrane such as its selectivity,
resistance and operating properties. The relative magnitudes of
these interactions differ from those observed in electrolyte
solutions due to the presence of fixed charges and polymer in the
membrane phase (19).

Ionic transport through these perfluorinated ionomers is considered now to be essentially a diffusional process whenever no
flow of current is imposed. This diffusion can be defined as a
rate process with an average energy barrier for diffusion that
must be exceeded before transport can occur. This approach is
useful because this activation energy provides a convenient inde-
cator of the minimum energy requirements for ion transport
through the membrane, and this provides a mechanism for diffusion
there. Ionic diffusion coefficients and the resultant activation
energies are thereby related to the operating characteristics of
the membrane under current flow or load conditions. A self-
diffusion coefficient can be obtained without imposing concentration gradients of water and ions across the membrane, and
so that it is an unambiguous measure of the diffusion ability of
an ion through a separator (18) (19) (20) (21) (22).

The thermodynamic diffusion coefficient, D_T, is defined as:

$$D = D_T(1 + d\ln\gamma/d\ln C) \tag{1.}$$

here γ is the activity coefficient, a/C,
and:

$$D_T = RTU \tag{2.}$$

where U is the mobility, and:

$$D = \lambda^2/2\tau \tag{3.}$$

and (3.) constitutes a new definition of the diffusivity in
terms of the mean molecular jump distance λ , and the mean
time per jump, τ , and (3.) also can be given as:

$$D = \lambda^2 k \tag{4.}$$

Here, (4.) gives the diffusivity in terms of molecular
properties. In this case, k is the Absolute Reaction Rate
constant given for a solution which is homogeneous, in which con-
ducting holes are distributed at random along with the solute
molecules across the thin film quasi-lattice. The specific rate

constant from Absolute Reaction Rate Theory (A.R.R.T) is given as:

$$k = k'T/h \ (F^*/F) \ exp \ (-E_A/k'T) \qquad (5.)$$

where F^* and F are the partition functions for the system and E_A the activation energy per molecule at $0°K$, and k' the normal Boltzmann's constant, so that:

$$D = D_o \ exp(-E_A/RT) \qquad (6.)$$

Given in terms of viscosity, experimental results confirm the theoretical Einstein formula relating the diffusion coefficient to viscosity:

$$D = \lambda^2/2\tau = RT/N(1/6\pi\eta \ r) \qquad (7.)$$

where λ is the average molecular displacement in time τ and η the viscosity (22) (23) and r the radius of the fine capillary across the thin film quasi-lattice.

Diffusion Related to Flux

Diffusivity is defined as the Fick's law coefficient which is based on an analogy with other physical phenomema, such as heat transfer and electrical conduction. The drag on the ions and molecules being driven through a solution and producing resistance to flow is caused by the viscosity of the medium. For diffusion rates which are not extermely high, the mean velocity of diffusing molecules is proportional to the force acting on them:

$$v \ (m/s) = U \ (m^2/V\text{-}s) \ f \ (V/m) \qquad (8.)$$

here v is the net velocity of the ion or molecule, U is the proportionality contant called the mobility, and f is the driving force acting on the particle, called the potential gradient or electric field strength. The concentration, C, times the velocity, v, gives the flux, J, as:

$$J = Cv = - \ RTU \ \{dC/dx\} \cong - \ D\{dC/dx\} \qquad (9.)$$

given as Fick's First law. The coefficient, D, is the diffusivity. It is more convenient to express the product Cv in terms of the molecular flux, J, and area of solution transferred:

$$J(moles/m^2s) = -D(m^2/s) \ \{dC/dx\} \ (moles/m^3)(1/m) \qquad (10.)$$

Experimentally, the quantity J is measured by the average time rate of change of concentration per unit area.
    In any case, diffusivity depends on the concentrations established across the barrier films and the diffusion coef-

ficient is very useful because it relates directly to the mobility, U, which can be determined from diffusion experiments(22)(23).

Diffusion through Quasi-Lattice Films

Crystal structures of many simple metallic oxides, including
strong alkalis, are considered to consist of hexagonal or cubic
close-packing structures of two types: 1.) Voids surrounded by
four oxygen ions, tetrahedral voids, and 2.) Voids surrounded by
six oxygen ions, the octahedral voids. In the close - packed
structures there are two tetrahedral sites and one octahedral site
per oxygen ion (24) (25).

Even though most of the simple MO oxides have the halide
structures where the metal ions are octahedrally coordinated by
the oxygen ions there are a few MO oxides where the metal ions
are tetrahedrally coordinated (26). The alkali metal oxides, Li_2O,
Na_2O, K_2O and Rb_2O possess the anti-fluorite structure with oxygen
ions considered as close-packed and cations occupying all of the
tetrahedral sites.

The structures of a number of concentrated aqueous solutions
have been examined by x-ray diffraction by Finbak and co-workers,
including nitric acid, sulfuric acid and sodium hydroxide.

The x-ray radial distribution curves obtained for a 38 weight
percent aqueous solution of NaOH is interpreted as indicating a
tetrahedral arrangement of water molecules that surrounds the

$Na+-OH_2$ at bond distances of 2.03 Å. At 38 weight percent caustic,
a $NaOH\cdot 3.6H_2O$ composition is found at 536 GPL (25). In these
very concentrated solutions, formation of $Na+OH^-$ ion - pairs is
assumed to exist. Even though it may not be generally agreed
that it is justifiable to draw detailed conclusions about structures of ionic solutions from their x-ray scattering patterns, it
is possible to obtain information about the structure and immediate environment of certain ions in this manner (25) (27).

Electrical Conductivity

Whenever an electric field, E, is applied across a thin film
quasi-crystal system, such as found in these metallic oxides, a
force is exerted on the charged particles in a quasi-crystal. If
an ion or a defect has a charge, Q_i, then the force, F_i, on this
ion or defect film is given as (21) (28):

$$F_i \ (joules/m) = Q_i(coul) \ E_i(V/m) \qquad (11.)$$

where: coulombs = amp-sec, and joules = watt-sec=coulomb-volt
This force causes a directional transport of the charged particles in the crystal, or quasi-crystal film, in addition to
their random thermal motion. In this case, Q_i is the net charge
contained within a mobile, collective Gaussian surface. Ions
cross interfacial boundaries such as membranes and create a net

transfer of charge to produce the steady-state potential surfaces observed. The resulting current density is given by (22) (26):

$$I \ (amps/m^2) \ = \ \sigma_i (ohm^{-1}m^{-1}) \ E(V/m) \tag{12.}$$

and where:

$$\sigma_i (ohm^{-1}m^{-1}) \ = \ \rho \ (amp\text{-}s/m^3) \ U(m2/V\text{-}s) \tag{13.}$$

and where σ_i is the conductivity, ρ is the charge density, and U is the ion mobility of ions within the Gaussian surface.

Here σ represents the total electrical conductivity, the ionic conductivity of ion i is given as σ_i and related to the total conductivity through a proportionality constant, t_i, or the transport, or transference number of species i (22) (26):

$$t_i \ = \ \sigma_i/\sigma \tag{14.}$$

In an ionizable compound of any geometry, the total electrical conductivity is given by the sum of anionic and cationic conductivities:

$$\sigma \ = \ \sigma_{\text{Cation}} \ + \ \sigma_{\text{Anion}} \ = \ \sigma(t_c + t_a) \tag{15.}$$

and, it follows that:

$$t_c + t_a \ = \ 1 \tag{16.}$$

The current density of the particles of type i, I_i, is related to their migration or drift velocity, v_i, through the relationships:

$$I_i = C_i Q_i v_i = C_i Z_i e v_i \ = \ \rho v_i \tag{17.}$$

where C_i is the ionic concentration, Q_i, the ionic charge and given as the product of charge e and number of charges, Z_i, and in terms of charge density, ρ, the current density is:

$$I(amp/m^2) \ = \ \sigma(ohm^{-1}m^{-1}) \ E(V/m) \ =$$

$$= \ \rho(amp\text{-}s/m^3) \ U(m^2/V\text{-}s) \ E(V/m) \tag{18.}$$

here with Z_i as ionic valence and C_i concentration of particles, the charge mobility U is defined as the velocity in a unit electric field (23).

Diffusion Related to Transport

Electrical transport through industrial membranes used in chlor-alkali cells is not shared equally among all of the mobile components within the conductive polymer film. In all cases, however, the sodium ion within the hydrated ionomeric phase of the polymeric film in the membrane is the major current carrier as these ions move through the membrane they drag along much water with them. A quantitative definition of the transport of water molecules and sodium ions through ion exchange membranes is thus found to be of fundamental importance in all phases of cell operation (16) (24).

Electroosmotic Coefficients

The electroosmotic transport coefficient for water through Nafion 295 and 1150 membranes is typical and is shown to be highly dependent on the anolyte concentration to the exclusion of all the other variables studied. The water transport coefficient varies almost linearly with anolyte concentration from 6 to 17 molar caustic, giving 2.9 to 0.8 moles/F, as shown in Figure 3. The sodium ion transport number goes through a maximum of 0.82 eq/F in the 7 to 13 molar caustic range (27).

The data shows that changes in water concentration and activity across the film, as controlled by anolyte concentration, regulates the close range ionic hydration of the hydrophylic macro-molecules making up the matrix of the membrane, by changing its phases. These phase changes induce changes in the diffusion mechanism for water molecules, sodium and hydroxide ions passing through the polymer film, especially at the thin film catholyte interface.

When one views the membrane as a multi-layer thin film device, he begins to understand how the interphases control its electrophysical properties and the reaction rates across its junctions (18) (29).

Voltage Profiles

Membrane voltages are plotted versus current density for Nafion 120 and 295 films and given in Figures 4 and 5. The curve in Figure 4 shows a greater slope change, dE/dI, for the Nafion 120 than for the Nafion 295. The essential difference between these membranes is the degree of hydration of the cathodic surface of the separator films. The Nafion 120 membrane material overall contains 30% water, and this is released in the presence of high concentrations of caustic or salt, thereby forming dynamic thin-film lattice barrier layers. The large variation in this slope implies existence of a selectivity that is strongly dependent on caustic concentration, while the Nafion 295 membrane contains a thin chemically modified layer of ethylenediamine, having 10-15% water (27).

Sodium Transport

The transport properties of the Nafion membranes can best be

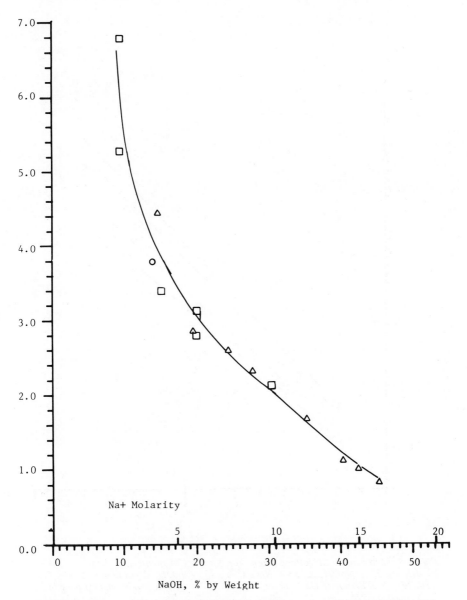

Figure 3. Water electro-osmotic coefficient vs. anolyte concentration for Nafion 295 membrane, 80° C, 2 kA/m². Key: ◯, measurements with new cell design, identical anolyte/catholytes; ☐, chlorate present in anolyte; △, measurements with old cell design, identical catholyte/anolytes.

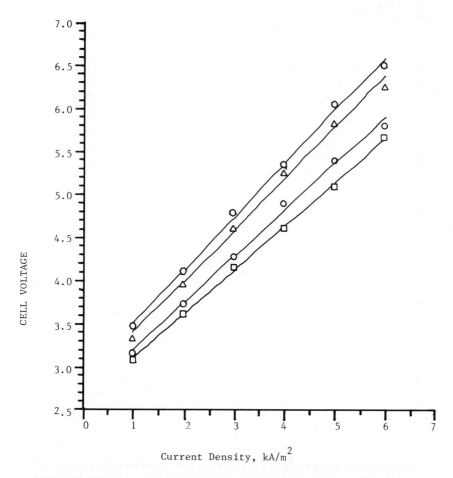

Figure 4. Membrane cell voltage profile for Nafion 120 membrane. Key: □, *25%*
NaOH; ○, *32% NaOH;* △, *33% NaOH;* ◐ , *35% NaOH. Conditions: 85–90° C,*
18–24% NaCl Anolyte DSA anode, Ni cathode, ⅛ inch electrode–membrane gap.

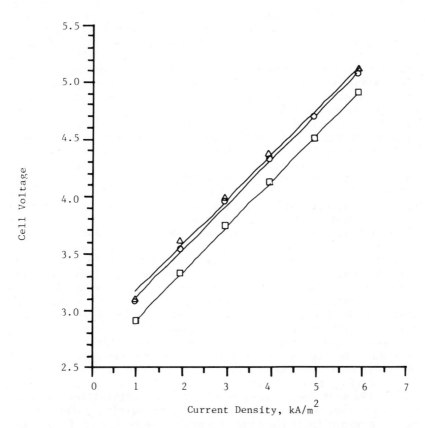

*Figure 5. Membrane cell voltage profile for Nafion 295 membrane. Key: □, 22%
NaOH; ○, 35% NaOH; △, 39% NaOH. Conditions: 85–90° C, 22–24% NaCl
anolyte DSA anode, Ni cathode, ⅛ inch electrode–membrane gap.*

explained based on the fine-capillary-pore membrane model. The membrane appears to be composed of separate phases formed as separate three-dimensional statistical networks generating phase boundary junctions at the membrane pore solution interface.

Few studies have been made on transport processes involving concentrated solutions. In the concentrated solutions, in the range of dehydrated melt formation, incompletely hydrated melts and anhydrous salt melts, various structural models are described to define their properties, i.e. the free-volume model, the lattice-model and the quasi-crystalline model. Measured and calculated transport phenomena do not always represent simple ion migration of individual particles, but instead we sometimes find them to be complicated cooperative effects (27).

At very high concentrations of the ionic solutions, the quasi-lattice model of Braunstein is useful wherein a highly concentrated electrolyte solution is considered to be a solution of water in fused salts (24) (26).

The data presented in Figure 6 gives a typical description of the transport properties of Nafion membranes within the framework of operating data for a chlor-alkali or a water electrolyzer system. In all cases, one finds better selectivity for discriminating against the back-flow of hydroxyl ions at or near the 33-38% caustic concentration. This gives a higher sodium to hydroxide transport number ratio. Since we know that in each case, the known structures of the corresponding crystalline hydrates are retained in concentrated solutions, then they may induce cation partial lattices within the membrane phase which in turn provides thin film anion partial lattices in the 35-36% caustic ranges. Whenever the concentrations go above or below this maximum in the curve, the short-range translational motion of ions, exemplified by activated jumps from one equilibrium position to the other, has great significance in relation to the kinetic transport properties of aqueous electrolyte solutions in the space charge region of these thin film separators.

The data on electrical conductance of aqueous salt solutions are of great interest for relating structural changes in the electrolyte solutions to the degree of swelling and variable pore structures. If the salt under consideration forms a crystalline hydrate, then for an isotherm not too far from the melting point of the hydrate a maximum in conductance occurs close to the melting point of the hydrate eutectic composition. The existence of these maxima is due to the fact that this viscosity increases rapidly at certain concentrations causing the hydroxyl ion mobility to decrease on the microscale within the micron thick solution quasi-lattice which forms in the pores. These conditions forming the maxima become imposed on the phase structure of the membrane and correspond to structural transformations within the solutions as shown(22)(24):

$$\eta_o/\eta = 1 + A\sqrt{C} \quad + \quad B\,C \qquad\qquad (19.)$$
$$\underset{\text{ion-ion}}{} \qquad \underset{\text{ion-solvent}}{}$$
$$\underset{\text{term}}{} \qquad\quad \underset{\text{term}}{}$$

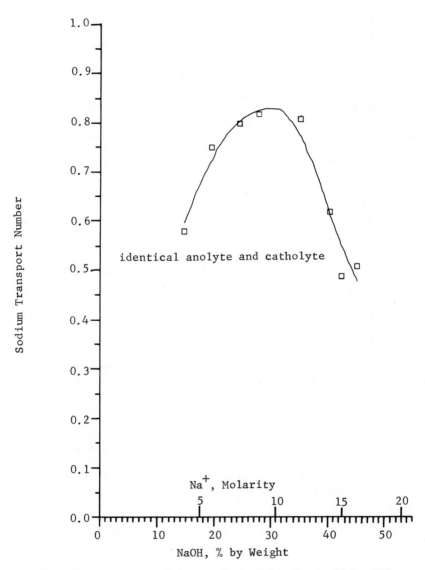

Figure 6. Na transport number vs. anolyte concentration for Nafion 295 membrane, 75–95° C, 1–3 kA/m².

Where η_0 is the solvent viscosity and η solution viscosity, C is the bulk concentration and A is a constant dependent on interionic attraction and B a constant dependent on ion-solvent interaction and is a function of the ionic mobility. The concentration of non-exchange electrolyte in the pore system of a membrane is determined by a distribution equilibrium, dependent on the width of the pore. The higher the concentration of the outside solution, the greater is the concentration of non-exchange electrolyte in the pore system of a poorly hydrated ionic membrane film.

Reactor Design

Electrochemical reactor design is ideally a good compromise between capital and power costs. The power consumption of a cell or reactor is the most important single factor needed to evaluate its performance. Both the power production and chemical process industries, (CPI), involve heat and electrical energy in a similar fundamental way, and so are governed by the second law of thermodynamics. The second law actually imposes an absolute natural limitation on the efficiency of any energy transformation, and therefore it provides a reliable standard with which to compare and control practical operations (30) (31) (32).

Power Costs

Power costs for aqueous chlor-alkali cells amount to about 50% of the total operating costs and almost 75% for water electrolysis. Molten salt sodium and aluminum processes are even more power intensive than chlor-alkali cells.

Energy Efficiency

Power consumption is thus seen to be the primary criterion of overall cell performance, where the energy consumption per mole of product, W_j is given (33):

$$W_j = (EIt)/N_j = watts/mole \qquad (20.)$$

where N_j is the moles of product, E the total cell potential, I is the amperage, and t the time of the current flow, so that the minimum electrical energy expended for the overall process in terms of watts (30):

$$W(watts) = N_j(moles)W_j(watts/mole) = N_j \Delta G_j = \text{Minimum electrical}$$
$$\text{energy expended in}$$
$$\text{the process.} \quad (21.)$$

From thermodynamics, at constant temperature we recall that the

$$\Delta G_j = \Delta H_j - T \Delta S_j \qquad (22.)$$

relationship exists,

where ΔG_j is the free energy available for a reversible process, ΔH_j the total enthalpy change and ΔS_j the entropy change, so that

$$W_j + N_j\ T\Delta S_j = N_j\Delta H_j \tag{23.}$$

and 23.) is obtained after subsituting 20.), 21.) and 22.), so that:

$$EI_j t = N_j\Delta H_j + Q \tag{24.}$$

where the energy balance for the condition $E > E°$ exists, and Q is the quantity of heat removed from the reactor at constant temperature, $N_j T\Delta S_j$, so that the total electrical pressure or voltage imposed across the electrodes in the reactor during the process shown in Figure 7, is given as (30) (34):

$$E = (N_j\Delta H_j + Q)/I_j t \tag{25.}$$

The general basis for calculating the power usage in electrochemical processes is the overall energy efficiency:

$$E.E. = V.E. \ X \ C.E. \tag{26.}$$

where E.E. is the total energy efficiency, and V.E. the voltage efficiency for the given process. The voltage efficiency portion of equation 26.) deals with the electrical pressure effects imposed across the electrodes, and is calculated by:

$$V.E. = (E°/E) \times 100 \tag{27.}$$

with E the actual bus to bus cell voltage and $E°$ the thermodynamic reversible cell potential. The current efficiency portion of the overall energy efficiency in 26.) above deals with the flow of current and is calculated:

$$C.E. = Eq.(Produced)/\ nF(Passed) \tag{28.}$$

where Eq.(Produced) is the equivalents of product produced and nF(passed) is the total Faradays of charge passed (34).

Voltage Balance

The total cell potential E for reaction can be resolved into the respective cell components, as given in Table 1., as follows (35) (36):

$$E\ =\ E° + \eta'_a + \eta'_c + E_{IR} + E_{IR} \tag{29.}$$
$$\text{(internal)}\quad\text{(external)}$$

As seen in Table 1., as above, I is the total net current flow, R_{ex} the external circuit resistance and R_{in} the internal cell

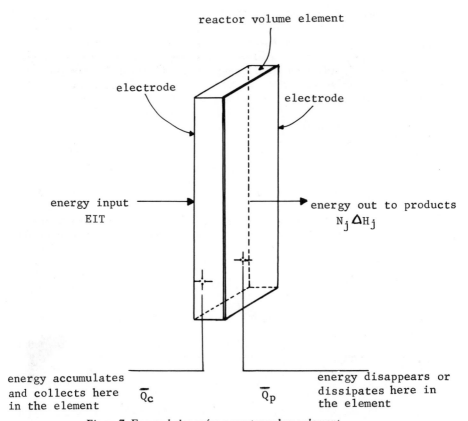

Figure 7. Energy balance for a reactor volume element.

TABLE I

Typical Voltage Breakdown
for Diaphragm, Membrane and Mercury Cells

	Voltage Component	Diaphragm Cell at 2.3 KA/m² 12%NaCl/16%NaOH	Mercury Cell at 10 KA/m² 50% Caustic: Graphite Anodes	Mercury Cell at 10 KA/m² 50% Caustic: Metal Anodes	Membrane Cell 2 KA/m² 37% Caustic (427 Membrane)
$E°$	$E°a$ (Anode Decomposition)	1.359	1.47	1.37	1.359
	$E°c$ (Cathode Decomposition)	0.930	1.74	1.74	0.950
η'	$\eta'a$ (Anode Overpotential)	0.160	0.18	0.12	0.150
	$\eta'c$ (Cathode Overpotential)	0.350	0.85	0.85	0.495
INTERNAL + EXTERNAL E_{IR}	E_{IRB} (Brine Gap plus Bubble Resistance-Internal)	0.390	0.35 (3mm)	0.30 (2mm)	0.29
	E_{IRC_1} (Caustic plus Bubble Resistance-Internal)	-	-		0.160
	E_{IRC_2} (Anode Contact to Base-External)	0.080	0.09	0.09	0.080
	E_{IRC_3} (Cathode Contact to Base-External)	-	0.05	0.05	-
	E_{IRC_4} (Base Plate Resistance-External)	0.060	0.06	0.06	0.040
	E_{IRC_5} (Separator Resistance-Internal)	0.520	-	-	0.650
	E_{Cell} Voltage-TOTAL	3.85	4.79	4.58	4.17

resistance. The internal resistance is made up of resistance in separator and electrolytes. The anode and cathode over potentials are η_a and η_c respectively. Table 1 gives the typical voltage breakdown for diaphragm, membrane and mercury cells, with $E°$ the total decomposition potential determined from equilibrium thermodynamics.

The electrode overpotential has several components possible:

$$\eta'_{Total} = \Sigma_i \eta'_i$$

These components of η'_i are: transition; concentration; diffusion; reaction; crystallization and resistance.

Process Development

Electrochemical processes are now becoming important in modern technology. There are certain basic fundamental design factors to consider for any electrolytic facility. These are comprised of four units: cell feed preparation, electrolytic cell reactions, electrical power supplies and finally product recovery equipment. The technology and costs for cell feed preparation and product recovery equipment are well established in process engineering and not considered further here. The focus of this section is the electrochemical reactor.

Reactor Design

A reactor of any type must be optimized in operating performance for best yields of products, and the minimum power and reactant consumption for any system can be evaluated using the basic law of transport as shown in Figure 7, and the following equation (40):

$$\Sigma \begin{bmatrix} Mass+ \\ Energy + \\ Momentum \end{bmatrix}_{input} - \Sigma \begin{bmatrix} Mass + \\ Energy + \\ Momentum \end{bmatrix}_{output} + \begin{bmatrix} Generation \\ or \\ Depletion \end{bmatrix} + \begin{bmatrix} External \\ Influences \end{bmatrix} =$$

$$\begin{bmatrix} Rate\ of\ Accumulation \end{bmatrix} \qquad (31.)$$

Critical Parameters

The design of electrochemical reactors impacts capital and production costs very significantly. Much effort has thus been expended in the electrochemical process industry during the past two decades toward reducing power consumption in order to meet production goals with much more expensive power and raw material costs.

Electrochemical processes are by their very nature more specific, but quite capital intensive and thus consume large amounts of energy in its most valuable form as electric current (7)(34)(40).

The essential task of the electrochemical engineer deals with the process optimization, that is the idea of defining the best economics in terms of compromises among the competing factors such as: space-time yield; energy consumption, product quality and materials of construction.

The removal of reaction products is the second factor that the engineer must consider in order to get the reactor scaled-up to the next pre-pilot stage of development successfully.

Product recycling must be examined at this stage after a suitable reactor system is selected, designed and proven (28)(30)(34).

Reactor scale-up is an extremely important step after bench-scale studies have been conducted.

In design development, reactants are charged into the batch-type reactor one at a time at the beginning of electrolysis while products are removed at the end of the run. A continuous flow system can be evaluated as a natural extension of the batch system.

Critical design parameters for these systems can be cast into idealized, quantitative design equations in order to define such factors as reactor volume-flow, electrode overpotential, and hold-up time, as functions of reactor design (40).

We realize that whenever the overpotential on an electrode is greater than ~60 millivolts, the reverse reaction can be neglected and the kinetic equation can be simplified as follows:

$$I = QA \{ k_f C_o - k_r C_r \} \cong QA \ k_f C_o =$$

$$= n \ F \ A \{ k_f C_o \exp (- \alpha \ n \ F n'_a /RT) \} \tag{32.}$$

Thus, the behavior at diffusion limited current flow conditions could be most simply represented as a function of the bulk concentration, C_o, and given as (5):

$$i_{max} \cong nFC_o(fr.) \tag{33.}$$

where (fr.) is the flow rate through the reactor, and r_{el} here the electrochemical rate:

$$r_{(electrochem)} = I/(nFA) = i_o/ \ nF \{ \exp (-\alpha n n'_a/RT) \} \tag{34.}$$

Considering the section of a reactor as shown in Figure 8, apart from its vertical orientation of the electrodes, the only other departure from previous nomenclature is the introduction of terms l_a and l_c which represent the thicknesses of the anode and cathode respectively. The electrical connections are located at the tops of the electrodes which are designated by the x = 0 positions.

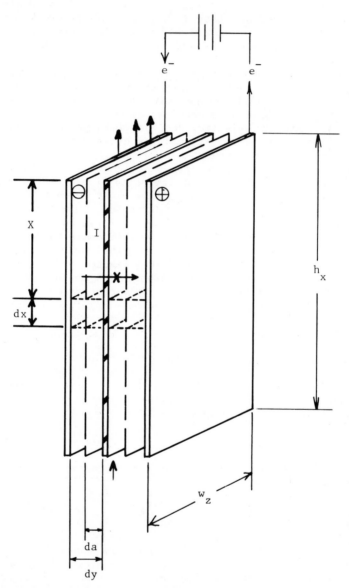

Figure 8. Design parameters for single-compartment parallel plate reactor (membrane chlor-alkali cell) with slow gas evolution in two dimensions.

The passage of current through an element of the electrolyte having dimensions $Bl_a dx$, is seen in Figure 8. If we assume that no current flows through the metal in the z direction, then conduction is one-dimensional only and takes place in the x direction. The metal current I_a, at a distance x from the top of the electrode is, from Ohm's law (30)(38):

$$I_a = - Bl_a \sigma_a (d\Phi_a/dx)_y \qquad (35.)$$

where Φ_a is the potential and σ_a is its specific conductance. A total voltage balance over any horizontal section must now include the extra potential contributions at anode and cathode due to mass and charge transfer limitations, $\Delta\Phi_a$ and $\Delta\Phi_c$, (5),(33) so that we find:

$$E = E^\circ + n_a' + n_c' + E_{IR} + (\Delta\Phi_a)_x + (\Delta\Phi_c)_x \qquad (36.)$$

Mass and Charge Transfer

For a general rate equation in a homogeneous back-mixed reactor under steady-state flow conditions, one finds (34) (40):

$$\text{Input} = \text{Output} + \text{Disappearance} + \text{Accumulation} \qquad (37.)$$

where:

$$\text{Input} = \text{Output} = 0 \qquad (38.)$$

so that neglecting convection and surface coverage, the rate of appearance of product is:

$$(dN_A/dt)_T = (dN_A/dt)_D + (dN_A/dt)_M + (dN_A/dt)_{ET} \qquad (39.)$$
$$\text{(TOTAL)} \quad \text{(DIFFUSION)} \quad \text{(ELECTROMIGRATION)} \quad \text{(ELECTRON TRANSFER)}$$

The general equation including all components is given:

$$(dN_A/dt)_T = ZFU_c \left[(RT/ZF)(dC/dx)_D + C^\circ(dE/dx)_M \right.$$
$$\left. + KC^\circ \exp(\alpha ZF n'/RT) \right]_{ET} \qquad (40.)$$

The first term, (dC/dx) represents the diffusion control, the second, (dE/dx), electromigration and the final terms, $\exp(\alpha ZF n'/RT)$, the charge transfer or electrode kinetics term.

Irreversible Liquid Hydrodynamics

Equilibrium properties of systems that are operating in a reversible way can be treated with steady state thermodynamics.

For the time dependent phenomena in a solution such as viscosity, diffusion and electrical conductivity, as noted, no fully developed theory exists. These are the irreversible processes, and because of their great importance, they have been studied and treated empirically by applied scientists for Newtonian and non-Newtonian fluids.

Convective Mass Transfer

Several methods are available to determine convective mass transfer rates, such as: solutions to the convective mass transfer Navier - Stokes equations, hydrodynamic mass transfer boundary layer equations, and correlation of experimental mass and momentum transfer data (36).

The basic aim of the method suggested here allows the engineer to express mass and momentum transfer data for a given system in terms of mass transfer coefficients that can be related to   measurable   physical   parameters   for   a   system.

The electrochemical aspects of mass transfer are examined here at the electrode-, or separator- solution interface.   In certain cases, such as with rotating disk electrodes, precise hydrodynamic analyses are possible and thus the diffusion equations   for   these   systems   can   be   solved   exactly. Electrochemical and non-electrochemical mass transfer phenomena can be correlated using analogies to heat transfer systems for both natural and forced flow convection, with reasonable success.

The   coupling   between   chemistry   and   mass   transport   and evaluated   roughly   by   external   macroscopic   measurements (36) using equations such as:

$$\delta_c = (\eta \, y/V_o)^{1/2} \, (D/\eta)^{1/3} \tag{41.}$$

where δ_c is the thickness of the diffusion layer, η the kinematic viscosity, V_o the liquid velocity at the surface and finally y is the distance from the given electroactive phase boundary to the diffusion layer in the solution.

The final general form of the flow-volume design equation with flow rate (fr.) is given as:

$$(fr.) = (nFC^\circ/I) \; \{ \; 2 \, nFD/\delta \; + K \exp \, (\alpha nF\eta'/RT)\} \tag{42.}$$

Commercial Reactors and Systems
Brine Process

Even though the main focus for the chlor-alkali engineers is the cell, and cell room, electrolysis is only one of many equally important operations involved in this process.   All electrosynthetic processes require the following ancillary processes: reactant feed, (or brine, here),   preparation, electrolysis, product recovery, and finally D.C. power.   These process units are related as shown in the typical membrane cell plant (6).

In all cases the depleted brine from the cells is dechlorinated before it is resaturated because it is difficult to remove iron in the presence of hypochlorite ions, and resaturation is less complicated when brine contains no chlorine. The cell effluent is acidified with HCl in sufficient amounts to react with all of the HOCl present. The remaining available chlorine can be removed by blowing compressed air in or by vacuum dechlorination. The tail gas from this process can be vented or killed in a milk of lime or caustic absorption tower. This dechlorinated brine contains less than 0.02 to 0.03 g/l available chlorine at this stage (6) (7). This dechlorinated strongly acid brine is next neutralized and made slightly alkaline before flowing to the saturator.

Depleted brine, dechlorinated and neutral to alkaline, containing 260 to 280 GPL NaCl, and at a temperature in the range of 50 to 60°C, is saturated by pumping it up through a bed of salt in dissolving tanks. Brine generally leaves the saturator hot and saturated; to prevent crystallization downstream it is generally diluted with a small bypass stream of weak brine.

Brine Purification - Chemical Treatment

If rock or solar salt is to be used, the full stream of brine is treated for the removal of Mg^{++}, Fe^{+++} and Ca^{++}, and other heavy metals and controlled to close tolerances. The $SO_4^{=}$ ion concentration exceeds the Ca^{++} ion levels when calcium chloride or soda ash is added. Heavy metals are usually sufficiently absorbed by the precipitates formed to remove them by coprecipitation. The reactions are:

$$CaSO_4 + BaCO_3 \rightleftarrows CaCO_3 \quad + BaSO_4$$

$$CaCl_2 + Na_2CO_3 \rightleftarrows CaCO_3 \quad + 2\ NaCl$$

$$MgCl_2 + 2NaOH \rightleftarrows Mg(OH)_2 \quad + 2\ NaCl$$

Treatment is carried out in a series of stirred tank reactors, the usual sequence of additions being Na_2CO_3 and then NaOH as given in Figure 9 (79). The brine is then subsequently clarified in a Dorr Oliver settler, and the overflow made sparkling clear by filtering through sand or "anthrafilt", or given a total filtration through a pressure leaf filter. The calcium levels at this stage are about 1 ppm. This is still too high for membrane cells (56).

As we see, it is most important that after a chemical treatment, insoluble materials are filtered by an appropriate method. These insoluble materials are filtered thoroughly before entering

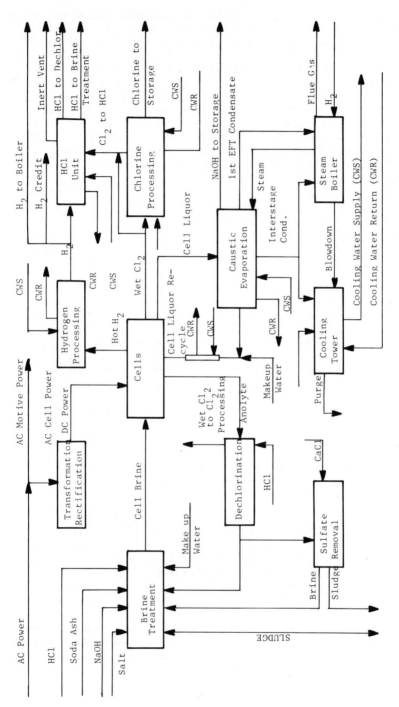

Figure 9. Process flow diagram for a 600 ton/day membrane cell Cl plant.

the process and the concentrations of the multivalent cations such as the calcium, magnesium, iron and alumimum ions have to be maintained as low as possible in the anolyte. For this purpose, the additional treatment of conventionally treated brine can be carried out by an ion exchange resin to remove such multivalent cations, or complexed with brine additives. However, the investment cost for the ion exchange column is relatively less due to the low concentration of such ions in the conventionally purified brine.

Brine Purification - Ion Exchange

The Ion Exchange Process

Whenever ion exchange resins are regarded as insoluble acids, bases or salts, the process of ion exchange can be regarded as salt formation and displacement. The interior of a water-swollen, strong acid cation exchange resin may be regarded as a concentrated acid solution.

Ion exchange is an equilibrium process, and the exchange reaction involving two cations, Na^+ and Ca^{++} can be written:

$$2\ R^-Na^+ + Ca^{++} \rightleftarrows R_2^= Ca^{++} + 2\ Na^+$$

The final position of the equilibrium is then given by the values of the equilibrium constant and the concentrations of reacting species (41).

Several new kinds of ion exchangers have been developed in recent years that give more specific and selective removal of divalent brine impurities such as calcium. One such resin was developed by Dow, Rohm and Haas and Mitsubishi, as a crosslinked styrene - divinylbenzene copolymer having iminodiacetate groups for joining fixed functional group sites to the metals by a chelate bond, (42) as:

The chelate resin selectivity for the heavy metals is similar to EDTA as it attaches preferentially to bivalent metal ions in the presence of monovalent metal ions.

Relations between pH and chelating ability of various metal ions are given in Figure 10. These plots show that this particular resin has a maximum rate for chelation above 2-5 for the bivalent metal ions, but depends strongly on the metal ion being chelated. Care must be taken that the metal ions not be precipitated as hydroxides and thereby increase metal ion leakage in the column effluent. The general selectivity for divalent and monovalent metal ions is:

$$Hg>Cu>Pb>Ni>Cd>Zn>Co>Mn>Ca>Mg>Ba>Sr>>Na$$

The operating cycle of such a column is:
1.) Removal; 2.) Backwash; 3.) Regenerate; 4.) Wash; 5.) NaOH treatment; 6.) Wash to clean effluent.

If the hardness is not to be removed from the brine system, then it must be sequestered by additives.

Brine Additives

The build-up of $Ca(OH)_2$ or $Mg(OH)_2$ at the anolyte interface of the membrane-brine system can be prevented by addition of certain sequestering-gelling agents into the brine such as phosphoric acid or phosphate salts (43). A strongly hydrogen bonded non-stoichiometric calcium orthophosphate gel is formed at pH's greater than 5 to sequester the divalent ions at the membrane-brine interface, but the gel dissolves at pHs of 2-3.5.

The pH sensitive reversible nature of the phosphate gel provides a continuously renewable surface for the entrapment of divalent impurities moving toward the membrane from the brine during operation and it eliminates the need for expensive ion exchange equipment for purifying the brine.

Certain membranes are more susceptible to damage by these impurities than others. The membrane structure thus determines the sensitivity of the films to contamination as well as its selectivity and ability to suppress free electrolyte diffusion (44).

Since the original development of low electrical resistance membranes by Walter Juda in the 1950's, (45), many other commercial membranes have been developed.

Commercial Membranes

The earliest commercial membranes tested for use in chloralkali cells were composed of ionomeric polymers having hydrocarbon backbones with attached carboxyl and sulfonate functional groups such as the polystyrene sulfonic or carboxylate materials, (46), (47), (48), (49).

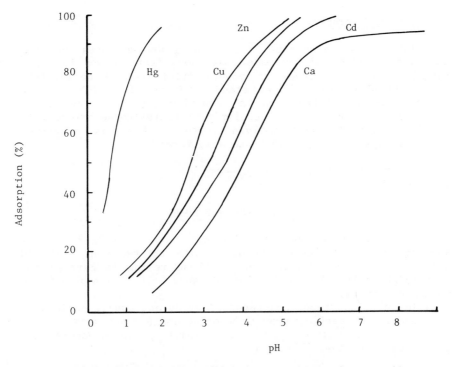

Figure 10. Relationships between pH and adsorption of various heavy metal ions.

The first successful chlor-alkali cell membrane materials (9)(10) tested in the 1970's were adopted from perflurosulfonic acid materials produced by duPont for GE for their fuel cell program in the 1960's. These materials had no hydrocarbon linkage in their backbone and thus could withstand the attack of the strongly oxidizing anolyte(11).

Remarkable advances in ion exchange membranes have been made since their inception and application to chlor-alkali cells in the 1970's, and since that time many patents have issued on their applications. Several companies besides duPont have developed proprietary membranes and electrolyzers for commercial application.

duPont Membranes

duPont has been active in developing new membranes over the past two decades and has made continuous improvements in their materials with reference to power efficiency and durability. Figure 12 shows the earlier standard Nafion 120, 295 and the newer Nafions plotted as a function of time on line.

The Nafion resin is a copolymer of tetrafluoroethylene with perfluorosulfonyl-ethoxypropylvinyl ether, which is converted from sulfonylfluoride to sulfonic acid form:

$$\sim\!\!-\!\!-(CF_2CF_2)_x\!-\!(CF_2\!-\!\underset{\displaystyle\overset{\displaystyle O}{|}}{CF})_y\!-\!\!\sim\!\!\sim$$

$$\underset{\displaystyle CF_3\!-\!CF\!-\!OCF_2CF_2SO_3H}{CF_2}$$

The membranes supplied by duPont have equivalent weights ranging from 1,100 to 1,500 meq/g with thickness of 5 to 10 mils, (2)(8). Mold processing of these resins is carried out only in the sulfonyl chloride form because it is thermoplastic in this case (50)(51).

The chemical reactions reported by duPont appear to involve the reaction of perfluoropropylene oxide and β - sultones to form sulfonic acid resins. The functional group monomers are generated as follows (52):

$$2\ CF_3\ -\ CFOCF_2\ +\ CF_2CF_2SO_2\ \xrightarrow[\text{Press}]{\Delta}$$

$$OHCCF(CF_3)OCF_2CF(CF_3)OCF_2CF_2SO_2F\ \xrightarrow[\text{Na}_2CO_3]{\Delta}$$

$$CF_2\!=\!CFOCF_2CF(CF_3)OCF_2CF_2SO_2F\ +\ CO_2^\uparrow\ +\ 2NaF$$

and then polymerized with tetrafluoroethylene in the ratio of m = 1 to n = 3-15:

$$n(CF_2CF_2) + m(CF_2CF)\ OCF_2(CF_3)\ OCF_2CF_2SO_2F \rightarrow$$

$$\sim\!\!\sim (CF_2CF_2)_n\!\!-\!(CF_2CF)_m\!\!\sim\!\!\sim \qquad\qquad KOH,\Delta$$
$$|$$
$$OCF_2CF(CF_3)OCF_2CF_2SO_2F \qquad\qquad \rightarrow$$

$$\sim\!\!\sim (CF_2CF_2)_{\overline{n}}\!\!\sim\!(CF_2\!-\!CF)_m\!\!\sim\!\!\sim$$
$$|$$
$$OCF_2CF(CF_3)OCF_2CF_2^-SO_3K^+$$

The Nafion membranes are produced in this way and with a fabric backing such as PTFE or mixed PTFE - rayon fabrics. These supporting materials improve the mechanical strength of the film and keep the dimensional changes in bounds. In general, for chlor-alkali electrolysis, the side of the membrane with the highest resistance, selectivity and charge density is preferred toward the cathode side to limit the undesirable effects of the back flow of hydroxide ions into the anode chamber. The anolyte side of the membrane polymer is thus less dense, less selective and more conductive than the catholyte side of the separator film.

The newer membranes provided by duPont have improved performance and Figure 11 shows the relative degradation rates for these materials used by GE in their fuel cells, (53).

Asahi Chemical Membranes

The Asahi Chemical Company of Japan has developed a perfluorocarboxylic acid membrane (54) (55) (56). It is reported to be formed from Nafion films wherein the SO_3H groups on the cathode surface are split off and the adjacent CF_2 groups thereafter oxidized to carboxylic acid groups.

$$\qquad\qquad\qquad\qquad\qquad\qquad\qquad\qquad [O]$$
$$\sim\!\!\sim (CF_2CF_2)_n\!\!\sim\!(CF_2CF)_m\!\!\sim\!\!\sim \qquad\qquad \rightarrow$$
$$OCF_2CF(CF_3)OCF_2C\overset{*}{F}_2SO_3H$$

$$\sim\!\!\sim (CF_2CF_2)_m\!\!\sim\!\!\sim(CF_2CF)_m\!\!\sim\!\!\sim$$
$$O\!-\!CF_2CF(CF_3)OCF_2\overset{*}{C}O_2H$$

These membranes are reported to achieve 93% cathode current efficiency at 21.6% caustic concentrations from the electrolysis process.

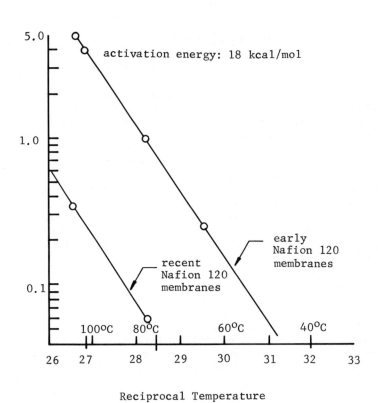

Figure 11. Degradation rates of perfluorinated sulfonic acid membranes.

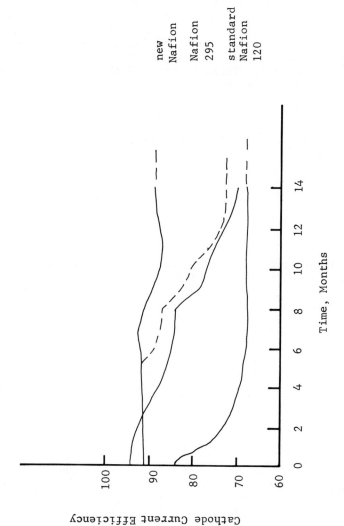

Figure 12. Performance of early standard DuPont Nafion 120, 295, and the newest membrane vs. time in a chlor-alkali cell.

Asahi Glass Co., Membranes

Asahi Glass Co. has recently disclosed their own perfluorocarbon membrane films for use in chlor-alkali cell electrolysis.
These membrane polymers have high molecular weights to
impart strong mechanical properties, and a tight structure to
supress unselective diffusion and resistance to swelling.
The perfluorinated structure imparts stability against
chlorine and strong caustic soda at high temperatures, copolymerization of tetrafluoroethylene with molecules such as
perfluoro-δ-butyrolactone and hexafluoropropylene oxide as:

$$CF_3CFOCF_2 + CF_2CF_2CF_2CO_2 \xrightarrow[\begin{subarray}{c}\text{Diglime}\\0-10°C\end{subarray}]{CH_3OH}$$

$$OHCCF(CF_3)OCF_2CF_2CF_2CO_2CH_3 \xrightarrow[\Delta]{Na_2CO_3}$$

$$2N_aF + CO_2 + CF_2CFOCF_2CF_2CF_2CO_2CH_3$$

next they polymerize with tetrafluoroethylene and a large monomer
molecule (57) (58) (59) (60) (61) (62).

$$n(CF_2CF_2) + m(CF_2\underset{\underset{\underset{CO_2CH_3}{|}}{\underset{(CF_2)_3}{|}}{\overset{|}{O}}}{CF}) + q(CF_2\underset{\underset{\underset{\underset{CO_2CH_3}{|}}{(CF_2)_3}}{\overset{|}{O}}}{\underset{\underset{CF(CF_3)}{|}}{\underset{CF_2}{|}}}{CF}) \xrightarrow{AIBN}$$

$$\sim\!\!-(CF_2CF_2)_{\overline{n}}\!\!-\!\!(CF_2\underset{\underset{\underset{CO_2CH_3}{|}}{\underset{(CF_2)_3}{|}}{\overset{|}{O}}}{CF})_{\overline{n}}\!\!-\!\!(CF_2\underset{\underset{\underset{\underset{CO_2CH_3}{|}}{(CF_2)_3}}{\overset{|}{O}}}{\underset{\underset{CF(CF_3)}{|}}{\underset{CF_2}{|}}}{CF})_{\overline{q}}\!\!\sim \xrightarrow[KOH]{\Delta}$$

$$\sim (CF_2CF_2)_n \sim (CF_2CF) \sim (CF_2CF)_q \sim$$

$$\begin{array}{cc}
O & O \\
| & | \\
(CF_2)_3 & CF_2 \\
| & | \\
C\bar{O}_2K^+ & CF(CF_3) \\
 & | \\
 & O \\
 & | \\
 & (CF_2)_3 \\
 & | \\
 & C\bar{O}_2K^+
\end{array}$$

with block perfluorinated divinyl monomer (63) copolymer structure of 10,000 to 100,000 molecular weights.

Copolymers containing up to 35 wt % carboxylated vinyl
ether were synthesized by regulating reaction pressure of the
tetrafluoroethylene in the polymerization. X-ray diffraction and
differential scanning calorimetry showed that the crystallinity
of the copolymer decreased with increasing vinyl ether content.
The ester groups in the film are quantitatively hydrolyzed in
caustic soda to yield sodium salt membranes. The incorporation
of ionic groups into the polymer greatly influences its mechanical properties. Conversion of the ester to the sodium salt is
accompanied by increased tensile strength and decreased elongation.

The water content of carboxylic-acid-type membranes
is low and does not change as much as that of a sulfonic-acid
type membrane. The low water content combined with high ion
exchange capacity results in a very high fixed ion concentration
within the membrane (63) (64).

Other companies have also developed ion exchange membranes
for alkali chloride electrolysis. These companies include:
Kureha Chemical, Maruzen Oil, Showa Denko, Tokuyama Soda, and
Toyo Soda.

Membrane Cell Electrolysis

Commercial Membrane Electrolysers

Electrolyzers for the membrane cell electrolysis process are
a filter press arrangement of cell units electrically arranged
in two different cell configurations, mono-polar and bipolar.
Figure 13 illustrates the examples of the electrical connection
of the cells (65). Figure 14 illustrates the electrolysis
circuit (66).

In the monopolar system, each cell unit in one electrolyzer
is connected in parallel and each electrolyzer is connected in
series. Voltage across one electrolyzer is low and current passing through one electrolyzer is high. Each electrolyzer unit is
connected to the next thru bolted bus bar connections which adds

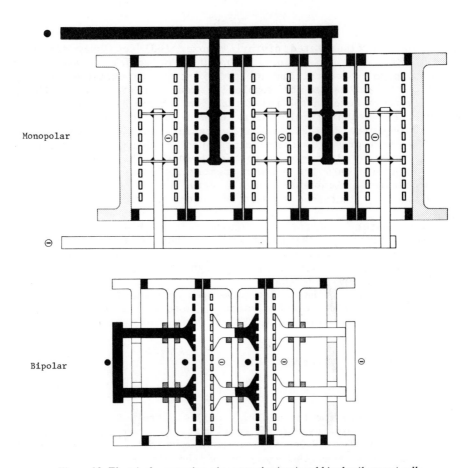

Monopolar

Bipolar

Figure 13. Electrical connection of monopolar (top) and bipolar (bottom) cells.

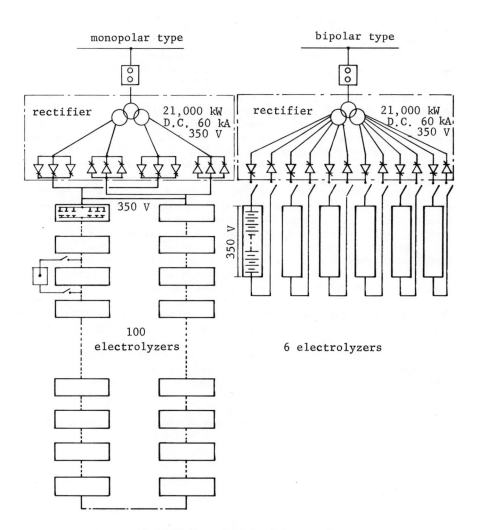

Figure 14. Example of electrical connection.

to the voltage loss of the electrolysis circuit. A portable switch allows for maintenance of an individual electrolyzer unit in the electrolysis circuit.

In the bipolar system, each cell unit in the electrolyzer is connected in series and each electrolyzer is connected in parallel. Voltage losses within the cell and the electrolysis circuit are minimized. Current leakage thru electrolyte paths between cells and corrosion of dissimilar metal joints (titanium/steel) can be encountered in the bipolar cell design.

Commercial electrolyzers have been developed using both the bipolar and monopolar cell configuration. The cell unit is generally made of metal or corrosion resistant plastics.

A large cell unit like the Asahi Chemical Co. bipolar cell Figure 15 is made of metal. The partition wall is an explosion bonded titanium to steel sandwich. The anode compartment is lined with titanium and the cathode compartment is steel. Metal components allow cell operation at 90°C while maintaining the high mechanical accuracy of the cell. Anodes and cathodes with an individual area of 2.7 m^2 are assembled into units containing up to 80 cells with a distance between the two electrodes in the range of 2-3 mm. Ohmic drops in the catholyte and anolyte are thereby minimized. Electrolytic power consumption of the Ashai Chemical Co. cell is shown in Figure 16 (67).

Tokuyama Soda Co., Ltd. has developed a large bipolar electrolyzer that has an electrolysis area of 2.7 m^2 per unit. The cell bodies are metallic. The anode chamber is lined with titanium plating and the cathode chamber and structural frame are carbon steel. The anode surface is titanium mesh with an electrocatalytic coating. The cathode surface can be either carbon steel or a low overvoltage cathode surface (LHOC) which Tokuyama Soda has developed. The hydrogen overvoltage of LHOC surface is 150-200 mV lower than carbon steel. Electrolysis power consumption of the 10,000 mt/year plant electrolysis is 2400-2500 KWH/mt NaOH at 2 KA/m^2 using steel cathodes. Using NEOSEPTA-F C-1000 membrane optimum catholyte concentration is about 20 wt% NaOH. NEOSEPTA-F C-2000 give a maximum value of current efficiency at nearly 30 wt% catholyte current ratio (68).

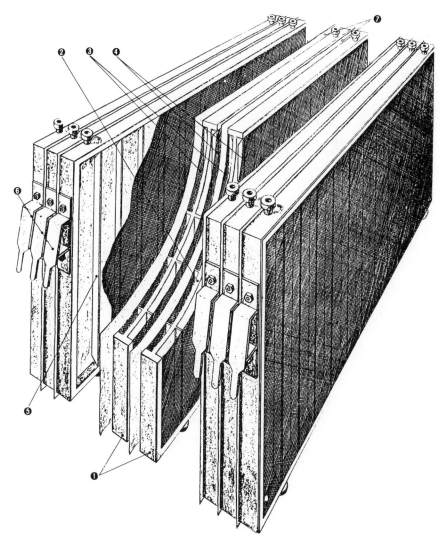

Figure 15. The Asahi Chemical Company bipolar electrolyzer.

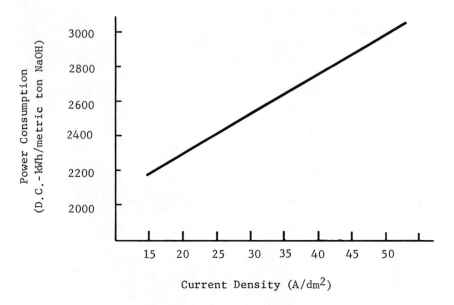

Figure 16. Power consumption of the Asahi Chemical Company membrane cell electrolyzer.

The Hooker MX, Allied PCF, and Ionics Chloromate electroly- zers are all bipolar designs that utilize plastic cell frames. The electrolytic area of these cells is typically 1.2-1.5 m^2 per cell unit.   The electrolyzers are typically composed of 10-50 cell units.   Power consumptions are reported to be 2750 KWH/metric ton of NaOH operating at 3KA/M^2 current density at 95% current efficiency producing 35% caustic soda (65,69,70).

The details of the monopolar electrolyzer of Diamond Shamrock are shown in Figure 17.   The electrolyzer is based on the monopolar design using DSA structures welded into titanium frames and the cathode structure welded into a steel frame.   Each cell has an active area of 1.41 m^2.   Expected performance of this cell is shown in Figure 18 (71)(72).

The Hooker-Uhde Monopolar Membrane electrolyzer is shown in Figure 19.   The active electrode surface of the HUMM electrolyzer is 1.7 m^2 per cell element.   The cell elements consist of anode frames of titanium and cathode frames of steel.   Anode-cathode gap is approximately 3 mm.   A separate frame is provided for holding the membrane.   The electrode block rests on a support structure that serves as the header system for electrolyte and electrolysis products.   Peformance of this electrolyzer is reported to be 2750 KWH/M ton NaOH at 35% caustic soda and 95% current efficiency at 3 KA/M^2 current density (65).

Asahi Glass Co. has developed a monopolar filter press type membrane electrolyzer that has an individual cell membrane area of 4 m^2. An electrolyzer may be composed of up to 32 cell units. Power consumption is reported to be 2500 KWH/M ton of NaOH at 94% current efficiency producing 35% caustic at 2 KA/M^2 current den- sity (73).

The Krebscosmo Bipolar Cell Electrolyzer BMZ 7.5 which is basically made from steel and titanium.   The partition wall of the bipolar electrode is a PTFE foil.   Anodes are expanded tita- nium sheet with noble metal coatings and the cathode structure is a perforated steel sheet.   The cell units can be arranged in parallel groups of bipolar elements.   Each cell element has 2.5 m^2 of membrane area that operate at a nominal amperage of 7.5 KA.   The cell block of the BMZ 7.5/64 electrolyzer consists of 64 elements with 16 series elements with 4 current paths with a nominal block amperage of 30 KA.   Electrolytic performance is not disclosed (74).

Commercial application of membrane cell technology began in 1975 with the installation of the Nobeoka No 1 (Japan) using Asahi Chemical Co. electrolyzers, Reed Paper (Canada) using Hooker MX electrolyzers and American Can of Canada (Canada) using Ionics Chloromate electrolyzers.   By the end of 1982 world capa-

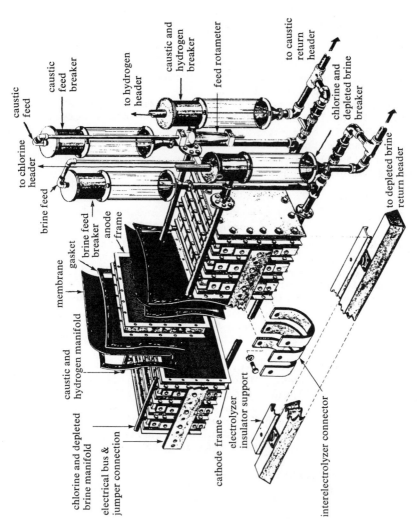

Figure 17. The Diamond Shamrock monoplar electrolyzer.

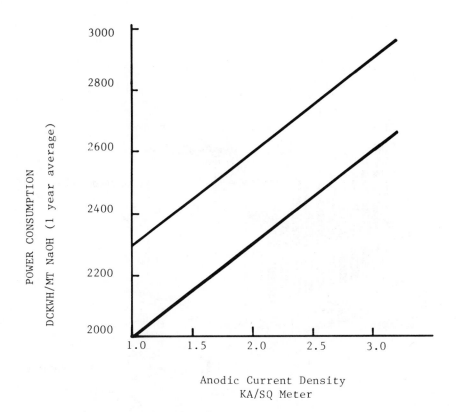

Figure 18. Expected membrane performance. Level 1: performance previously demonstrated during long term commercial plant operation. Level 2: performance demonstrated in laboratory cells. Membranes in commercial cells are expected to achieve this level of performance in the future.

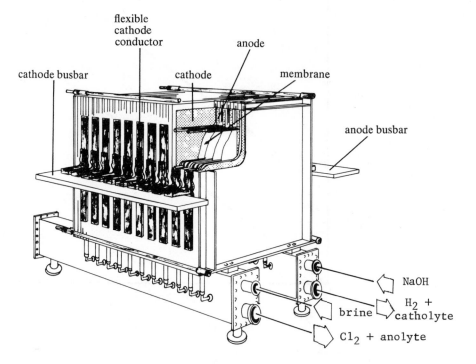

Figure 19. Principle of the Hooker–Uhde cell.

city for the membrane cell process is projected to be approximately 600,000 metric tons of NaOH per year. The installed capacity by electrolyzer system is shown:

World Membrane Electrolyzer Capacity (66)

Electrolyzer System	Estimated 1982 Installed Capacity
Asahi Chemical Co.	*470,000 MT/yr NaOH
Diamond Shamrock Co.	41,000
Hooker	34,000
Asahi Glass Co.	20,000
Tokuyama Soda	10,000
Ionics	10,600

*Assumes AKZO will be installed.

Membrane cell plants show considerable capital cost advantages against diaphragm cell plants for all capacities.
Investment costs for a membrane plant are 15-20% lower than the
diaphragm cell plants. Production cost comparison shows that the
membrane cell plant has a 10-15% lower net production cost per
ton NaOH (72) (79).

Membrane chlor-alkali cells represent a very successful,
commercially verified, economically competitive technology with a
short histroy of very rapidly advancing technology.

Membrane Developments

The Nafion membranes utilized in the early 1970's produced
caustic soda concentrations of 10-15wt% at electrolytic power
consumptions of approximately 3450 KWH/MT NaOH. Advancements in
the technology of membranes by duPont, Asahi Glass Co., and Asahi
Chemical Co., Tokuyama Soda Co., have achieved membranes that
today can produce caustic soda concentrations of 28-40wt% with
caustic current efficiency well over 90% for long term operations.

Asahi Glass Co., has announced the improvement of its
Flemion membrane, Flemion 723, which reduces electrolysis power
from 2500 to 2300 KWH/M Ton NaOH, operating at 35% NaOH and 2.0
KA/M^2 current density.   Cell voltage is 3.23 volts at 2
KA/M^2 with 94% current efficiency (75).

DuPont has recently announced the development of a new high
performance chlor-alkali membrane Nafion 901X. Caustic soda is
produced at 33 wt% with over 94% current efficiency. The Nafion
901X is capable of operating at minimum voltage and high current
efficiency for extended periods estimated to be in excess of two
years (76).

Further rapid developments of the membrane technology are to be expected, which are certain to further decrease electrolytic power requirements.

Membrane Electrolyzer Developments

Advancement of the membrane cell technology by Oronzo De Nora Impianti Elettrochimici S.p.A., utilizing the solid polymer electrolyte (SPE) brine electrolyzer, was first reported in May of 1979. This new technology utilizing specially activated perm-selective membranes is reported to have achieved cell voltages of less than 3.15 v at 3.3 KA/^{2} current density producing 28-30% caustic soda at 94% current efficiency, equivalent to 2200 KWH/MT NaOH. This technology is reported to be operating in excess of one year in a commercial prototype cell having an electrode area of 0.5 X 1.7 meters (77).

Further development of the "Zero Gap" membrane cell technology by Asahi Glass Co., called AZEC, is reported to have achieved at laboratory scale an electrolytic power consumption of 1950 KWH/M ton NaOH at 2 KA/M^2 current density and 35% NaOH and at a current density of 4 KA/M^2 the power consumption is 2140 KWH/M ton NaOH (78).

Development of these new zero gap membrane cell electroly-zers represents a major new approach in the membrane cell tech-nology and promises to provide even more rapid development in this quiet revolution of the membrane cell chlor alkali process.

Bibliography

(1) Coulter, M.V., "Modern Chlor-Alkali Technology," Society of Chemical Industry, Ellis Horwood Limited, Chichester, England, 1980.

(2) O'Leary, K. J., "Membrane Chlorine Cell Design and Technology," Lecture, The Electrochemical Technology Group of the SCI, London, June 16-17, 1976.

(3) Bergner, D., "Current Efficiencies, Chlorate and Oxygen Formation in Alkali Chloride Electrolysis According to the Membrane Method," Chemiker Zeitung, Vol.104, No 7-8, pp 215-224, 1980.

(4) ORONZIO DE NORA SYMPOSIUM, ORONZIO dE Nora Impianti, Elettrochimicia S.P.A. - Milano, 1979.

(5) Dotson, R. L., "Modern Electrochemical Technology," Chem. Eng. (Feature Article), July, 106-118, 1978.

(6) Sconce, J. S., "Chlorine, ACS Monograph Series, No.154," Reinhold Publishing Co., New York, 1962.

(7) Mantell, C. L., "Electrochemical Engineering," McGraw-Hill, New York, 1960.

(8) Bergner, D., "Electrolytic Chlorine Generation by the Membrane Process," (Elektrolytische Chlorerzeugung nach dem Membranverfahren), Chemikerzeitung, Vol. 101, No.10., pp 433-447, 1977.

(9) Dotson, R. L. and O'Leary, K. J., "Electrolytic Production of High Purity Alkali Metal Hydroxide," USP 4,025,405, German P 2,251,660, 1977.

(10) Stacy, A. J. and Dotson, R. L., "Control of Anolyte-Catholyte Concentrations in Membrane Chlor-Alkali Cells," USP 3,773,634; British P. 1,369,579, and German P 2,311,556, 1973.

(11) Flory, P. J., " Principles of Polymer Chemistry," Cornell University Press, Ithaca, New York, 1953.

(12) Kesting, R.E., "Synthetic Polymeric Membranes," McGraw-Hill Book Company, New York, 1971.

(13) Houwink, R. and Burgers, W. G., "Elasticity, Plasticity and the Structure of Matter," Cambridge, At the University Press, 1937.

(14) Ferry, J. D., "Viscoelastic Properties of Polymers," John Wiley & Sons, Inc., New York, 1970.

(15) Eisenberg, A., and King, M., "Ion-Containing Polymers, (Physical Properties and Structure)," Vol. 2. Academic Press, New York, 1977.

(16) Skelland, A.H.P., "Diffusion and Mass Transfer," John Wiley & Sons, New York, 1974.

(17) Dotson, R. L. and Yeager, H. L., "Fundamentals of Transport and Diffusion through Chlor-Alkali Cell Membranes", Presented at the Symposium of the Advances in Chlor-Alkali Technology in London, June, 1979, Sponsored by SCI.

(18) Hwang, Sun-Tak, and Kammermeyer, Karl, "Membranes in
     Separations," John Wiley & Sons, 1975.
(19) Yeager, H. L., Kipling, B. and Dotson, R. L., "Sodium Ion
     Diffusion in Nafion® Ion Exchange Membranes", J. of the
     Electrochemical Society, 127, No. 2., 303-307, Feb. 1980,
     Olin.
(20) Dotson, R. L., Yeager, H. L., Ford, J. M., and Bennion, D. N.,
     "Parameter Correlations for A Multicomponent Transport Model
     for Chlor-Alkali Membrane Cells," Presention at the 157th
     Meeting of the Electrochemical Society, St. Louis, Mo. 5-11,
     16, 1980, Olin.
(21) Helfferich, F., "Ion Exchange," McGraw-Hill, New York, 1962.
(22) Tuwiner, S. B., Miller, F.P., and Brown, W. E., "Diffusion and
     Membrane Technology," Reinhold Publishing Corp., 1962.
(23) Lakshminarayanaiah, N., "Transport Phenomena in Membranes,"
     Academic Press, New York, 1969.
(24) "Chemical Physics of Ionic Solutions," John Wiley & Sons, Inc.
     1966.
(25) Wells, A. F., "Structural Inorganic Chemistry," Oxford
     University Press, 1962.
(26) Per Kofstad, "Nonstoichiometry, Diffusion, and Electrical
     Conductivity in Binary Metal Oxides," Wiley-Interscience A
     Division of John Wiley & Sons, Inc., New York, 1972.
(27) Dotson, R. L., Lynch, R. W., and Hilliard, G. E., Transport of
     Water Molecules and Sodium Ions through Nafion Ion Exchange
     Membranes," Presentation at the 158th Meeting of The
     Electrochemical Society, Hollywood, Florida, 10-5,10, 1980,
     Olin.
(28) Bird, R. D., Stewart, W. E., and Lightfoot, E. N., "Transport
     Phenomena," John Wiley & Sons, Inc., New York, 1960.
(29) "Perfluorocarbon Ion Exchange Membranes," 152nd National
     Meeting of The Electrochemical Society, Atlanta, Georgia,
     Oct. 10-14, 1977.
(30) Pickett, D. J., "Electrochemical Reactor Design", Elsevier
     Scientific Publishing Co., New York, 1977.
(31) Chem. Techn. 32(3), 119-122 (1980).
(32) Chem. Techn. 31 140(1979).
(33) Dotson, R. L., "The Electron as Reagent," Chem. Tech., Vol.
     8., No. 1, 1978.
(34) Gerischer, H., and Tobias, C.W.,"Advances in Electrochemistry
     and Electrochemical Engineering," Vol.11., John Wiley & Sons,
     New York, 1978.
(35) Bard, A. J., and Faulkner, L. R., "Electrochemical Methods,
     Fundementals and Applications," John Wiley & Sons, New York,
     1980.
(36) Erdey-Gruz, T., "Kinetics of Electrode Processes,"
     Wiley-Interscience, New York, 1972.
(37) Bockris, J. O'M., "Overpotential," J. Electrochem. Soc.
     Vol. 98. No. 12., 1951.
(38) Kortum, G., "Treatise on Electrochemistry," Elsevier Pub. Co.,
     New York, 1965.

(39) Hampel, C.A., "The Encyclopedia of Electrochemistry," Robert E. Kreiger Publishing Co., Huntington, N.Y. 1972.

(40) Levenspiel, O., "Chemical Reaction Engineering," John Wiley & Sons, Inc., New York, 1972.

(41) Coulson, E. H., et.al., "Chemistry of Ion Exchange a Special Study, Nuffield Advanced Science, C. Tinling and Co., Ltd. London and Prescot, England, (1970).

(42) Diaion Manual of Ion Exchange Resins (1)-(2), Mitsubishi Chemical Industries, Ltd., Tokyo, Japan.

(43) Dotson, "USP 3,793,163 and British P 1,375,126, (1974).

(44) C. J. Molner and M. M. Dorio "Perfluorocarbon Ion Exchange Membranes," 152nd National Meeting of the Electrochemical Society Atlanta, Ga., Oct 10-14, 1977.

(45) Juda, W. and McRae, W. A., USP 2,636,851 and U.K.P. 720,002.

(46) Bergsma, F., Chem. Woekbl., Vol. 48, 361, 1952.

(47) Juda, W., Marinsky, J. A. and Rosenberg, N. W., Ann. Rev. Physic. Chem., Vol. 4, P. 373, 1953.

(48) Chrysikopoulos, S., Tombalakian, A. S. and Graydon, W. F., Canad. J. Chem. Engng., Vol. 6. p. 91, 1963.

(49) Kaden, H. and Schwabe, K., Chem. Techn., Vol. 19, P. 87, 1967

(50) Grot, W., Chemie Ing. Techn. 44. Jahry . Nr. 4, 1972.

(51) Nafion Products, E. I. duPont de Nemours & Co. (Inc.), Plastics Products & Resins Department, Wilmington, DE 19898

(52) Grot, W., Chemie Ing. Techn., 47, 617., 1975.

(53) Russell, J. H., "An Update on Solid Polymer Electrolyte Electrolysis Programs at G.E.," 3rd World Hydrogen Energy Conference, Tokyo, Japan, June 23-26, (1980).

(54) Seko, M., "Commercial Operation of Ion Exchange Membrane Chlor-Alkali Process," ACS Meeting New York, April 4-9, 1976.

(55) Seko, M., "The Asahi Chemical Membrane Chlor-Alkali Process," The Chlorine Institute, New Orleans, Feb., 1977.

(56) Seko, M., "New Development of the Asahi Chemical Membrane Chlor-Alkali Process", Oronzio De Nora Symposium on Chlorine Technology 15-18 May 1979, Venice, Italy.

(57) US 4,138,373

(58) Jap 116,790 (1977)

(59) Jap 81,485

(60) Jap Kokai 76,282

(61) British 1,522,877

(62) British 1,523,047

(63) Ukihashi, H., "A Membrane for Electrolysis," CHEMTECH, Feb, 1980.

(64) Suhara, M. and Oda, Y., "Transport Number through the Perfluorinated Cation Membrane, Flemion," 158th Meeting of the Electrochemical Society, Hollywood, FL., Oct. 5-10, 1980.

(65) UHDE, "Alkaline Chloride Electrolysis by the Membrane Process,"
      Uhde Gmb H.
(66) Ogawa, Shinsaku, "Asahi Chemical Membrane Chlor Alkali Process,"
      Chemical Age of India Vol 31, No. 5, May, 1980, pg 451.
(67) Coulter, M.O., "Modern Chlor Alkali Technology," Ellis Howard
      Limited, 1980, pgs 195-210.
(68) Coulter, M.O., "Modern Chlor Alkali Technology," Ellis
      Howard Limited, 1980, pg 223-234.
(69) Celleco, "Chlor-alkali Plants, with Ion Selective Membrane
      Cells."
(70) Wadsworth, A.C. 3rd; "Captive or Over-Fence Chlorine &
      Caustic Soda for Bleached Pulp Production, Advantages of
      Membrane Cells," Allied Chemical Corp. 1979.
(71) Diamond Shamrock, DM-14 Membrane Electrolyzer, ES-ECL-4A
      Diamond Shamrock Corp.
(72) Klamp, K. Lohrberg G. "Membrane Cell Technology-View of an
      Engineering Co.," Chemical Age of India Vol. 31, No. 5,
      May, 1980, pgs. 463-470.
(73) Asahi Glass, "The Flemion Membrane Chlor-Alkali Process,"
      Asahi Glass Co., Ltd. Sept. '78.
(74) Hausmann, E.; Will, H.; Belloni, A.;"Plate Type Cells Ion
      Brine Electrolysis," Chemical Age of India, Vol 31, No.5,
      May, 1980, pg 433-440.
(75) Ukihashi, H.; Oda, Y.; Asawa, T.; Morimoto, T.;"Progress of
      Electrolysis Technology with Flemion Membrane" Kyoto
      Symposium of Japanese Soda Industrial Association, 1980.
(76) duPont, NEWS CONFERENCE ANNOUNCEMENT, Japan, 1980.
(77) Nidola, A; "Brine Electrolysis with an Ironzio DeNora Design
      for the SPE Cell;" 24th Chlorine Plant Operations Seminar,
      Chlorine Institute, Feb. 1981.
(78) "New Caustic Soda Process is Devised by Asahi Glass," The
      Japan Economic Journal, November 25, 1980.
(79) Abam Engineers, Inc. "Process Engineering and Economic
      Evaluations of Diaphragm and Membrane Chloride Cell
      Technologies", ANL/OEPM-80-9, December, 1980, Argonne National
      Laboratory, Argonne, Ill. Page 82.

RECEIVED August 7, 1981.

Perfluorocarboxylic Acid Membrane and Membrane Chlor-Alkali Process Developed by Asahi Chemical Industry

MAOMI SEKO, SHINSAKU OGAWA, and KYOJI KIMOTO

Asahi Chemical Industry Co., Ltd., 1-2, Yurakucho 1-chome, Chiyoda-ku, Tokyo, Japan

Asahi Chemical started research and development of the ion-exchange membrane chlor-alkali process in 1966. Research was carried out on the effects of the type of ion-exchange group, ion exchange capacity, degree of crosslinking, membrane structure, caustic concentration, and many other parameters on current efficiency, operation voltage, etc. In 1969, a bench-scale plant started operation based on a three compartment process using hydrocarbon membrane. Further study on fluorinated monomers and polymers started in 1970, to improve the chemical stability of the membrane.

After intensive research and development work, Asahi Chemical filed the basic patents of fluorinated carboxylic acid membrane and carboxylic and sulfonic acid membrane and the related electrolysis processes in 1974 ($\underline{1}$ - $\underline{8}$).

In April 1975, Asahi Chemical started operation of a membrane chlor-alkali plant with a capacity of 40,000 MT/Y of caustic soda using Nafion perfluorosulfonic acid membrane. In 1976, this membrane was replaced by perfluorocarboxylic acid membrane developed by Asahi Chemical. The total caustic production capacity of plants based on Asahi Chemical's membrane chlor-alkali technology using perfluorocarboxylic acid membrane will reach 520,000 MT/Y in 1982, at seven locations in various countries.

General Requirements for Membranes for Chlor-Alkali Process

Ion-exchange membranes for the chlor-alkali process should satisfy the following requirements, some of which tend to be mutually contradictory.
- Chemical stability
- Physical stability
- Uniform strength and flexibility
- High current efficiency
- Low electric resistance
- Low electrolyte diffusion

In order to satisfy these requirements and to optimize the electrochemical properties of the membrane, the following factors must be considered in relation to specific electrolysis conditions.

- Water content
- Type of ion-exchange group
- Ion exchange capacity
- Polymer structure
- Polymer composition
- Physical structure of membrane
- Distribution of ion-exchange groups in the membrane
- Membrane thickness

The typical membranes are homogeneous, and are preferably reinforced with an inert material. The chemical composition of the membranes are hydrolyzed copolymers of tetrafluoroethylene (TFE) and perfluoro vinyl ether monomer containing an ion-exchange group or its precursor (PVEX), represented by the following general formula (1 - 21).

$$\text{PVEX:} \quad CF_2 = CFO-(CF_2\overset{\overset{\displaystyle CF_3}{|}}{C}FO)_m-(CF_2)_n-X \quad \ldots\ldots \quad (1)$$

Where: m = 0 or 1
        n = 2 - 12
        X = ion-exchange group or its precursor such as SO_2F, SR, SO_2R, COOR, COF or CN

In preparing the membranes, the following steps must be carefully designed to control the above factors.

- Monomer synthesis
- Polymerization
- Membrane fabrication
- Treatment to improve the electrochemical properties of the membrane
- Reinforcement

Classification of Membranes

The membranes for the chlor-alkali process are classified by chemical structure of ion-exchange group, number and type of membrane layers, and polymer structure.

Ion-Exchange Group. The following five ion-exchange groups have been reported in the literature.
  a. sulfonic acid group $-SO_3H$ (9, 11, 12, 13)
  b. sulfonamide group $-SO_2NHR$ (14, 15)
  c. carboxylic acid group $-COOH$ (1-8, 10, 16-21)
  d. phosphoric acid group $-PO_3H_2$ (22, 23)
  e. quaternary alcohol group $-\overset{|}{\underset{|}{C}}-OH$ (24, 25)

Groups d and e are not yet utilized in commercial membranes, probably because of difficulties in the monomer synthesis. Table I shows a comparison of the membranes currently in use which contain sulfonic acid, sulfonamide, carboxylic acid, and both carboxylic and sulfonic acid groups.

The strong acidity and high hydrophilicity of the perfluorosulfonic acid group result in a membrane of high water content and low electric resistance. Since the fixed-ion concentration in the sulfonic acid membrane is also low, current efficiency is less than 80% with caustic concentrations of 17% or more (26). The chemical stability of perfluorosulfonic acid group is excellent. Because of its low pKa value, the membrane can be exposed to solutions of pH 1.

The weak acidity and relatively low hydrophilicity of the perfluorocarboxylic acid group results in a very high current efficiency of over 96%, although its electric resistance is high (1, 3, 5, 7, 10, 26-29). The membrane can be exposed to fairly acidic solution as the pKa value is around 2. Its chemical stability is quite good under electrolysis conditions.

Perfluorosulfonamide has also been proposed as ion-exchange group having very weak acidity (27, 30, 31). It is necessary to keep the membrane of this type in an alkaline solution in order to maintain dissociation of the ion-exchange group. Another drawback of this membrane is its rather poor chemical stability due to its tendency to be hydrolyzed during electrolysis.

Some patent applications report membranes which contain ion-exchange groups of different types or exchange capacities, thus achieving better performance. These membranes can be classified into those composed of a homogeneous mixture of different ion-exchange groups and those with a multilayer structure of ion-exchange groups differing in type or in ion exchange capacity.

To obtain the former, ion-exchange groups of different types can be incorporated by terpolymerization, blending (mixing), or impregnation (2, 3, 4, 6, 8, 19, 32, 33). These membranes show fairly high current efficiency and fairly low electric resistance but do not fully utilize the merits of each of the groups, and thus do not exhibit the highly superior characteristics of multilayer membranes.

The latter are obtained by lamination, chemical treatment, or coating to incorporate two or more layers, with each layer containing an ion-exchange group of a specific type or exchange capacity (2, 3, 4, 6, 8, 11, 13, 14, 15, 18-22, 25, 34).

Figure 1 shows the general methods for preparation of perfluorocarboxylic acid monolayer and multilayer membranes.

Multilayer Membranes. Various types of multilayer membranes have been developed to obtain a combination of high current efficiency and low electric resistance. For high current

Table I. Membrane Properties Classified by Ion Exchange Group and Membrane Structure.

Ion-exchange group	R_f-SO_3H	R_f-SO_2NHR	R_f-COOH	R_f-COOH/R_f-SO_3H
pKa	<1	8–9	2–3	2–3/ <1
Hydrophilicity	high	very low	low	low/high
Water content	high	very low	low	low/high
Current efficiency % (8N NaOH)	75	88	96	96
Electric resistance	low	very high	high	low
Chemical stability	very high	low	high	high
Handling condition (pH)	>1	>10	>3	>3
pH of anolyte	>1	>10	>3	>1
Neutralization of OH⁻ by HCl	applicable	impossible	impossible	applicable
O₂ in product Cl₂	$<0.5\%$	$>2\%$	$>2\%$	$<0.5\%$
Life of anode	long	short	short	long
Current density	high	low	low	high
Necessary number of cells	large	large	large	small

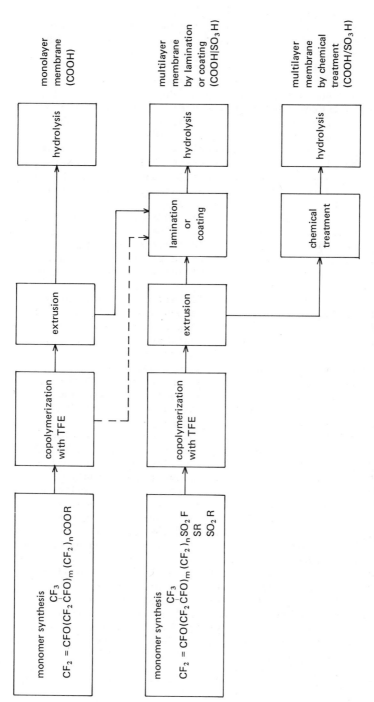

Figure 1. Process for the preparation of perfluorocarboxylic acid membrane.

efficiency, the layer facing the catholyte should contain an
ion-exchange group with an associated water content lower than
that of the rest of the membrane. Since the thickness of this
layer can be much less than the whole membrane thickness, it is
possible to obtain both higher current efficiency and lower
electric resistance than that of monolayer membrane.

Figure 2 shows the influence of the carboxylic acid layer
thickness on current efficiency and electric resistance for a
multilayer perfluorocarboxylic and sulfonic acid membrane
prepared by chemical treatment (COOH/SO$_3$H). The electric
resistance increases almost linearly with increasing thickness
of the carboxylic acid layer. Current efficiency, however,
increases rapidly with increasing thickness of the carboxylic
acid layer, and reaches substantial saturation near a thickness
of about 7 microns. This indicates that a carboxylic acid
layer thickness in the range of 2 to 10 microns is the optimum
to reduce power consumption in electrolysis. This optimization
of the thickness of carboxylic acid layer allows high current
efficiency without increasing electric resistance, a major
advantage of the chemical treatment method for preparing multilayer membrane.

Comparison is made in Table II between a monolayer membrane
containing only a carboxylic acid group and multilayer membranes
containing both carboxylic and sulfonic acid groups prepared by
lamination, coating or chemical treatment. From the table, it
is clear that chemical treatment is most desirable, because of
the low electric resistance. Moreover, only one type of PVEX
monomer containing sulfonic acid group or its precursor is
required for the preparation of the multilayer membrane. It is
necessary, on the other hand, to find an efficient method for
replacing the sulfonic acid group on one surface of the membrane
with the carboxylic acid group by mild chemical reaction (18,
20, 21, 35-38). The monolayer perfluorocarboxylic acid membrane
can be prepared, if necessary, by chemical treatment which
causes the chemical reaction throughout the membrane.

The use of membrane containing only the carboxylic acid
group in an acidic anolyte is rather difficult, because the
carboxylic acid group tends to remain undissociated, thus
increasing the electric resistance of the membrane. This makes
it impossible to neutralize migrating hydroxyl ion by adding
acid to the anolyte. The migrating hydroxyl ion is converted
to oxygen at the anode, resulting in a shorten anode life.
With a multilayer membrane containing both carboxylic and
sulfonic acid groups, however, it is possible to neutralize
proton from the anolyte with hydroxyl anion at the surface of
or in the membrane before the proton reaches the carboxylic
acid layer facing the catholyte, and thus achieve high chlorine
purity and longer anode life, as indicated in Table I. This is
one of the essential features of the Asahi Chemical process
patented in various countries (39). Another advantage of the

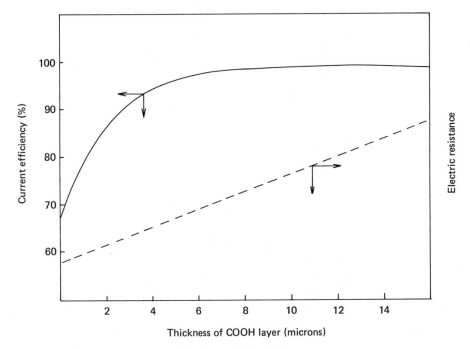

Figure 2. The thickness of COOH layer, current efficiency, and electric resistance of a multilayer perfluorocarboxylic and sulfonic acid membrane prepared by chemical treatment (COOH/SO₃H).

Table II. Comparison of Monolayer and Multilayer Perfluoro Ionomer Membranes.

Type of membrane	Monolayer perfluorocarboxylic acid membrane (COOH)	Multilayer membrane of perfluorocarboxylic and sulfonic acid by lamination or coating (COOH|SO$_3$H)	Multilayer membrane of perfluorocarboxylic and sulfonic acid by chemical treatment (COOH/SO$_3$H)
Thickness of COOH layer (micron)	> 125	25–50	$\leq$ 10
Current efficiency % (8N NaOH)	96	96	96
Electric resistance	high	medium	low
Necessary number of PVEX monomers	1	2	1
Membrane fabrication	extrusion	extrusion and lamination or coating	extrusion and chemical treatment

multilayer membrane is the higher current density which can be
applied because of its lower electric resistance as well as the
higher current efficiency, which allows a reduction in the
number of cells needed.

Relation between Ion Exchange Capacity and Pendant Structure
of PVEX Monomer. Generally speaking, the physical strength of
the membrane depends on various factors such as the ratio of
copolymers having ion-exchange groups in the membrane, the TFE
content of the copolymers, the molecular weight and molecular
weight distribution of the copolymers, and the type of rein-
forcing material used. The desired ion exchange capacity of
the membrane is obtained by adjusting the ratio of the copolymers
in the membrane and the TFE content of each copolymer. The
most important factor for adjusting both physical strength and
ion exchange capacity is the TFE content.

The ion exchange capacity of perfluorocarboxylic acid
membranes reportedly falls in the range from 0.5 to 4 meq/gram
dry resin (1-8, 10, 16-21). At a given ion exchange capacity,
a greater molecular weight in the PVEX monomer results in a
decrease in the TFE content of the copolymer, and a lowering of
the physical strength of the membrane. Conversely, the attainable
ion exchange capacity decreases when the molecular weight of
PVEX monomer is increased in order to obtain sufficient physical
strength. In cases where PVEX monomer with m = 1, n = 3 and
X = COOCH3 in formula (1) is utilized for perfluorocarboxylic
acid membrane, the highest ion exchange capacity is reported to
be approximately 1.3 meq/gram dry resin (10). A much higher
ion exchange capacity can be attained in cases where m = 0
(1-8, 10).

Membranes with a high ion exchange capacity are desirable
for the production of concentrated caustic soda, and it is
therefore essential to use PVEX monomer with m = 0 as a comonomer.

Although it is desirable to use the PVEX monomer with
n = 2 and X = SO2F in formula (1) as a comonomer to obtain
membrane containing sulfonic acid group (9, 11, 12, 18, 21,
35 - 38), which is useful for the preparation of multilayer
membrane with both carboxylic and sulfonic acid groups by
chemical treatment, it is reported that when m = 0 this PVEX
monomer undergoes the cyclization reaction shown below during
vinylization, and also undergoes a cyclization reaction during
polymerization under certain conditions (13, 20, 40). This
makes its synthesis impractical, and may cause a low molecular
weight of the resultant polymer.

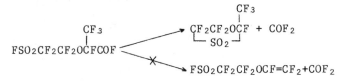

For these reasons, the PVEX monomer with m = 1 is generally
utilized in commercial applications, and the ion exchange
capacity is therefore limited to around 0.9 meq/gram dry resin
($\underline{9}$, $\underline{41}$).

Asahi Chemical's Japanese patent applications claim a
method to overcome these difficulties, in which PVEX monomer
represented by the following general formula are utilized as a
comonomer ($\underline{13}$, $\underline{20}$).

$$CF_2 = CFO-(CF_2CFO)_m-(CF_2)_n-X \quad \ldots\ldots\ldots\ldots \quad (2)$$
$$\overset{\displaystyle CF_3}{\underset{\displaystyle |}{}}$$

Where: m = 0 or 1
         n = 3 - 5
         X = precursor of sulfonic acid group such as
             SO$_2$F, SR or SO$_2$R

Under suitable conditions, even with m = 0 these PVEX
monomers do not cyclize during monomer synthesis nor during
polymerization, because the functional end group or the size of
the ring which would form upon cylization is different from
that in the case of n = 2. This facilitates monomer synthesis
and the formation of a polymer with high molecular weight, and
allows the use of PVEX monomer with m = 0 as the main functional
comonomer in the membrane preparation. Consequently, physically
strong membrane with high ion exchange capacity can be prepared
since the polymer contains sufficient amount of TFE.

Figure 3 shows the relation between electric conductivity
and ion exchange capacity of membranes produced from the PVEX
monomers with m = 0 and 1 which are indicated by formula (2).

General Properties of Perfluorocarboxylic Acid Membranes

As membranes employed in the chlor-alkali industry are
generally of the non-crosslinked type, their properties are
influenced significantly by the conditions in which they are
utilized. The extreme temperature, concentration and current
density to which they are subjected in the chlor-alkali process
are not encountered in other applications such as electrodialysis.
Clarification of the membrane properties is therefore both
practically necessary and theoretically interesting, and application of ion cluster theory has been attempted ($\underline{42}$, $\underline{43}$).

The water content, electric resistance, current efficiency
and mechanical properties are influenced by various factors.

Water Content. Figure 4 shows the relation between the
water content of perfluorocarboxylic acid membrane prepared by
chemical treatment and the ion exchange capacity with varying
external solution concentration. As the concentration of the
external solution increases, the membrane shrinkage increases
and the water content is therefore decreased. The influence of

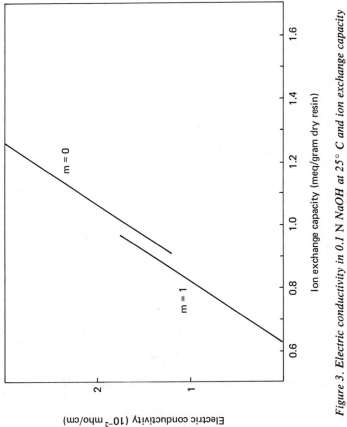

Figure 3. Electric conductivity in 0.1 N NaOH at 25° C and ion exchange capacity

$$CF_3$$

for perfluorosulfonic acid $\}-O(CF_2\overset{|}{C}FO)_m(CF_2)_3\text{-}SO_3$ *H membrane.*

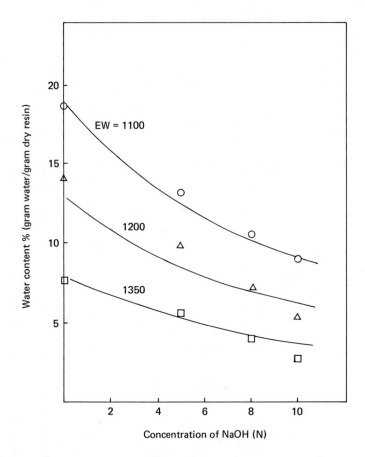

Figure 4. *Water content, equivalent weight and concentration of NaOH, for per-*

$$\text{fluorocarboxylic acid. } \{-OCF_2 \overset{\overset{\displaystyle CF_3}{|}}{C}FOCF_2 COOH \text{ membrane prepared by chemical}$$

treatment. Line represents: $W = \dfrac{0.001855}{1 + 0.1065C} \times exp\,(\dfrac{5104}{EW})$
$\cdot (1100 \leq EW \leq 1400).$

concentration is much greater for the membrane with higher ion
exchange capacities. This can be interpreted to indicate that
crystallized domains originating from TFE behave like cross-
linked points, and thus suppress swelling at low external
concentrations.

Although the copolymer constituting the membrane can be
considered basically random in structure in view of the monomer
reactivity ratios, some crystallinity occurs when the TFE
content is high, similar to that of PTFE. For perfluorosulfonic
acid membrane, a nearly linear relationship is found between
the TFE content and the degree of crystallinity as measured by
X-ray diffraction, as shown in Figure 5. Figure 6 shows the
melting point of the copolymer measured by DSC (Differential
Scanning Calorimetry) as a function of TFE content. In this
figure, extrapolation to a TFE content of 100 mole percent
yields a melting point almost equal to 326°C, the melting point
of PTFE. A similar relation between TFE content and degree of
crystallinity occurs in perfluorocarboxylic acid membrane (10).

Water content in Figure 4 can be expressed as a function
of ion exchange capacity and external solution concentration by
the following empirical equation, which is similar to that
proposed for perfluorosulfonic acid membranes by W.G.F. Grot in
1972 (44). The water content of perfluorocarboxylic acid
membrane is much lower than that of perfluorosulfonic acid
membrane.

$$W = \frac{0.001855}{1 + 0.1065C} \times \exp\left(\frac{5104}{EW}\right)$$

$$(1100 \leq EW \leq 1400)$$

where, W: water content % (gram water/gram dry resin)
        C: concentration of NaOH (N)
        EW: equivalent weight defined by

$$\frac{1000}{\text{ion exchange capacity (meq/gram dry resin)}}$$

Electric Resistance. Figures 7, 8 and 9 show the dependence
of electric resistance on ion exchange capacity and external
solution concentration for perfluorosulfonic acid membrane,
multilayer perfluorocarboxylic and sulfonic acid membrane
prepared by chemical treatment (COOH/SO$_3$H), and perfluorocarboxylic acid membrane prepared by chemical treatment. The
electric resistance dependent on type of ion-exchange group is
lowest for the first and highest for the third of these three
membranes. With increasing concentration of the external
solution, the concentration of sorbed electrolyte in the membrane
increases, and the electric resistance might therefore be
expected to decrease because of the increasing number of mobile
ions present in the membrane. However, the electric resistance
actually increases due to the decreased water content in the

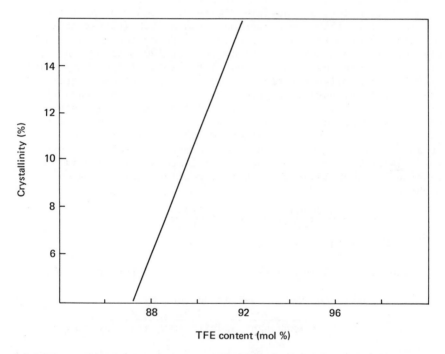

Figure 5. TFE content and degree of crystallinity measured by x-ray diffraction for perfluorosulfonic acid membrane.

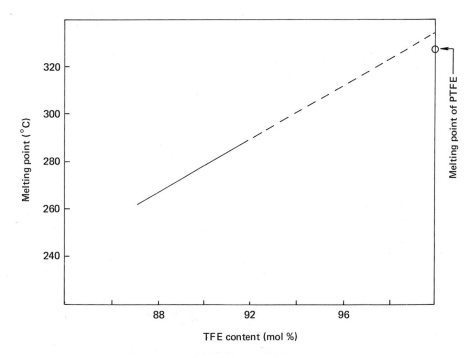

Figure 6. TFE content and melting point measured by DSC for perfluorosulfonic acid membrane.

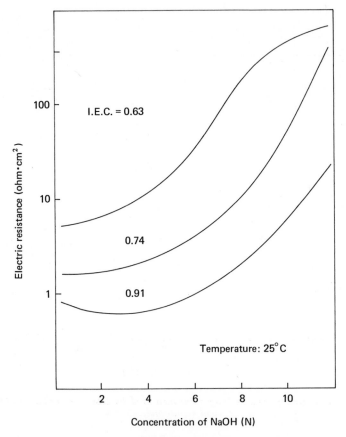

Figure 7. Electric resistance, ion exchange capacity (meq/g dry resin) and concentration of NaOH, for perfluorosulfonic acid membrane.

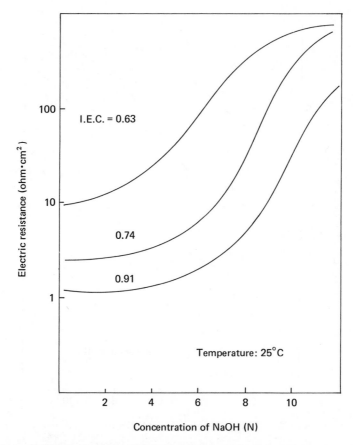

Figure 8. Electric resistance, ion exchange capacity (meq/g dry resin) and concentration of NaOH, for multilayer perfluorocarboxylic and sulfonic acid membrane prepared by chemical treatment (COOH/SO₃H).

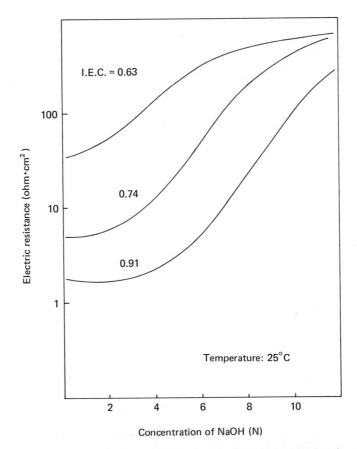

Figure 9. Electric resistance, ion exchange capacity (meq/g dry resin) and concentration of NaOH, for perfluorocarboxylic acid membrane prepared by chemical treatment.

membrane and the consequent sharp reduction in the diffusion
coefficients of the ions which occurs with increasing membrane
shrinkage when the external solution concentration increases.
The minimum electric resistance which occurs at the concentration
of about 2N in Figures 7, 8 and 9 is considered to be the
result of the interacting effect of concentration of the sorbed
electrolyte and diffusion coefficients of the ions in the
membrane.

Figure 3 shows the relation between the pendant structure
of PVEX monomer and the electric conductivity of the membrane.
As shown in Figure 3, electric conductivity is higher for the
membrane from PVEX monomer containing longer pendant (m = 1) at
a given ion exchange capacity. The same pattern was observed
in the case of PVEX monomer with $COOCH_3$ (10). This is caused
by the occurrence of higher water content with lower TFE content
when m=1.

Current Efficiency. Current efficiency is plotted against
the external solution concentration in Figure 10 for a multilayer
perfluorocarboxylic and sulfonic acid membrane prepared by
chemical treatment ($COOH/SO_3H$). For this membrane, a maximum
current efficiency occurs at a certain external solution
concentration. The maximum current efficiency tends to occur
at higher external concentration when the ion exchange capacity
is relatively high, thus indicating that membranes with higher
ion exchange capacity are preferable for the production of
alkali of higher concentrations.

According to the Donnan equilibrium theory, the concentration of sorbed electrolyte in the membrane increases with
increasing external solution concentration which tends to
result in lower current efficiency at higher external solution
concentrations (45).

However, occurrence of maximum current efficiency indicates
that a region exists in which the Donnan exclusion effectively
overcomes this general tendency, probably because the fixed-ion
concentration increases rapidly with the decreasing water
content in highly concentrated external solution. This can be
considered the general characteristic of membranes for the
chlor-alkali process, which are non-crosslinked and are utilized
for high concentration alkali electrolysis. The same phenomenon
has also been observed with membranes containing only carboxylic
acid group (46), sulfonamide group (15), and sulfonic acid
group (47-50).

Mechanical Properties. The mechanical properties of
membranes for industrial applications are important, because of
their influence on the membrane's resistance to wrinkling and
pinhole formation.

Table III shows the mechanical properties of perfluorocarboxylic acid membrane prepared by chemical treatment and
perfluorosulfonic acid membrane.

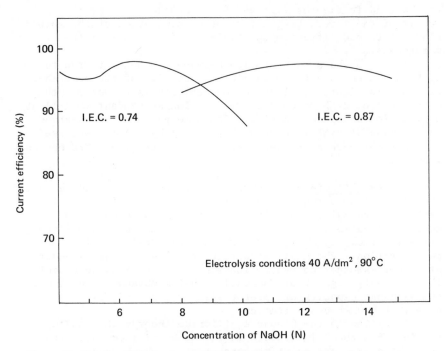

Figure 10. Current efficiency, ion exchange capacity (meq/g dry resin) and concentration of NaOH, for multilayer perfluorocarboxylic and sulfonic acid membrane prepared by chemical treatment (COOH/SO₃ H).

Table III. Mechanical Properties of Perfluoro Ionomer Membrane. (Ion Exchange Capacity: 0.03 meq/g Dry Resin).

Membrane	Perfluorocarboxylic acid membrane prepared by chemical treatment	Perfluorosulfonic acid membrane
Conditions	25°C in water	25°C in water
Tensile strength (kg/cm²)	236	224
Elongation (%)	253	155
Tensile modulus (kg/cm²)	1,560	791

Table IV indicates the degree of elongation found for membranes reinforced with PTFE fabric.

Membrane Life in Industrial Applications

The life of the membrane is governed by its characteristics and the conditions under which it is utilized. For long life, it is essential to select the ion-exchange group, polymer structure and composition, layer structure, and reinforcement appropriate to the conditions in which it will be employed, and also to establish and maintain the necessary electrolysis and handling conditions.

Optimum Ion-exchange Group. The perfluorocarboxylic acid group has been found to best satisfy the dual requirements of long membrane life and high current efficiency. Although perfluorocarbon copolymer — the main constituent of the membrane — is completely inert to chlorine and concentrated caustic soda under electrolysis conditions, ion-exchange groups are reactive in varying degrees. Perfluorosulfonic acid group has excellent chemical stability even at caustic soda and chlorine concentrations and temperatures much higher than those normally encountered in the electrolysis cell, but results in low current efficiency in the membrane. Perfluorosulfonamide group tends to undergo hydrolysis under electrolysis conditions. Although perfluorocarboxylic acid group may be decarboxylated under extreme conditions (34), it is stable under normal electrolysis conditions. Since membranes with this group also exhibit high current efficiency, it is therefore the best choice for use in electrolysis.

Influence of Polymer Pendant Structure. The polymer pendant structure generally has a significant effect on the life of membrane. For a multilayer perfluorocarboxylic and sulfonic acid membrane prepared by chemical treatment (COOH/SO$_3$H), a great difference in the long-term stability of current efficiency is observed depending on the polymer pendant structure of the perfluorosulfonic acid membrane utilized as the starting material. When the PVEX monomer represented by the general formula (2) is utilized in preparing perfluorosulfonic acid membrane, the resultant multilayer membrane shows a current efficiency with much greater long-term stability than that obtained when the conventional PVEX monomer with m=1, n=2 and X=SO$_2$F in formula (1) is utilized (20). Although the mechanism is not yet clear, both physical and chemical factors are believed to be involved in this difference.

Influence of Polymer Composition and Layer Structure. Under electrolysis conditions, the copolymer constituting the membrane tends to creep due to the violent mass transfer of

Table IV. Elongation of Perfluoro Ionomer Membrane Reinforced with PTFE Fabric.

			(basis: dry H form)
Conditions	25°C in water	90°C in 5 N NaOH	90°C in 10 N NaOH
Multilayer perfluoro-carboxylic and sulfonic acid membrane by chemical treatment ($COOH/SO_3H$)	6.0%	4.5%	2.9%
Perfluorosulfonic acid membrane (SO_3H)	8.5%	7.5%	6.1%

Table V. PVEX Monomer Appropriate for Perfluorocarboxylic Acid Membrane.

Type of membrane	PVEX monomer with carboxylic acid group or its precursor	PVEX monomer with sulfonic acid group or its precursor
Monolayer perfluorocarboxylic acid membrane (COOH)	yes	no
Multilayer perfluorocarboxylic and sulfonic acid membrane by lamination or coating ($COOH|SO_3H$)	yes	yes
Multilayer perfluorocarboxylic and sulfonic acid membrane by chemical treatment ($COOH/SO_3H$)	no	yes

Table VI. Synthesis of PVEX Monomer (Scheme 1).

$$X - (CF_2)_{n-1} - COF \xrightarrow{(m+1)\ HFPO} X - (CF_2)_n - (OCFCF_2)_m O - \overset{\overset{\displaystyle CF_3}{|}}{C}FCOF$$

with CF_3 groups

$$\downarrow - COF_2$$

$$X - (CF_2)_n - (OCFCF_2)_m O - \overset{\overset{\displaystyle CF_3}{|}}{C}F = CF_2$$

ions and water molecules through the membrane, thus causing a decrease in current efficiency. This tendency can be avoided by including sufficient amount of TFE in the polymer composition, since the crystallized domains originating from TFE behave like crosslinked points and effectively resist the forces caused by the mass transfer. For a multilayer membrane, particularly that prepared by lamination, the same forces tend to cause the layers to peel apart. It is therefore essential that the layers be strongly bonded, and for this purpose the mutual solubility of the copolymers in the multilayer membrane is of primary importance.

Necessity for Reinforcement. Handling of the membranes may tend to cause pinholes or tears. Adequate reinforcement of the membrane is therefore necessary to ensure long service life. It is also important to establish methods for repair of the damaged membranes.

Influence of Electrolysis Conditions. Among the various electrolysis conditions, brine purity has the most significant effect on the life of the membranes. The presence of a small amount of multivalent cations leads to formation of metal hydroxide deposits in the membrane, and thus causes a decrease in current efficiency, an increase in cell voltage, and damage to the polymer structure of the membrane. With perfluoro-carboxylic acid membrane, the presence of more than 1 ppm of calcium ion will begin to cause these problems in a very short period (1 - 8). To obtain stable current efficiency and cell voltage, it is therefore essential to establish effective brine purification methods.

Preparation of Membrane

The three main steps of membrane preparation are monomer synthesis, polymerization and membrane fabrication.

Synthesis of Monomers. The appropriate selection of PVEX monomer structure and a high yield in monomer synthesis are essential to the successful membrane preparation. The type of PVEX monomer to be utilized is determined by the required membrane structure, as indicated in Table V.
These monomers are generally produced by the addition of hexafluoropropylene oxide (HFPO) to a bifunctional compound represented by the general formula $X-(CF_2)_{n-1}-COF$, followed by vinylization by heating with a catalyst such as sodium carbonate, as indicated in Table VI (13, 20, 51-63).
The bifunctional compounds $X-(CF_2)_{n-1}-COF$ are key inter-mediates and the method of synthesis of these intermediates is the major determinant of the monomer synthesis, as indicated in Tables VII (51-56), VIII (57, 58), and IX (13, 20, 59-63).

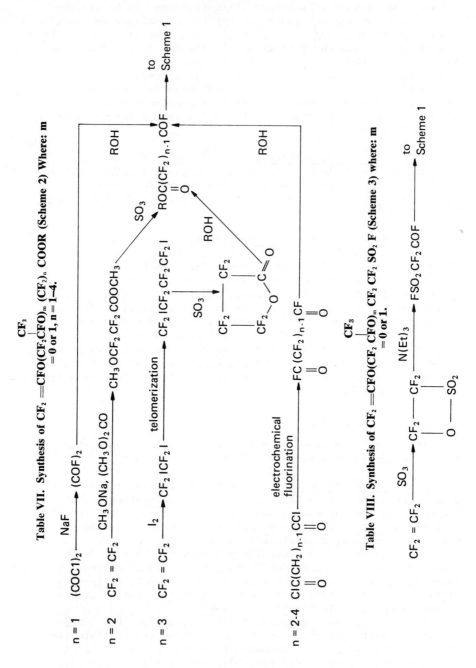

Table VII. Synthesis of $CF_2 = CFO(CF_2CFO)_m(CF_2)_n COOR$ (Scheme 2) Where: **m = 0 or 1, n = 1–4.**

Table VIII. Synthesis of $CF_2 = CFO(CF_2CFO)_m CF_2 CF_2 SO_2 F$ (Scheme 3) where: **m = 0 or 1.**

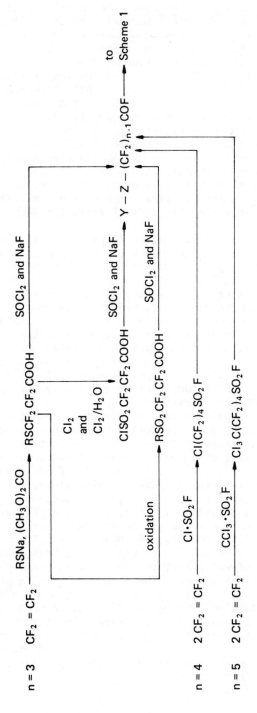

Table IX. Synthesis of $CF_2 = CFO(CF_2 CFO)_m (CF_2)_n -Z-Y$ (Scheme 4) where: $m = 0$ or 1, $n = 3$–5; and when $Z = -S-$, $Y = R$, R_f, or Cl; when $Z = -SO_2-$, $Y = R$, R_f, F, Cl, or OM (M = H, metal or ammonium ion).

Polymerization Process. The copolymer of TFE and PVEX monomer may be prepared according to the well known methods for homopolymerization and copolymerization of a fluorinated ethylene (10, 17, 59, 60, 64). The methods may be broadly classified as polymerization in a non-aqueous system and polymerization in an aqueous system. The polymerization temperature is generally from 5 to 100°C depending on the half-life temperature of the initiator. The pressure may be varied from 0 to 30 kg/cm^2 to adjust the composition of the copolymer.

Non-aqueous polymerization is frequently carried out in an inert fluorinated solvent such as 1,1,2-trichloro 1,2,2-trifluoro ethane, perfluoro methyl cyclohexane, perfluoro dimethyl cyclobutane, perfluoro octane or perfluoro benzene. As the free radical initiator, a perfluorocarboxylic acid peroxide (R$_f$COO)$_2$, an azo bis type compound such as azobisisobutylo-nitrile, or a fluorine radical initiator such as N$_2$F$_2$ may be used.

Aqueous polymerization is carried out in either an emulsion or a suspension. In the emulsion polymerization, monomers are polymerized in an aqueous medium containing a water soluble free radical initiator and an emulsifier to obtain a polymer particle slurry. In the suspension polymerization, monomers are polymerized in an aqueous medium containing both a free radical initiator which is soluble in the monomer and a dispersion stabilizer inert to telomerization to achieve the dispersion of polymer particles, followed by precipitation of the dispersion. The free radical initiator may be a redox catalyst such as ammonium persulfate-sodium hydrogen sulfite, a perfluorocarboxylic acid peroxide (R$_f$COO)$_2$, or an azo bis type compound such as azobisisobutylonitrile.

During polymerization, the formation of end groups of carboxylic acid or its derivatives may occur, depending on the type of initiator. This causes difficulties in melt fabrication of the resultant polymer because of the relatively low thermal stability of these groups. Methods of avoiding this problem have been reported in patent applications, such as converting these groups into ester groups with methanol (65), into a CF$_3$ group with fluorine gas (66), or into a CF$_2$H group by heating.

The molecular weight of the copolymer is generally from 8,000 to 1,000,000 (17, 59, 60). It is of course necessary to produce a polymer of molecular weight high enough to obtain membrane of sufficient physical strength and good electro-chemical performance. The melt flow rate is preferably in the range of 0.1 gram/10 minutes to 500 gram/10 minutes, as measured under a load of 2.16 kg at 270°C according to ASTM D1238.

The monomer reactivity ratios for copolymers of TFE and PVEX monomer containing COOCH$_3$ have been reported in the literature (10). The following is an example of monomer reactivity ratios obtained for the copolymer of TFE and PVEX monomer containing SO$_2$F.

$$CF_2 = CF_2 \qquad\qquad r_1 = 8$$

$$CF_2 = CFOCF_2\overset{\displaystyle\underset{|}{CF_3}}{C}FO(CF_2)_3SO_2F \qquad\qquad r_2 = 0.08$$

Membrane Fabrication, Lamination and Reinforcement. The thermoplastic copolymer of TFE and PVEX monomer can be extruded or molded into a thin film by conventional methods. Extrusion is most appropriate for continuous production of large membranes because it facilitates control of film thickness. Lamination can be performed with a roll press to obtain multilayer membrane. To improve their mechanical properties, membranes are generally reinforced with an inert material such as fabric made of PTFE (67, 68), fibriled PTFE (69), or wire netting (70).

Incorporation of carboxylic acid group

Perfluorocarboxylic acid membrane is prepared by hydro-lyzing the film of copolymer of PVEX monomer with carboxylic acid group or its precursor and TFE by conventional method. For membrane prepared from PVEX monomer with sulfonic acid group or its precursor, the carboxylic acid group is incorpo-rated by one of the following methods.

Chemical Treatment. Sulfonic acid groups on one surface of the membrane can be replaced by carboxylic acid groups through one of the following chemical reactions, most of which were first reported in Asahi Chemical patents and patent appli-cations.

A. Chemical reaction via sulfinic acid group (18, 20, 35). As indicated by the following scheme, perfluorosulfonyl halide group is reduced to sulfinic acid group. The resulting perfluorosulfinic acid group is unstable particularly in its H form and can be easily replaced by carboxylic acid group.

$$-O(CF_2)_n SO_2X \xrightarrow{\text{reduction}} -O(CF_2)_n SO_2H$$

$$(X=Cl, \ F) \qquad\qquad -SO_2 \downarrow \begin{array}{l}\text{reducing agent,}\\ \text{oxidizing agent,}\\ \text{or heat}\end{array}$$

$$-O(CF_2)_{n-1}COOH$$

Reducing agents such as hydroiodic acid, hydrobromic acid, hypophosphorus acid, and hydrazine can be utilized in this reaction. Sulfonyl chloride group (SO_2Cl), the most common starting chemical moiety, can be prepared from the sulfonic acid group by reaction with phosphorus pentachloride or phosphorus trichloride and chlorine (18, 20, 21, 36, 37, 38, 71) or from a sulfide group by reaction with chlorine and water (20).

B. Chemical reaction via $-CF_2X$ (36, 37)
The functional group $-CF_2X$ (X=I or Br) can be readily prepared by contacting the membrane having the sulfonyl chloride group with a halogen such as iodine or bromine. When exposed to ultraviolet light or SO_3, these groups ($-CF_2X$) can be replaced by a carboxylic acid group, as indicated in the following scheme.

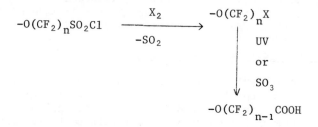

C. Desulfonylation reaction (18, 35, 38)
Perfluorosulfonyl chloride or sulfonic acid group can be desulfonylated under various conditions, to prepare a carboxylic acid group.

$$-O(CF_2)_nSO_2Cl \xrightarrow{\text{UV, heat, peroxides, or oxidizing agent}} -O(CF_2)_{n-1}COOH$$

$$-O(CF_2)_nSO_3H \xrightarrow{F_2 + O_2} -O(CF_2)_{n-1}COOH$$

D. Addition reaction (21, 37)
In accordance with the above reactions, pendant carboxylic acid has one less carbon atom than the original pendant. Some patent applications report a method to prepare the pendant containing carboxylic acid in which the number of carbon atoms equal to or greater than that in the original, by application of an addition reaction with the sulfonyl chloride group or $-CF_2X$ (X=I or Br).

Reaction A is most appropriate for the preparation of multi-layer membrane because of its extremely high selectivity.

Lamination and Coating (2, 3, 4, 6, 8, 19, 33).
Multilayer perfluorocarboxylic and sulfonic acid membrane
(COOH|SO$_3$H) is prepared by hydrolyzing a laminated or coated
film containing a carboxylic acid ester group in one layer of
the membrane and a sulfonyl fluoride or sulfonic acid group in
the other layer in the membrane. Multilayer ((COOH + SO$_3$H)|SO$_3$H)
membrane can be obtained by hydrolyzing a laminated or coated
film containing a mixture or blend of copolymers with carboxylic
acid ester and sulfonyl fluoride.

Terpolymerization (2, 3, 4, 6, 8, 33). Membranes containing both perfluorocarboxylic and sulfonic acid groups can be
prepared by hydrolyzing a film formed by terpolymerization of
TFE and two PVEX monomers containing a carboxylic ester group
and sulfonyl fluoride group.

Impregnation and Blending (2, 3, 4, 6, 8, 19). Membrane
with mixed perfluorocarboxylic and sulfonic acid groups can be
prepared by impregnating perfluorosulfonic acid membrane with a
PVEX monomer containing a carboxylic acid ester group, and
polymerizing the monomer in the membrane, and then hydrolyzing
the resultant film. Porous film of PTFE or Teflon PFA can be
utilized as a base material in place of perfluorosulfonic acid
membrane. The same membrane with mixed carboxylic and sulfonic
acid groups can be obtained by hydrolyzing the film formed by
blending the two copolymers, in which one is a copolymer of TFE
and PVEX monomer containing a carboxylic acid ester group and
the other is a copolymer of TFE and PVEX monomer containing a
sulfonyl fluoride group.

Asahi Chemical's Membranes

The typical perfluorocarboxylic acid membrane developed by
Asahi Chemical is a multilayer membrane prepared by chemical
treatment. The structure of the membrane is optimized for high
current efficiency and low electric resistance. The thickness
of the carboxylic acid layer is in the range of 2 to 10 microns.
The chemical structure of the membrane is as follows (72).

$$-(CF_2-CF_2)_x-(CF_2-CF)_y-$$
$$O-(CF_2CFO)_m-(CF_2)_n-COOH$$

with CF_3 branch

$$-(CF_2-CF_2)_z-(CF_2-CF)_w-$$
$$O-(CF_2CFO)_m-(CF_2)_\ell-SO_3H$$

with CF_3 branch

(m = 0 or 1, n = 1-4, ℓ = 2-5)

The performance and life of the membrane are excellent.
The following are typical performance data for Asahi Chemical's
membrane.

membrane	current efficiency	electric resistance 25°C in 0.1N NaOH	cell voltage*
membrane for 21% NaOH	96%	4.2 ohm	2.98 V
membrane for 30% NaOH	96%	3.9 ohm	3.12 V

* in laboratory cell at 40 Amperes/dm^2 and 90°C

Membrane Chlor-Alkali Process Developed by Asahi Chemical

Perfluorocarboxylic acid membrane differs greatly from
conventional asbestos diaphragm in many respects, the most
important of which are its high cation permselectivity, effective
gas impermeability, low water transport, small pore diameter,
high mechanical strength, ease of handling, and relatively high
membrane cost. It also differs from perfluorosulfonic acid and
sulfonamide membrane in current efficiency, electric resistance,
and chemical stability and other characteristics. The overall
membrane process should be specifically designed for optimum
utilization of the carboxylic acid membrane characteristics in
industrial applications. In particular, the basic electrolysis
conditions described below are specific to the membrane process,
and must be considered in the design and selection of the brine
purification, electrolyzer, anode, cathode, and evaporation
process.

Limiting Current Density. In the membrane process, boundary
layers form at both sides of the membrane due to its cation
permselectivity. Such boundary layers do not occur in the
diaphragm process. For the boundary layer at the surface of
the membrane facing the anolyte, the following basic equation
is established (73).

$$\frac{I}{F}(T_{Na}{}^+ - t_{Na}{}^+) = \frac{D}{\delta}(C - Co)$$

Where:

I = current density (Ampere/cm^2)

$T_{Na}{}^+$ = transport number of Na$^+$ in the membrane (Na$^+$ current efficiency)

tNa^+ = transport number of Na$^+$ in the anolyte

F = Faraday constant (96500 Ampere.sec/ equivalent)

D = diffusion coefficient of sodium chloride at the boundary layer (cm^2/sec)

C =   concentration of sodium chloride in the
bulk phase of anolyte (equivalent/cm^3)
Co = concentration of sodium chloride at the
surface of membrane (equivalent/cm^3)
δ =   thickness of the boundary layer (cm)

The left side of the equation represents the rate of
sodium chloride removal from the boundary layer due to the
difference between the transport number of Na^+ ion in the
membrane and that in the anolyte. The right side of the
equation represents the rate of sodium chloride supply to the
boundary layer caused by diffusion from the bulk phase. At a
certain current density (I = Io), Co approaches zero and the
following equation is established.

$$Io = \frac{C}{\delta} \cdot D \cdot F \cdot \frac{1}{T_{Na^+} - t_{Na^+}}$$

At a higher current density than this limiting current
density (Io), the supply of Na^+ ion to the boundary layer
becomes insufficient for the transport of electric current, and
water therefore decomposes to hydrogen ion and hydroxyl ion for
the transport of electric current in the boundary layer. Under
these conditions, electric current is carried through the
membrane not only by Na^+ ion but also by this hydrogen ion.
This results in higher operating voltage and lower current
efficiency.

The operating current density must therefore be lower than
the limiting current density over the entire membrane surface.
This requires careful design of the cell to ensure uniform
current distribution throughout the membrane and uniform
distribution of the electrolyte concentration throughout the
cell.

A bipolar configuration, in which the electric current in
the individual cell is unaffected by the number of cells in the
electrolyzer, is preferable for this purpose (26-29). The
bipolar configuration, moreover, is highly preferable for
minimization of the electrode gap, facilitates detection of any
variation in performance due to membrane manufacturing or cell
construction through measurement of the operating voltage of
each cell, and allows automatic trip-off of one electolyzer
independently of others (74).

For uniform distribution of electrolyte concentration in
and among cells, forced, continuous circulation of a large
amount of electrolyte is highly preferable to the conventional
drop-wise supply used in the mercury and diaphragm process.
Forced circulation also allows effective removal of the heat
generated by electrolysis.

If the boundary layer (δ) is narrow, a lower sodium
chloride concentration in the anolyte can be used at a given
current density. This results in a high rate of sodium chloride

utilization in the anolyte stream, and a consequent reduction
in brine purification cost. It has been reported that δ can
be reduced by locating the membrane near the anode, with
appropriate turbulence at the boundary layer thus provided by
the evolving chlorine gas (75). The anode configuration must
promote both effective diffusion of sodium chloride into the
boundary layer and uniform distribution of current through the
membrane.

Caustic soda of high purity can be obtained by operation
at a current density slightly below Io, since this results in a
very low sodium chloride concentration near the membrane (Co)
and thus effectively prevents diffusion of sodium chloride
through the membrane and into the catholyte.

A boundary layer with a gradient in caustic soda concentration also forms at the surface of the membrane facing the
catholyte based on a similar principle, resulting in a caustic
soda concentration on the membrane surface which is higher than
that in the bulk phase. Since this tends to reduce the current
efficiency and electric conductivity of the membrane, it is
necessary to minimize the boundary layer thickness or reduce
the caustic soda concentration in the bulk phase. It is also
essential to purify the brine with ion-exchange resin of high
selectivity, in order to prevent precipitation of metal ions as
hydroxides in the membrane and the boundary layer (74).

In the diaphragm process, these phenomena do not occur
because the diaphragm has no cation permselectivity.

In the solid polymer electrolyte (SPE) cell process, the
membrane and the electrode are bonded together, and it is
difficult to reduce the boundary layer thickness on both surfaces
of the membrane. This process also requires very thin electrodes
which must be made highly porous without increasing their ohmic
resistance or reducing their mechanical strength.

The optimum current density of the membrane process
(including the SPE cell process) is higher than that of the
diaphragm process, because of the relatively high cost of the
perfluoro ionomer membrane and its greater sensitivity to
impurities, which requires the use of more expensive material
for equipment.

Caustic Soda Concentration. The maximum electric conductivity of caustic soda solution occurs at a concentration of
about 20% at the ordinary electrolysis temperature, and the
membrane conductivity tends to decline sharply with caustic
soda concentration in the catholyte exceeding 20% (26). The
boundary layer effect described in the previous section also
makes relatively low concentrations preferable. With increasing
concentration of caustic soda, moreover, the allowable concentration of multivalent cation in the brine must be decreased
exponentially because the solubility products of multivalent
cation hydroxides are constant, and operational difficulties

occur as a result. The advanced perfluorocarboxylic acid
membrane now available has a current efficiency of over 90% in
a broad range of caustic soda concentrations, and minimum power
consumption is achieved at a caustic soda concentration of
approximately 20%-30%, where the electrolysis voltage is lowest.
    The high purity of the caustic soda obtained by the membrane
process eliminates the need for a caustic soda evaporator in
cases where it is to be supplied to customers such as pulp
mills which utilize a dilute caustic soda. This is in marked
contrast to the diaphragm process which inevitably requires
evaporation to separate sodium chloride. For the general
trade, in which caustic soda at 50% concentration is required,
a conventional multiple effect evaporator is generally utilized
to concentrate the catholyte. Caustic soda from the membrane
process contains a very slight amount of sodium chloride which
does not cause corrosion of the evaporator materials or
precipitation of sodium chloride, and thus allows easier and
more stable evaporator operation than in the diaphragm process.
    Although as previously described it is preferable to
operate the cell at around 20% - 30% of caustic soda concentration
to minimize electrolysis power consumption, higher concentrations
of catholyte are generally preferable to minimize steam
consumption in the evaporation process. However, Asahi Chemical
has developed a heat recovery evaporator which greatly reduces
the need for external stream supply and therefore permits a
significant reduction in the total energy consumption of
electricity and steam (74,76).
    The heat recovery evaporator is a multistage, multi-effect
evaporator which is different from conventional multiple effect
evaporator or multistage flash evaporator. Asahi Chemical's
heat recovery evaporator can concentrate the catholyte from 21%
to about 40% without steam by utilizing heat generated during
electrolysis. To obtain product caustic soda of 50% concentration, a small amount of steam is supplied to the finishing
evaporator.

    Operating Pressure. In the diaphragm process, a small
difference in hydraulic pressure is applied between anolyte and
catholyte to transport anolyte through the diaphragm to the
catholyte compartment. It is therefore impractical to pressurize
the anolyte, since this would cause chlorine gas to dissolve in
the anolyte and thus mix with the catholyte.
    In the membrane process, however, pressurized operation
does not cause mixing of anolyte and catholyte due to the dense
structure of the membrane, and various advantages such as
reduced operating voltage and membrane vibration can be gained
by pressurizing both chlorine gas and hydrogen gas (77).
Figure 11 shows the relation between operating voltage and
operating temperature at various operating pressures (77).

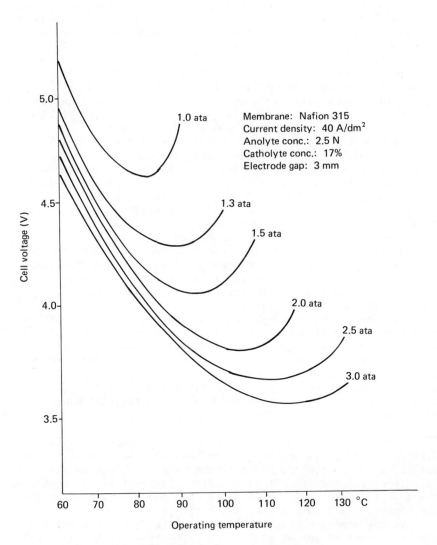

Figure 11. Cell voltage vs. operating pressure.

With increasing temperature, electrolyte and membrane
resistance tend to decrease while water vapor pressure tends to
increase, and minimum operating voltage occurs at a specific
temperature.

The use of metal cells is justified by the advantages of
pressurized operation, as well as by their other operational
and mechanical advantages (74).

Anode. Metal electrodes of high dimensional precision are
essential for maintaining a uniform distance between the electrode
and a membrane of large size, and an electrode distance as
small as possible in the membrane cell. It is preferable to
locate the membrane near the anode, to minimize the thickness
of the boundary layer at the membrane during electrolysis.
During operation, alkali migrates from the catholyte to the
anode through the membrane. If the alkali is not neutralized,
oxygen is generated at the anode at a considerable rate and
tends to shorten the anode life. It is therefore preferable to
neutralize this alkali with hydrochloric acid, and to utilize
multilayer R_f-COOH/R_f-SO$_3$H membrane with high current efficiency.

For the same reason, the anode coating for the membrane
process must possess superior alkali resistance, while fulfilling
the requirements for low chlorine overvoltage and high oxygen
overvoltage.

Asahi Chemical has developed an unique anode coating which
satisfies the requirements of the membrane process, and has
used it industrially since 1975. This coating is a completely
solid solution of ruthenium, titanium and oxygen, in which the
molar percentage of ruthenium is at least 50% of the total
metal content and various other metal components are
incorporated to provide high oxygen overvoltage (26, 78).

Cathode. Because iron is an inexpensive material and has
a rather small hydrogen overvoltage, iron cathode is generally
utilized, in the form of a mesh, perforated plate, or expandable
metal sheet.

In a discussion on the hydrogen overvoltage of various
metal wires, it has been shown that although iron cathode
having a flat surface has a relatively high hydrogen voltage,
about 0.45 volt at the current density of 25 Amperes/dm^2, the
hydrogen overvoltage of the iron cathode utilized in an industrial
cell for hydrogen generation is about 0.2 volt due to the fact
that the surface of the cathode, which is flat in the initial
period of operation, is converted to a porous surface of large
effective area by the deposition of iron from the catholyte
(79).

The increasing cost of energy has also stimulated research
on the reduction of hydrogen overvoltage by utilizing various
metal coatings.

The performance of an activated cathode having a low

hydrogen overvoltage tends to degrade gradually, probably due
to the precipitation of iron on the cathode. In the study of
this problem, it is useful to refer to the Pourbaix diagram,
which is generally used in consideration of corrosion problem.

Figure 12 shows the Pourbaix diagram for iron calculated
for solution of caustic soda at 90°C. The horizontal axis
represents pOH calculated by the equation $pKw = pH + pOH$ at
90°C. The thermodynamic parameters at 90°C shown in Table X
are calculated from the figures at 25°C and at 100°C (By the
courtesy of Professor M. Takahashi, Yokohama National University).

Because the potential difference between point (A) and (D)
is 0.19 volt, the equilibrium concentration of $HFeO_2^-$ on the
cathode should be about 10^{-5} mol/1. This means that the iron
cathode with a hydrogen overvoltage of about 0.2 volt is in an
immunity state. If the iron cathode is not polarized, the
equilibrium concentration of $HFeO_2^-$ on the iron is about 10^{-2}
mol/1 as determined at point (B), and the iron therefore corrodes.
With an activated cathode of a small hydrogen overvoltage such
as 0.12 volt, which corresponds to the difference between the
potentials at (A) and (E), it becomes necessary to reduce
the $HFeO_2^-$ concentration to less than 10^{-4} mol/1.

The above calculations are based on equilibrium
considerations and suggest the following.
1.   The hydrogen overvoltage of iron is favorable for the
      protection of the iron cathode.
2.   If an active cathode with a low hydrogen overvoltage is
      used, the $HFeO_2^-$ concentration in the catholyte must be
      kept low as determined by equilibrium calculation.
      If the $HFeO_2^-$ concentration is not low, iron will precipitate
      on the active cathode, resulting in a hydrogen overvoltage
      as high that of iron cathode.
      Asahi Chemical has developed an catalytic cathode for
industrial applications which has been tested in commercial
operation since 1980.

Oxygen Depolarized Cathode. It was proposed by Juda in
1964 to use oxygen at the cathode in the membrane chlor-alkali
process in order to reduce electrolysis voltage without generating
hydrogen at the cathode (80). The cell voltage is theoretically
reduced by 1.23 volt by using the oxygen depolarized cathode,
but the actual reduction in electrolysis is reported to be
about 0.6 volt (81).

In industrial application of the oxygen depolarized cathode,
air is the preferred oxygen source. However, air will cause
higher electrolysis voltage than pure oxygen, and nitrogen from
the air together with excess oxygen will remove water and heat
from the cathode area. This causes local deposition of sodium
chloride, sodium carbonate and other compounds.

Practical application of the oxygen depolarized electrode
has been limited to fuel cells with pure oxygen for special

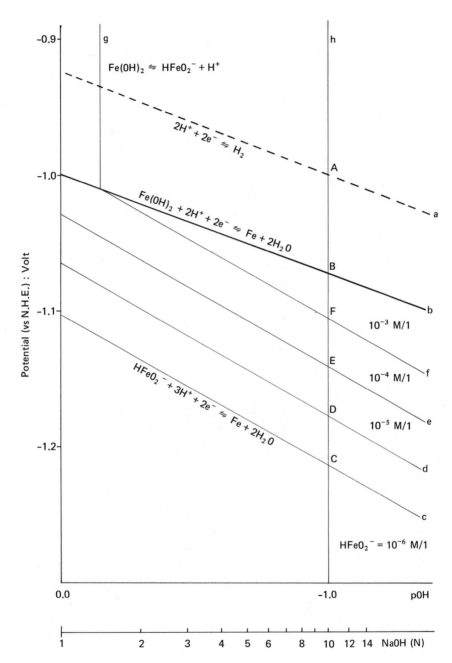

Figure 12. Pourbaix diagram for Fe in caustic soda at 90° C.

Table X. Thermodynamic Parameters (79).

		H_2O (liq)	$HFeO_2^-$ (aq)	$Fe(OH)_2$ (S)	pKw
$\mu^\Phi{}_{25^\circ C}$	kcal/mol	-56.69	-90.60	-117.56	14.00
$\mu^\Phi{}_{100^\circ C}$	kcal/mol	-53.84	-84.31	-110.95	12.26
$\mu^\Phi{}_{90^\circ C}$	kcal/mol	-54.22	-85.15	-111.83	12.49

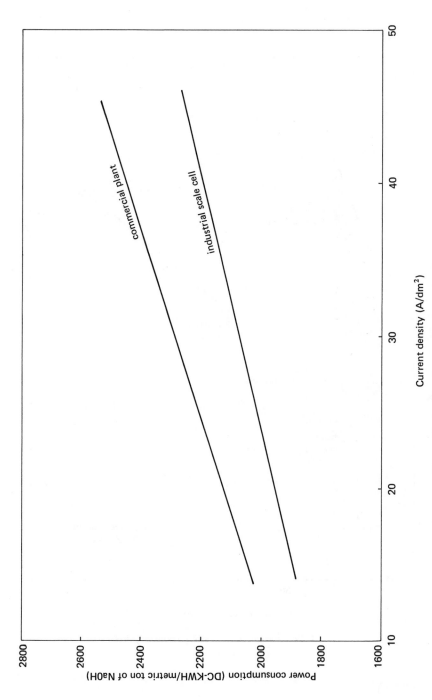

Figure 13. Power consumption vs. current density.

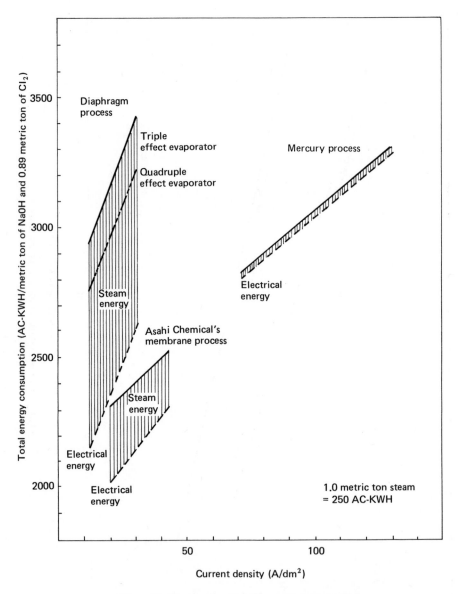

Figure 14. Comparison of total energy consumptions.

purposes such as spacecraft. No fuel cell in which air is utilized as the oxygen source has yet been applied practically or industrially. Some years will be required for development of the industrial application of oxygen depolarized cathode for the membrane chlor-alkali process, because of the difficulties in manufacturing cathodes of large size and in preventing contamination of the cathode and the resulting decline in catalytic activity, and because of the danger of the formation of an explosive mixture of oxygen and hydrogen in the event of a failure in the catalytic activity of the cathode. This danger is not present in fuel cell applications.

Energy Consumption. Electric power consumption of electrolysis is the major part of the energy consumption in a chlor-alkali process. The power consumption of the membrane process has recently been greatly reduced by various improvements. The latest performance of Asahi Chemical's membrane process realized at a commercial plant and also in an industrial scale cell is shown in relation to current density in Figure 13 (82).
Because steam is consumed for caustic soda evaporation in the diaphragm and the membrane processes to obtain 50% caustic soda, it is also important to compare the combined energy consumption of electric power for electrolysis and steam for caustic soda evaporation. Figure 14 shows the total energy consumption of the modern asbestos diaphragm process with metal anode and modified asbestos, the modern mercury process with metal anode and Asahi Chemical's membrane process. In this comparison, it is assumed that one metric ton of steam is equivalent to 250 KWH of electric power based on the rate of fuel consumption in the modern power generation plants. In this figure, the broken lines represent the power consumption for electrolysis and the solid lines represent the total energy consumption. In calculation of steam consumption, the diaphragm process is assumed to utilize triple and quadruple effect evaporators. As the optimum current density is 20 - 25 Amperes/dm^2 for the diaphragm process, 100 - 130 Amperes/dm^2 for the mercury process, and 30 - 50 Amperes/dm^2 for the membrane process, reference to Figure 14 clearly shows that Asahi Chemical's membrane process is the most economical in terms of total energy consumption.

Literature Cited

1. Seko, M., (to Asahi Chemical Ind. Co., Ltd.) Japanese Patent Publication 55-1351 (Jan. 12, 1980).
2. Seko, M., (to Asahi Chemical Ind. Co., Ltd.) Japanese Patent Publication 55-14 148 (Apr. 14, 1980).
3. Seko, M., (to Asahi Chemical Ind. Co., Ltd.) French Patent 2 263 312 (May 11, 1979).

4. Seko, M., (to Asahi Chemical Ind. Co., Ltd.) U.S. Patent
    4 178 218   (Dec. 11, 1979).
5. Seko, M., (to Asahi Chemical Ind. Co., Ltd.) German Patent
    Publication 2 510 071 (Nov. 27, 1980).
6. Seko, M., (to Asahi Chemical Ind. Co., Ltd.) German Patent
    Publication 2 560 151 (Nov. 27, 1980).
7. Seko, M., (to Asahi Chemical Ind. Co., Ltd.) U.K.
    Patent 1 497 748 (Jan. 12, 1978).
8. Seko, M., (to Asahi Chemical Ind. Co., Ltd.) U.K. Patent
    1 497 749 (Jan. 12, 1978).
9. Grot, W.G. Chem. Ing. Tech. 1972, 44(4), 167.
10. Ukihashi, H. CHEMTECH 1980 (February), 118.
11. Walmsley, P.N., (to E.I. du Pont de Nemours and Co.) U.S.
    Patent 3 909 378 (Sep. 30, 1975).
12. Yamabe, M.; Miyake, H.; Arai, K., (to Asahi Glass Co.,
    Ltd.) Laid Open Japanese Patent Application 52-28 588
    (Mar. 3, 1977).
13. Kimoto K.; Miyauchi, H.; Ohmura, J.; Ebisawa, M.;
    Hane, T., (to Asahi Chemical Ind. Co., Ltd.) Laid Open
    Japanese Patent Application 55-160 030 (Dec. 12, 1980).
14. Grot, W.G., (to E.I. du Pont de Nemours and Co.) U.S.
    Patent 3 969 235 (Jul. 13, 1976).
15. Hora, C.J.; Maloney, D.E., Nafion membranes structured for
    high efficiency chlor-alkali cell, presented at
    Electrochem. Soc. Meeting, Atlanta Georgia U.S.A.
    Oct. 1977.
16. Oda, Y.; Suhara, M.; Endo, E., (to Asahi Glass Co., Ltd.)
    U.S. Patent 4 065 366 (Dec. 27, 1977).
17. Ukihashi, H.; Asawa, T.; Yamabe, M.; Gunjima, T.; Miyake,
    H., (to Asahi Glass Co., Ltd.) U.S. Patent 4 126 588
    (Nov. 21, 1978).
18. Seko, M.; Yamakoshi, Y.; Miyauchi, H.; Fukumoto, M.;
    Kimoto, K.; Watanabe, I.; Hane, T.; Tsushima, S., (to
    Asahi Chemical Ind. Co., Ltd.) U.S. Patent 4 151 053
    (Apr. 24, 1979).
19. Molnar, C.J.; Price E.H.; Resnick, P.R., (to E.I. du Pont
    de Nemours and Co.) U.S. Patent 4 176 215 (Nov. 27, 1979).
20. Kimoto, K.; Miyauchi, H.; Ohmura, J.; Ebisawa, M.;
    Hane, T., (to Asahi Chemical Ind. Co., Ltd.) Laid Open
    Japanese Patent Application 55-160 029 (Dec. 12, 1980).
21. Seko, M.; Yamakoshi, Y.; Miyauchi, H.; Fukumoto, M.;
    Kimoto, K., (to Asahi Chemical Ind. Co., Ltd.) Laid Open
    Japanese Patent Application 54-6887 (Jan. 16, 1979).
22. Hane, T.; Kimoto, K.; Yamakoshi, Y.; Miyauchi, H., (to
    Asahi Chemical Ind. Co., Ltd.) Laid Open Japanese Patent
    Application 53-82 684 (Jul. 28, 1978).
23. Ukihashi, H.; Asawa, T.; Yamabe, M.; Miyake, H., (to Asahi
    Glass Co., Ltd.) Laid Open Japanese Patent Application
    52-22 599 (Feb. 19, 1977).

24. Asawa, T.; Oda, Y.; Yamabe, M., (to Asahi Glass Co., Ltd.) Laid Open Japanese Patent Application 52-76 299 (Jun. 27, 1977).
25. Seko, M.; Yamakoshi, Y.; Miyauchi, H.; Kimoto, K.; Hane, T., (to Asahi Chemical Ind. Co., Ltd.) Laid Open Japanese Patent Application 53-125 283 (Nov. 1, 1978).
26. Seko, M., Commercial operation of the ion exchange membrane chlor-alkali process, presented at Centennial Meeting of Amer. Chem. Soc., New York U.S.A. Apr. 4-9, 1976.
27. Seko, M., New development of the Asahi Chemical membrane chlor-alkali process, presented at the Chlorine Institute Inc., 22nd Chlorine Plant Managers Seminar, Atlanta Geogia U.S.A. Feb. 7, 1979.
28. Seko, M., New development of the Asahi Chemical membrane chlor-alkali process, presented at Oronzio De Nora Symposium, Venice Italy May 15-18, 1979.
29. Seko, M., New development of the Asahi Chemical membrane chlor-alkali process, presented at Electro. Chem. Tech. Group, London U.K. June 13-15, 1979.
30. Peters, E.J.; Pulver, D.R., The commercialization of membrane cells to produce chlorine and caustic soda, presented at Electrochem. Soc. Fall Meeting, Atlanta Georgia Oct. 1978.
31. Pulver D.R., The commercial use of membrane cells in chlorine-caustic plants, presented at Chlorine Institute's 21st Plant Manager's Seminar, Houston Texas Feb. 15, 1978.
32. Asawa, T.; Oda, Y.; Yamabe, M., (to Asahi Glass Co., Ltd.) Laid Open Japanese Patent Application 52-23 192 (Feb. 21, 1977).
33. Asawa, T.; Oda, Y.; Gunjima, T., (to Asahi Glass Co., Ltd.) Laid Open Japanese Patent Application 52-36 589 (Mar. 19, 1977).
34. Suhara, M.; Arai, K., (to Asahi Glass Co., Ltd.) U.S. Patent 4 212 713 (Jul. 15, 1980).
35. Grot, W.G.; Molnar, C.J.; Resnick, P.R., (to E.I. du Pont de Nemours and Co.) Belgian Patent 866 122 (Oct. 19, 1978).
36. Seko, M.; Yamakoshi, Y.; Miyauchi, H.; Fukumoto, M.; Kimoto, K.; Hane, T.; Hamada, M., (to Asahi Chemical Ind. Co., Ltd.) Laid Open Japanese Patent Application 53-125 986 (Nov. 2, 1978).
37. Sata, T.; Nakahara, A.; Ito, J., (to Tokuyama Soda Co., Ltd.) Laid Open Japanese Patent Application 53-137 888 (Dec. 1, 1978).
38. Onoue, K.; Sata, T.; Nakahara, A.; Ito, J., (to Tokuyama Soda Co., Ltd.) U.S. Patent 4 200 711 (Apr. 29, 1980).
39. Seko, M.; Yamakoshi, Y.; Miyauchi, H.; Kimoto, K.; Masuda, Y., (to Asahi Chemical Ind. Co., Ltd.) U.S. Patent 4 123 336 (Oct. 31, 1978).

40.   Resnick, P.R., (to E.I. du Pont de Nemours Co.)
      U.S. Patent 3 560 568 (Feb. 2, 1971).
41.   Anderson, A.W., Fluoropolymer containing sulfonic acid
      group, presented at 67th National Meeting of American Inc.
      of Chem. Eng., Atlanta Georgia U.S.A. Feb. 15-18, 1970.
42.   Yeager, H.L.; Kipling, B. J. Phys. Chem. 1979,
      83(14), 1836.
43.   Hopfinger, A.J.; Mauritz, K.A.; Hora, C.J., Prediction of
      the molecular structure of Nafion under different physico-
      chemical conditions, presented at the Electrochem. Soc.
      Fall Meeting, Atlanta Georgia U.S.A. Oct. 1977.
44.   Grot, W.G.F.; Munn, G.E.; Walmsley, P.N.,
      Perfluorinated ion exchange membranes, presented at 141st
      National Meeting of Electrochem. Soc., Houston Texas U.S.A.
      May 7-11, 1972.
45.   Helfferich, F. "Ion Exchange"; McGraw-Hill: New York,
      1962; P.140.
46.   Ukihashi, H.; Shiragami, O.; Oda, Y.; Asawa, T.,
      3rd Soda Kogyo Gijutsu Toron Kai Koen Yoshi-shu, P.5,
      Kyoto, Japan, Nov. 21-22, 1979.
47.   Stanley, A. J.; Dotson, R. L., (to Diamond Shamrock Corp.)
      U.S. Patent 3 773 634 (Nov. 20, 1973).
48   Oleary, K., Membrane chlorine cell design and technology,
      presented at Electrochem. Tech. Group of Soc. of Chem.
      Ind. London U.K.   June 16-17, 1976.
49.   Burkhardt, S.F., Radioactive tracer measurement of sodium
      transport efficiency in membrane cell, presented at
      Electrochem. Soc. Meeting, Atlanta Georgia U.S.A.
      Oct. 1977.
50.   Bergner, V.D. Chemiker-Zeitung 1977, 10, 433.
51.   England D.C., (to E.I. du Pont de Nemours and Co.) Laid
      Open Japanese Patent Application 53-132 519 (Nov. 18,
      1978).
52.   Fritz, C.G.; Moore, Jr., E.P.; Selman, S., (to E.I.
      du Pont de Nemours and Co.) U.S. Patent 3 114 778 (Dec. 17,
      1963).
53.   Anderson, D.G.; Gladding, E.K.; Sullivan, R., (to E.I.
      du Pont de Nemours and Co.) U.K. Patent 1 145 445
      (Mar. 12, 1969).
54.   Yamabe, M.; Munakata, S.; Sugaya, Y.; Jitsugiri, Y., (to
      Asahi Glass Co., Ltd.) Laid Open Japanese Patent
      Application 52-59 111 (May 16, 1977).
55.   Yamabe, M.; Munakata, S.; Kumai, K.; Akatsuka, Y., (to
      Asahi Glass Co., Ltd.) Laid Open Japanese Patent
      Application 52-83 417 (Jul. 12, 1977).
56.   Yamabe, M.; Munakata, S.; Samejima, S., (to Asahi Glass
      Co., Ltd.) Laid Open Japanese Patent Application 52-78
      826 (Jul. 2, 1977).
57.   Putnam, P.E.; Nicoll, W.D., (to E.I. du Pont de Nemours
      and Co.) U.S. Patent 3 301 893 (Jan. 31, 1967).

58. Connally, D.J.; Gresham W.F., (to E.I. du Pont de Nemours
     and Co.) U.S. Patent 3 282 875 (Nov. 1, 1966).
59. Kimoto, K.; Miyauchi, H.; Ohmura, J.; Ebisawa, M.; Hane,
     T., (to Asahi Chemical Ind. Co., Ltd.) Laid Open Japanese
     patent Application 55-160 007 (Dec, 12, 1980).
60. Kimoto, K.; Miyauchi, H.; Ohmura, J.; Ebisawa, M.;
     Hane, T., (to Asahi Chemical Ind. Co., Ltd.) Laid Open
     Japanese Patent Application 55-160 008 (Dec. 12, 1980).
61. Kimoto, K.; Miyauchi, H.; Ohmura, J.; Ebisawa, M.;
     Hane, T., (to Asahi Chemical Ind., Co., Ltd.) Laid Open
     Japanese Patent Application 56-12 362 (Feb. 6, 1981)
62. Kimoto, K.; Miyauchi, H.; Ohmura, J.; Ebisawa, M.;
     Hane, T., (to Asahi Chemical Ind. Co., Ltd.) Laid Open
     Japanese Patent Application 56-15 260 (Feb. 14, 1981).
63. Kimoto, K.; Miyauchi, H.; Ohmura, J.; Ebisawa, M.; Hane,
     T., (to Asahi Chemical Ind. Co., Ltd.) Laid Open Japanse
     Patent Application 56-16 460 (Feb. 17, 1981).
64. Carlson, D.P., (to E.I. du Pont de Nemours and Co.)
     U.S. Patent 3 528 954 (Sep. 15, 1970).
65. Carlson, D.P., (to E.I. du Pont de Nemours and Co.)
     U.S. Patent 3 674 758 (Jul. 4, 1972).
66. Manwheeler, C.H., (to E.I. du Pont de Nemours and Co.)
     Japanese Patent 632 937 (Jan. 29, 1972).
67. Grot W.G., (to E.I. du Pont de Nemours and Co.) U.S.
     Patent 3 770 567 (Nov. 6, 1973).
68. Watanabe, I.; Yamakoshi, Y.; Miyauchi, H.; Tsushima, S.;
     Fukumo, M., (to Asahi Chemical Ind. Co., Ltd.)
     U.S. Patent 4 072 793 (Feb. 7, 1978).
69. Ukihashi, H.; Asawa, T.; Gunjima, T., (to Asahi Glass Co.,
     Ltd.) U.S. Patent 4 218 542 (Aug. 19, 1980).
70. Oda, Y.; Asawa, T.; Gunjima, T., (to Asahi Glass Co.,
     Ltd.) Laid Open Japanese Patent Application 55-139 842
     (Nov. 1, 1982).
71. Yokoyama, S.; Kimoto, K.; Muranaka, F., (to Asahi Chemical
     Ind. Co., Ltd.) Laid Open Japanese Patent Application
     52-134 888 (Nov. 11, 1977).
72. Seko, M., Ion-Exchange Membrane For the Chlor-Alkali Process,
     presented to 159th Meeting The Electrochemical Society,
     Minneapolis, Minnesota May 13th, 1981.
73. Seko, M.; Ogawa, S.; Takemura, R., (to Asahi Chemical Ind.
     Co., Ltd.) British Patent 1 543 249 (Mar. 28, 1979).
74. Ogawa, S., Asahi Chemical Membrane Chlor-alkali Process,
     presented at Seminar on Developments in Chlor-alkali
     Industry, New Delhi India Mar. 7-8, 1980.
75. Seko, M.; Ogawa, S.; Yoshida, M., (to Asahi Chemical Ind.
     Co., Ltd. U.S. Patent 4 108 742 (Aug. 22, 1978).
76. Ogawa, S., (to Asahi Chemical Ind. Co., Ltd.) U.S. Patent
     4 132 588 (Jan. 2, 1979).
77. Ogawa, S.; Yoshida, M., (to Asahi Chemical Ind. Co., Ltd.)
     U.S. Patent 4 105 515 (Aug. 8, 1978).

78.   Seko, M.; Ogawa, S.; Yoshida, M.; Nakamura, A., (to Asahi
      Chemical Ind. Co., Ltd.)   U.S. Patent 4 005 004 (Jan. 25,
      1977).
79.   Takahashi, M. Soda and Chlorine (in Japanese) 1978, 11,
      511-520.
80.   Juda, W., (to Ionics Incorp.)   U.S. Patent 3 124 520
      (Mar. 10, 1964).
81.   Coker, T.G.; Dempsey, R.M.; La Conti, A.B., (to General
      Electric Co.)   U.S. Patent 4 191 618   (Mar. 4, 1980).
82.   Yomiyama, A.,   Energy reduction in a membrane chlor-alkali
      process (in Japanese), presented at 4th Meeting on
      Industrial Chlor-alkali Technology of the Electrochem.
      Soc. of Japan, Kyoto Japan   Nov. 21, 1980.

RECEIVED November 9, 1981.

Perfluorinated Ion Exchange Membranes

TOSHIKATSU SATA and YASUHARU ONOUE

Research and Development Division, Tokuyama Soda Co., Limited, Mikage-cho 1-1, Tokuyama City, 745 Yamaguchi Prefecture, Japan

The requirements for ion-exchange membranes in membrane cell caustic-chlorine process are : high permselectivity, low electric resistance, excellent chemical resistance to oxidants and alkali, good heat resistance, low diffusion of salt and low permeability to water and good mechanical properties. Chemical resistance to oxidants in particular could not be achieved in conventional hydrocarbon type ion-exchange membranes, although various electrolytic methods were tried to prevent ion-exchange membranes of that type from deteriorating by oxidants. In 1972, the difficulty in chemical resistivity was overcome by perfluorocarbon ion-exchange membrane made by du Pont de Nemours and Co. (1), but this membrane was still too low in permselectivity while its electric resistance was sufficiently low. The low electric resistance and high permselectivity are factors generally forced to be mutually inconsistent. However, the use of an anisotropic structure for ion-exchange membranes enables both of these requirements to be achieved together (2,3). Anisotropic ion-exchange membranes like reverse osmosis membrane are well known, i.e., monovalent cation and anion permselective ion-exchange membranes in electrodialytic concentration of sea water to make edible salt (4,5). In the case of ion-exchange membranes for the caustic-chlorine process, concentration of fixed ions in the membrane should be kept high to prevent permeation of hydroxide ions through the membrane. In general, however, a membrane which has a high concentration of fixed ions throughout also shows high electric resistance. Therefore, it is desirable that the ion-exchange membrane is composed of a thin layer having high concentration of fixed ions and a thick layer of low electric resistance. Various methods can be used to achieve this purpose, such as to differentiate ion exchange capacity of strongly acidic ion-exchange groups along the cross-section of the membrane, or to stratify weakly acidic ion-exchange groups over the surface of an ion-exchange membrane which has strongly acidic ion exchange groups. Various attempts were made by us to reduce the ion exchange capacity of perfluorocarbon sulfonic acid type membrane, i.e., decomposition or inactivation of ion exchange groups by chemical

0097-6156/82/0180-0411$05.00/0

reaction, impregnation of hydrophobic materials on the membrane
surface and so on (6). However it was found that these methods
could not achieve adequate performance in terms of electrical resistance and sodium ion permselectivity.

Therefore further efforts were directed to prepare an anisotropic membrane using some weakly acidic cation exchange groups.
We selected carboxylic acid groups as the main component of weakly
acidic cation exchange groups, rather than sulfonic acid amide
with a dissociable hydrogen, phosphonic acid, phenolic hydroxide
or perfluoro-tert-alcohol exchange sites, from the viewpoint of
ease of preparation, stability and good performance in electrolysis.

NEOSEPTA-F and its Preparation Methods

The chemical structure of NEOSEPTA-F made by us is basically
as follows :

$$-(CF_2-CF_2)_l \quad -(CF_2-CF)_m \quad -(CF_2-CF)_n$$

with side chains:

$$\begin{array}{ccc} & O & O \\ & CF_2 & CF_2 \\ CF_3CF & CF_3CF \\ & O & O \\ & (CF_2)_p & (CF_2)_q \\ & SO_3Na & COONa \end{array}$$

$$1/(m + n) = 6 - 8$$
$$m/n = 5 - 20$$
$$p, q = 1 - 2$$

Although perfluorocarbon sulfonic acid groups are very stable
chemically as well as thermally, perfluorocarbon sulfonyl halide,
especially sulfonyl chloride groups, are quite reactive.   For
example, sulfonyl chloride groups react with oxidants, reductants,
various amines, phenol compounds, iodine compounds, etc. and give
carboxylic acid, sulfinic acid, sulfonic acid amide, $-CF_2I$ and so
forth. Some examples of how this feature can be used to generate
various kinds of membranes will next be described
1) (7) A membranous material having a thickness of 0.2 mm
and composed of a copolymer of tetrafluoroethylene (monomer A)
and perfluoro (3,6-dioxa-4-methyl-7-octene-sulfonyl fluoride)
(monomer B) in a mole ratio of about 7 : 1, which had an ion
exchange capacity upon hydrolysis of 0.91 milliequivalent/gram of
dry membrane (meq./g.dry membrane of H^+-form) was hydrolyzed an
aqueous solution of dimethyl sulfoxide and potassium hydroxide to
afford an ion-exchange membrane having sodium sulfonate groups.
Sulfonate groups of the membrane was converted to sulfonic
acid form completely by nitric acid. The membrane was then dried,
and reacted at 130 °C for 75 hours in a bath consisting of phosphorous pentachloride and phosphorous oxychloride. After the reaction, the product was washed with carbon tetrachloride and dried.

To examine the resulting membrane, the reflective infrared spectrum of this membrane was measured. It was found that the absorption band at 1060 cm^{-1} observed in the sulfonic acid-type membrane disappeared, and a strong absorption band corresponding to the sulfonyl chloride group was observed at 580 and 1420 cm^{-1}.

The membrane having sulfonyl chloride groups and the membrane having sulfonic acid groups were dipped in n-butyl alcohol. Air was introduced in the medium, and applied uniformly to the surface of the membrane for oxidation reaction (at 110 $^{\circ}$C for 3 hours). The membrane were then washed with methanol and water, and dried. To examine their surface structures, the reflective infrared spectrum of the treated membranes were measured. No appreciable difference was seen in the sulfonic acid-type membrane before and after the treatment. In the sulfonyl chloride-type membrane, the absorption band at 580 and 1420 cm^{-1} ascribable to sulfonyl chloride totally disappeared, and a new absorption band at 1790 cm^{-1} was observed. This absorption band is assigned to carboxylic acid group. These membranes were treated for hydrolysis with methanol solution containing 10 % of sodium hydroxide (for 16 hours at 60 $^{\circ}$C), washed with water, and dried. The reflective infrared spectrum of the treated membranes were measured. The absorption band at 1790 cm^{-1} disappeared which had been observed on sulfonyl chloride-type membrane. Instead, a new absorption band was observed at 1680 cm^{-1}.

These membranes were each dipped in a dye solution of 1 % crystal violet and 10 % ethanol in a 0.5 N HCl aqueous solution. Then the membranes were washed with water and cut by a microtome. A microscopic examination indicated that the membrane derived from sulfonic acid type membrane was uniformly dyed deep green throughout, whereas the membrane derived from sulfonyl chloride-type membrane was dyed deep green only in its inner part leaving the outer layers of 20 μ each from its both surfaces free from dyeing. This dyeing test shows that in the sulfonyl chloride type membrane, the oxidation treatment has converted the sulfonyl chloride groups at the outer layers into carboxyl groups to the extent of 20 μ from respective surfaces.

The properties of the membrane hydrolyzed with 10 % sodium hyroxide were measured, and a saturated sodium chloride solution was electrolyzed using this membrane. The results are given in Table I. For comparison, the sulfonyl chloride-type membrane was treated in n-butyl alcohol at 110 $^{\circ}$C for 3 hours without introducing air. The treated membrane was subjected to hydrolysis treatment in a methanol solution containing 10 % of sodium hydroxide. Electric resistance of the membrane was 450 Ω $-cm^2$, and the current efficiency could not be measured.

Electric resistance was measured on the membrane which was placed partitioning 3.5 N NaCl solution to its one side and 6.0 N NaOH solution to the other side at 85 $^{\circ}$C and the solutions were electrolyzed. The electrolysis was carried out by using a saturated solution of sodium chloride as an anolyte, a titanium lath

Table I

Properties	Membrane having sulfonic acid groups (blank)	Membrane with oxidation reaction
Electric Resistance ($\Omega - cm^2$)	1.95	2.25
Ion Exchange Capacity (Meq./g.dry membrane of H^{+}-form)	0.91	0.85
Water Content (%)	17	12
Catholyte Concn. (N)	7.50	7.50
Current Efficiency (%)	49	93
NaCl in Catholyte (ppm) (as 48 % NaOH)	238	18

material coated with ruthenium oxide and titanium oxide as an anode, a mesh-like mild iron as a cathode. Water was added to the cathode compartment, and aqueous solution of sodium hydroxide was obtained in a certain concentration. The current density was 30 A/dm^2 and the temperature of the electrolytic solution was 80 to 90°C.

2) (8) The same membranous material as mentioned in 1) having a thickness of 100 μ was set in a reactor of the design which would allow only one surface of the membrane to contact with reaction reagents. Thereafter, the reactor compartment was filled with vapour of phosphorous pentachloride (at 170°C for an hour) to have one surface of the membrane reacted. The reflective infrared spectrum and dyeing test respectively showed that the membrane had sulfonyl chloride groups and that approximately 5 μ of non-dyed layer was stratified at the membrane surface where phosphorous pentachloride had contacted. The electric resistance of this membrane was about 1500 $\Omega - cm^2$ in a 1.0 N hydrochloric acid solution at 25°C when measured by 1000 cycle A.C. The electric resistance of the same membrane before the reaction with phosphorous pentachloride was only 0.38 $\Omega - cm^2$ under the same conditions.

The membrane which had a thin layer of sulfonyl chloride groups was treated by triethylamine at room temperature for 16 hours, washed with water and then heated at 170°C. Thereafter, the membrane was also immersed in the same mixed solution composed of water, dimethyl sulfoxide and potassium hydroxide as mentioned before. Electric resistance of the membrane was 1.5 $\Omega - cm^2$ when measured in the environment of 3.5 N sodium chloride solution to its one side and 6.0 N sodium hydroxide solution to the PCl5-reacted side of the membrane at 85°C. The measurement in 3.5 N sodium chloride solution of pH 0.5 (adjusted by hydrochloric acid) showed the electric resistance of 432Ω - cm^2 at 25.0°C. On the other hand, the electric resistance of the membrane before the reaction with phosphorous pentachloride and triethylamine was 1.1Ω - cm^2 and 1.0Ω - cm^2 respectively when measured in the environment of 3.5 N sodium chloride solution to its one side and 6.0 N sodium hydroxide solution to the other side, and in the environment of 3.5 N sodium chloride solution of pH 0.5. According to reflective infrared spectrum, the absorption bands observed were different between the surfaces reacted with phosphorous pentachloride and non-reacted. Namely the absorption band at 1680cm^{-1} corresponding to carboxyl groups was observed, and the absorption band at 1060cm^{-1} observed in the sulfonic acid type membrane disappeared on the surface which had been reacted with phosphorous pentachloride.

Using the treated membrane electrolysis of sodium chloride solution was carried out under the same electrolysis conditions as 1). The treated surface of the membrane was faced to the cathode side in the electrolyzer. When 6.5 N sodium hydroxide solution was obtained as catholyte, the current efficiency was 93% and the cell voltage was 3.85v. On the other hand, the ion exchange membrane not treated by phosphorous pentachloride and triethylamine showed the current efficiency of 52% and the cell voltage of 3.68v when 6.5 N sodium hydroxide was obtained as catholyte.

3) (9) Similar membranous copolymer as mentioned in 1) having a thickness of 150 μ (The ion exchange capacity of this membrane was 0.83 meq./q.dry membrane of H$^+$-form.) was set in a horizontal reactor of the design which would allow only one surface of the membrane to contact with reaction reagents. Then fine crystal powder of phosphorous pentachloride was uniformly placed to cover one surface of the membrane and heated (at 155°C for 40 min.).

The membrane having the sulfonyl chloride groups was treated by an aqueous n-butyl amine solution for 2 hours, washed with water, heated in air for 24 hours at 90°C and then dipped in 10% methanol solution of sodium hydroxide. The reflective infrared spectrum showed that absorption band ascribable to the sulfonyl chloride disappeared and new absorption bands appeared at 1620, 1680 and 3400 cm^{-1}. The electric resistance and the electrolysis results of both of the treated and untreated membranes are shown in Table II respectively.

4) (10) The membranous copolymer as mentioned before reinforced by a plain woven cloth of polytetrafluoroethylene was reacted with vapour of phosphorous pentachloride to form a membrane having sulfonyl chloride groups on its only one side. The membrane having sulfonyl chloride groups on its only one surface was treated by 29 % aqueous ammonia solution for 30 min. at 25 °C. After the ammonia treatment, the absorption bands ascribable to

Table II

Properties	Untreated Membrane	Treated Membrane
Electric Resistance in 6.0 N NaOH (Ω-cm^2)	3.6	3.8
Electric Resistance in 1 N HCl (Ω-cm^2)	0.43	125
Catholyte Concn. (N)	6.5	6.5
Current Efficiency (%)	58	94
NaCl in Catholyte (ppm) (as 50 % NaOH)	143	12
Cell Voltage (v)	3.55	3.65

sulfonyl chloride disappeared and new absorption bands were observed at 940, 1010, 1410, 1680 and 3400 cm^{-1}. The absorption bands
of 940 and 1010 cm^{-1} disappeared after the membrane was heated in
oxygen atmosphere. Thereafter, the membrane was treated by nitric
acid. The absorption band of 1680 cm^{-1} disappeared and a new absorption band was observed. It was deduced that sulfonic acid
amide groups and carboxylic acid groups had been introduced by
these treatments.

In electrolysis of sodium chloride solution under the same
conditions as mentioned before, 8.0 N sodium hydroxide was
obtained as catholyte at the current efficiency of 95 % and the
cell voltage of 4.1 V.

5) (11) Sulfonic acid groups of cation exchange membrane,
Nafion 110 (trademark for products of E. I. du Pont de Nemours
& Co.) were converted to sulfonyl chloride by a mixture of phosphorus pentachloride and phosphorus oxychloride. The resulting
membrane having sulfonyl chloride groups was set in an oxidation
device which allows a uniform air circulation. Air saturated with
n-butanol vapor was introduced into the oxidation device at 110 °C
to allow reaction on one side only. It was found that in the
spectrum of the treated surface, the absorption band ascribable
to the sulfonyl chloride groups disappeared, and a new absorption
band ascribable to the carboxylic acid groups appeared at 1790
cm^{-1}. In the spectrum of the other surface, the absorption band
of sulfonyl chloride groups remained likewise as before the treatment, and no absorption of the carboxyl acid groups was observed.

As shown in the above, sulfonic acid groups of perfluorocarbon polymer can be easily changed to weakly acidic cation exchange
groups. There are various other methods to change sulfonic acid
groups to weakly acidic cation exchange groups, i.e., contacting
the membrane having sulfonyl chloride groups with aromatic compounds with phenolic hyroxide groups, various amines, ammonium ions
and so on (12).

When these bilayer or multilayer ion-exchange membranes were used in the electrolysis of alkali metal salt solution, performance of the electrolysis was excellent. Sulfonic acid groups and carboxylic acid groups were selected for NEOSEPTA-F as the main ion exchange groups.

General Properties

The NEOSEPTA-F membrane properties examined are mainly of those relating to the electrolysis of sodium chloride solution. Table III shows characteristics of typical grades of NEOSEPTA-F. These membranes are chemically stable, i.e., against acid, base, oxidants and reductants because the membranes have perfluorocarbon backbone. And also the membranes have strong mechanical strength because of reinforcement with the fabric of polytetrafluoroethylene. Used in the electrolysis of sodium chloride solution, no deterioration of performance or mechanical strength was observed in continuous service for 2 years under appropriate electrolysis conditions. NEOSEPTA-F membranes are always improved to get better performance in the electrolysis and various grades which show better performance are developed.

Electric Resistance of the Membranes.   Figure 1 shows the relationship between the electric resistance of NEOSEPTA-F C-1000 and pH value of 3.5 N sodium chloride solution (pH was adjusted by adding hydrochloric acid). The measurements were carried out at 25.0 ^{o}C using 1000 cycle A.C. NEOSEPTA-F C-2000 also shows the similar relationship between the electric resistance and pH value of sodium chloride solution. It is recognized that these NEOSEPTA-F ion-exchange membranes have weakly acidic cation exchange groups which are dissociable in the range between pH 2 and 3.

Generally, when an ion-exchange membrane contacts with highly concentrated solution, it shrinks and then the electric resistance increases remarkably. These phenomena were observed in the case of NEOSEPTA-F also. Figure 2 shows the electric resistance of the membrane measured with 1000 cycle A.C. at 80 ^{o}C in sodium hydroxide solution of various concentrations. Figure 3 shows the electric resistance of the membrane measured with direct current under the same conditions as the electrolysis was to be carried out. (The membrane was placed partitioning 3.5 N sodium chloride solution to one side and sodium hydroxide solution of varied concentration to the other side and the direct current was passed at the current density of 30 A/cm^2.) The electric resistance measured with direct current was considerably lower than that measured with alternating current, while the difference should normally be very minute if measured under exactly the same conditions. The above significant difference is seemingly attributable to that in case of direct current measurement anolyte is sodium chloride solution kept at constant concentration of 3.5 N, which would increase water content of the membrane due to highly hydrated sodium ions passing through, thus lowering the electric resistance.

Table III   Characteristics of NEOSEPTA-F C-1000 and C-2000

Name	NEOSEPTA-F C-1000	NEOSEPTA-F C-2000
Backing	Polytetrafluoro-ethylene fabric	Polytetrafluoro-ethylene fabric
Resin	Perfluorocarbon	Perfluorocarbon
Main Ion Exchange Groups	$-SO_3Na$ $-COONa$	$-SO_3Na$ $-COONa$
Ion Exchange Capacity[*]	0.83	0.91
Electric Resistance[**] ($\Omega -cm^2$)[***]	2.0	1.7
Water Content	10.3	11.7
Tensile Strength[****] (Kg/cm)	10.6	10.6

[*]. Meq./g. dry resin of H^+-form.

[**]. Measured by electrolysis of 3.5 N NaCl ┃ 6.0 N NaOH (C-1000) and 3.5 N NaCl ┃ 9.0 N NaOH (C-2000) at the current density of 30 A/dm^2 at 80 ^{O}C.

[***]. g.H$_2$0/ g. dry resin of H^+-form. Measured in atmosphere after the membrane was equilibrated with 6.0 N NaOH (C-1000) and 9.0 N NaOH (C-2000) at room temperature.

[****]. Both wet and dry.

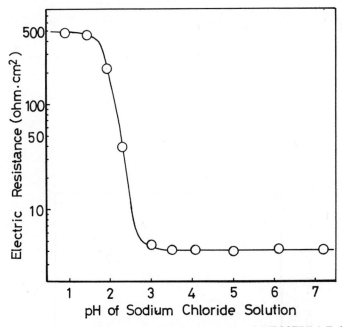

Figure 1. The relationship between electric resistance of NEOSEPTA-F C-1000 and pH value of 3.5 N NaCl solution.

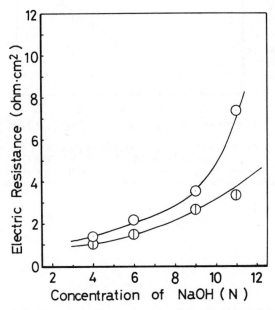

Figure 2. Relationship between electric resistance of NEOSEPTA-F and concentration of NaOH. Key: ○, NEOSEPTA-F C-1000; ⊕, NEOSEPTA-F C-2000. Measured by 1000 cycle alternating current at 80° C.

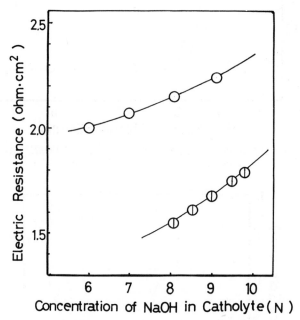

*Figure 3. Relationship between electric resistance of NEOSEPTA-F measured by
direct current and concentration of catholyte. Key: ◯, NEOSEPTA-F C-1000; ⦶,
NEOSEPTA-F C-2000. Measurements were made with direct current under the
same conditions as the electrolysis of NaCl solution was to be carried out.*

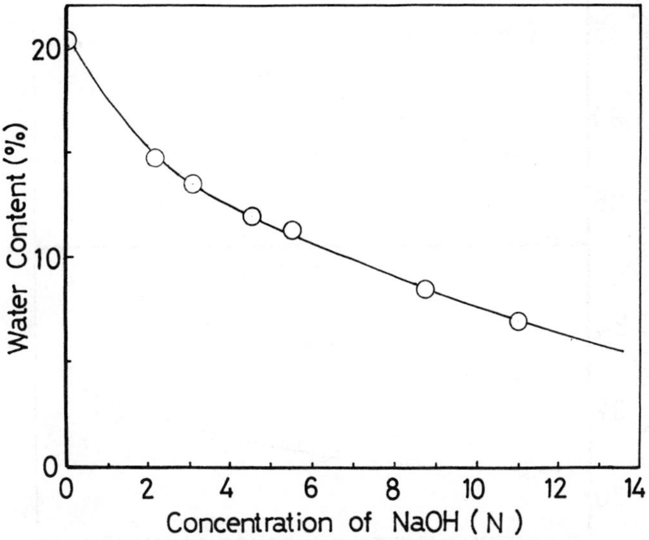

Figure 4. Change of H₂O content of NEOSEPTA-F C-1000 with concentration of NaOH solution. The membrane was immersed in NaOH solution of various concentration after boiling for 1 h in pure H₂O. Measurement was made at 20° C after the membrane had been immersed in NaOH solution of various concentration for 4 days at room temperature.

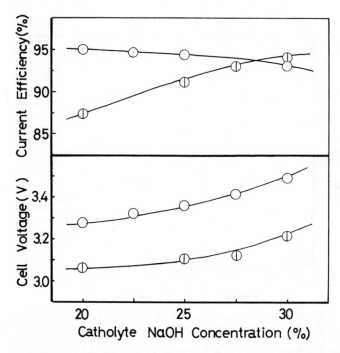

Figure 5. Relationship of current efficiency and cell voltage to NaOH concentration of catholyte. Key: ○, *NEOSEPTA-F C-1000;* ⊕, *NEOSEPTA-F C-2000. Current density is 20 A/dm².*

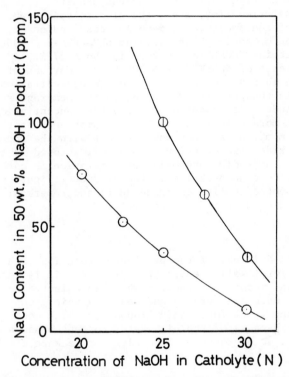

Figure 6. Relationship between NaCl in caustic product and concentration of NaOH in catholyte. Key: ◯, NEOSEPTA-F C-1000; ⊕, NEOSEPTA-F C-2000.

Figure 4 shows the water content of the membrane corresponding to various concentrations of sodium hydroxide. The extent of decrease of the water content with increase of concentration of the external solution is remarkable in comparison with the case of crosslinked hydrocarbon type cation exchange membranes.

Performance of NEOSEPTA-F in Sodium Chloride Solution Electrolysis. Figure 5 shows the relationship of the cell voltage and the current efficiency respectively with the concentration of sodium hydroxide in catholyte when electrolysis of sodium chloride solution was carried out at the current density of 30 A/cm^2. From the economical viewpoint, i.e., the electrolysis power cost, depreciation of equipment cost, membrane cost and so on, the optimum concentration of sodium hydroxide for NEOSEPTA-F C-1000 is about 20 % and that for NEOSEPTA-F C-2000 is about 27 %.

In the case of NEOSEPTA-F C-2000, the current efficiency increases with increase of sodium hydroxide concentration in catholyte. It is thought that the water in the membrane surface portion of cathode side is dehydrated and the concentration of fixed ion in the membrane increases. The presumption that the cathode side of the membrane surface would shrink with the increase of sodium hydroxide concentration is obviously proved in the relationship between the sodium hydroxide concentration in catholyte and sodium chloride concentration in the product (Figure 6). The diffused amount of sodium chloride decreased remarkably with increase of sodium hydroxide concentration.

Conclusion

NEOSEPTA-F is one of the perfluorocarbon ion exchange membranes for Chlor-Alkali electrolytic process. It is considered to be of an ideal membrane structure which is anisotropically composed of sulfonic acid groups and weakly acidic groups as ion exchange groups. Sulfonic acid groups give high conductance to the membrane because of the high water content. And a thin layer of carboxylic acid groups is a barrier for leakage of hydroxide ions.

Literature Cited

1. Grot, W. G. ; Munn, G. E. ; Walmsley, P. N. "Perfluorinated Ion Exchange Membranes", presented at the 141 st National Meeting , the Electrochemical Society, Houston, Texas, May (1972).
2. Sata, T. ; Murakami, S. ; Murata, Y. Japan. Pat. Application Publication No. 14595/1979, U.S.Pat. 4166014.
3. Sata, T. , Murakami, S. , Murata, Y. Japan. Pat. Application Publication No. 14596/1979, U.S.Pat. 4169023, Brit. Pat. 1493164, Ger. Pat. 2504622.
4. Sata, T. J. Colloid Interface Sci. 1973, 44 393 ; Sata, T.;

Izuo, R. Colloid and Interface Sci. 1978, <u>256</u>, 757. ;
Sata, T. ; Yamane, R. ; Mizutani, Y. J. Polymer Sci. Polymer Chem. 1979, <u>17</u> 2071, etc.

5. Mizutani, Y. ; Yamane, R. ; Sata, T. Japan. Pat. Application Publication No. 23607/1971, U.S.Pat. 3510417, 3510418, Brit. Pat. 1238656, ; Mizutani, Y. ; Yamane, R. : Sata, T. ; Izuo, R., Japan. Pat. Application Publication No. 3801/1972, 3802/1972, U.S.Pat. 3647086, Brit. Pat. 1251550, etc.

6. Sata, T. ; Nakahara, A. ; Murata, Y. ; Ito, J. Japan. Pat. Open Publication No. 26284/1978, 26285/1978, 26286/1978, 138489/1977, ; Sata, T. ; Nakahara, A. ; Murata, Y. ; Ito, J. ; Shirouzu, M. Japan. Pat. Open Publication No. 55383/ 1978, 58493/1978.

7. Onoue, Y. ; Sata, T. ; Nakahara, A. ; Ito, J. Japan. Pat. Open Publication No. 132069/1978, U.S.Pat. 4200711 (1980).

8. Sata, T. ; Nakahara, A. ; Ito, J. ; Shirouzu, M. Japan. Pat. Open Publication No. 64090/1979.

9. Sata, T. ; Nakahara, A. ; Ito, J. ; Shirouzu, M. Japan. Pat. Open publication No. 41287/1979.

10. Sata, T. ; Nakahara, A. ; Ito, J. ; Shirouzu, M. Japan. Pat. Open Publication No. 21478/1979.

11. Onoue, Y. ; Sata, T. ; Nakahara, A. ; Ito, J. Japan. Pat. Open Publication No. 83982/1979.

12. Sata, T. ; Nakahara, A. ; Ito, J. Japan. Pat. Open Publication No. 125974/1978, 137888/1978, ; Sata, T. ; Nakahara, A. ; Ito, J. ; Shirouzu, M. Japan. Pat. Open Publication No. 20981/1979.

RECEIVED August 26, 1981.

Perfluorocarboxylate Polymer Membranes

HIROSHI UKIHASHI and MASAAKI YAMABE

Research Laboratory, Asahi Glass Company, Limited, Yokohama, Japan

A new process using ion exchange membranes has recently gained wide acceptance in the chlor-alkali industry from the viewpoint of energy saving and environmental control. One of the important breakthroughs for this process was to develop a membrane of high performance.

A novel perfluorocarboxylate membrane named Flemion was developed by Asahi Glass Co., Ltd. in 1975, resulting from its long experience both with manufacture of various fluorochemicals and with electrodialysis using membranes.

Flemion is quite different from prior membranes in that it is based on specific perfluorinated copolymers with pendant carboxylic acid as a functional group. The introduction of carboxylic functions in the polymer has realized high permselectivity in cation transport with high conductivity, which is indispensable to electrochemical application of ion exchange membranes.

This chapter summarizes the preparation and the fabrication of perfluorocarboxylate polymers and their fundamental properties including those of the ionized salt-type membranes. The application of Flemion in chlor-alkali electrolysis is also described.

Preparation of Perfluorocarboxylate Polymer

Perfluorocarboxylate polymers were prepared by copolymerization of tetrafluoroethylene and carboxylated perfluorovinyl ether. The general formula of copolymers are shown as follows.

$$-(CF_2CF_2)_x-(CF_2CF)_y-$$
$$(OCF_2CF)_m-O(CF_2)_n-C-O-R$$
$$CF_3 \qquad \qquad O$$

(where $m = 0$ or 1, $n = 1 \sim 5$, $R = $ alkyl)

The ratio of carboxylated perfluorovinyl ether to tetrafluoroethylene in copolymers (y/x) can be controlled by the monomer feed
ratio and the values of m and n can be determined by choice of
monomer structures.

The synthesis of carboxylated perfluorovinyl ether was very
difficult and no preparative method had been known before. Recently, however, several synthetic methods have been proposed. An example of the synthetic route to methyl perfluoro-5-oxa-6-heptenoate
(M_1) and methyl perfluoro-5,8-dioxa-6-methyl-9-decenoate (M_2) is
shown in the following scheme.

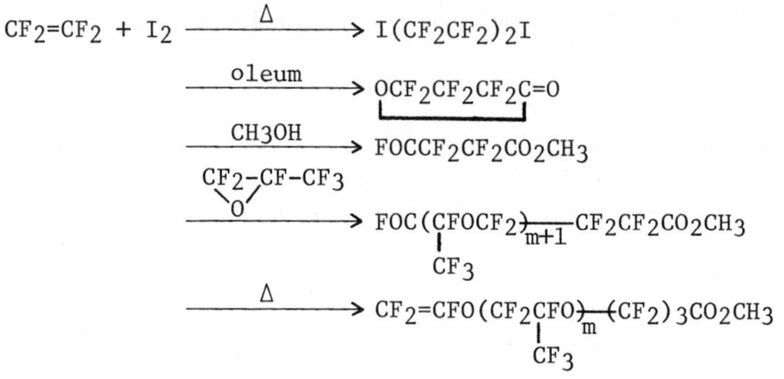

$$CF_2=CF_2 + I_2 \xrightarrow{\Delta} I(CF_2CF_2)_2I$$

$$\xrightarrow{oleum} OCF_2CF_2CF_2C=O$$

$$\xrightarrow{CH_3OH} FOCCF_2CF_2CO_2CH_3$$

$$\xrightarrow{CF_2-CF-CF_3 \atop O} FOC(CFOCF_2)_{\overline{m+1}}CF_2CF_2CO_2CH_3 \atop CF_3$$

$$\xrightarrow{\Delta} CF_2=CFO(CF_2CFO)_m(CF_2)_3CO_2CH_3 \atop CF_3$$

$$M_1 \ (m = 0), \ M_2 \ (m = 1)$$

Copolymerization of tetrafluoroethylene and carboxylated perfluorovinyl ether is carried out either in solution, bulk or emulsion system with a radical initiator. A typical copolymer composition curve is given in Figure 1, where M_1 or M_2 was copolymerized
with tetrafluoroethylene in bulk system at 70°C. The monomer reactivity ratios of tetrafluoroethylene and each vinyl ether are
calculated as 7.0 and 0.14, respectively.

Copolymers containing up to 35 mole% of carboxylated vinyl
ether were synthesized by regulating the reaction pressure of
tetrafluoroethylene in the copolymerization procedure.

The studies of the copolymer with X-ray diffraction and differential scanning calorimetry revealed that the crystallinity of
the copolymer decreased with increasing vinyl ether content as
shown in Figure 2. The copolymer became amorphous at the vinyl
ether content of 20 mole%. The glass transition temperature of
the amorphous copolymer lies around 10°C.

Copolymers having high molecular weight of more than 3×10^5
can be obtained by adopting bulk or emulsion system with extremely
purified functional comonomers.

Fabrication

The ester type copolymer as polymerized has a non-crosslinked
linear structure and is melt processable.

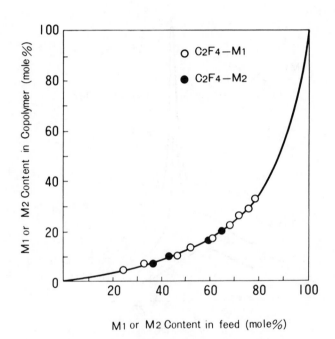

Figure 1. Copolymer composition curve.

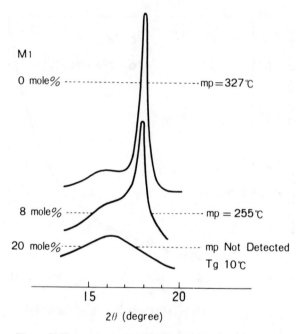

Figure 2. *X-ray spectra and melting point of polymer.*

Figure 3 shows the dependence of apparent viscosity of the copolymer upon shear rate. Since apparent viscosity decreases with increasing shear rate, the copolymer is regarded as a kind of pseudoplastics. Apparent viscosity of ca. 10^4 poise is obtained around the shear rate of 10^2 sec^{-1} in the temperature range of 230 ~ 250°C.

Figure 4 plots logarithm of apparent viscosity of the copolymer against reciprocal temperature. Apparent viscosity decreases with increasing temperature with the apparent activation energy of ca. 12 kcal/mole.

In addition, the copolymer is thermally stable as is represented by the thermal decomposition temperature of 320°C measured by thermogravimetry as shown in Figure 5.

Accordingly, the copolymer can be molded by press or extrusion method into films of arbitrary thickness under proper conditions.

After fabrication, the ester type films are hydrolyzed in a caustic solution to be converted to carboxylic acid type membranes. Figure 6 shows the change of infrared spectrum by hydrolysis of the ester type film in 25 wt% caustic solution at 90°C for 16 hrs. Complete hydrolysis is indicated by the fact that the absorption at 1780 cm^{-1} due to $\nu_c=0$ (-COOCH_3) is wholly shifted to 1680 cm^{-1} of $\nu_c=0$ (-COONa).

Fundamental Properties

Essential properties are described mainly for the sodium type membrane made of copolymers of tetrafluoroethylene and carboxylated vinyl ether (M_1), soaked in caustic solutions. The content of carboxylated vinyl ether in the copolymer determines ion exchange capacity of the resulting membrane, which is expressed as milliequivalent of carboxylic acid group per gram weight of dry sodium type membrane.

Physical properties
tensile properties   Sodium type membranes behave quite differently from ester type ones under tensile stress. In Figure 7, stress-strain curves are compared between both types of membranes. By incorporation of ions into the membrane, significant increase of tensile modulus and decrease of elongation, which are often acknowledged in high crosslinking of polymers, are brought about. Such a change in nature is due to the structural change in the membrane, which is suggested by the advant of a new peak in the small angle X-ray scattering pattern of the sodium type membrane as shown in Figure 8. It appears that ions form some kind of aggregations, which presumably act as transient crosslinks.

The presence of ions imparts specific features on the mechanical properties of the membrane; being relatively little dependent upon temperature and being affected by the concentration of caustic solution in which it is soaked. The sodium type membrane exhibits higher tensile strength than the ester type one. Its strength is

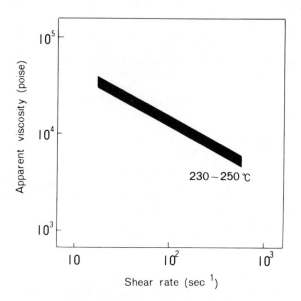

Figure 3. Melt flow behavior of polymer.

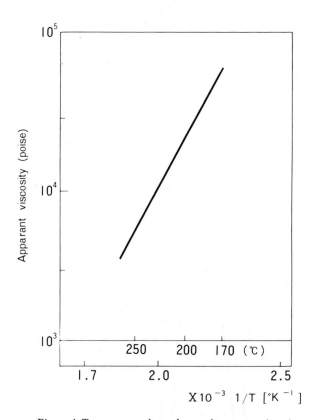

Figure 4. Temperature dependence of apparent viscosity.

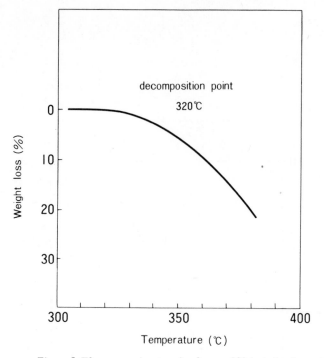

Figure 5. Thermogravimetry of polymer, 10°/min in air.

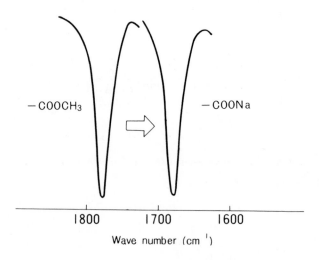

Figure 6. Characteristic absorption in IR spectra.

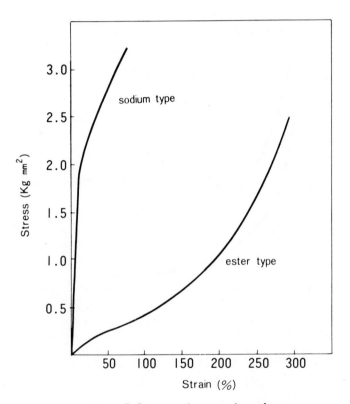

Figure 7. Stress–strain curve of membrane.

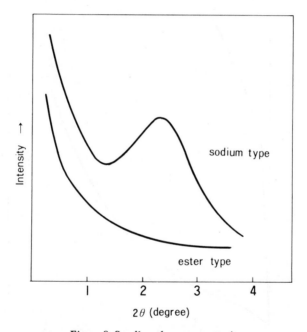

Figure 8. Small angle x-ray scattering.

retained even at high temperature of 90°C as shown in Table I, while ester type membrane loses its strength drastically with the rise of temperature.

Tensile strength of sodium type membrane in caustic solutions of various concentration are exhibited in Figure 9. Tensile strength increases with the increase of caustic concentration, which corresponds to the decrease of water content of the membrane. The relationship between water content of the membrane and caustic concentration at which it is soaked is shown in the following section.

Creep behavior of the sodium type membrane is given in Figure 10, where deformation is seen to be highly dependent upon caustic concentration. Such a phenomenon also reflects the effect of water content of the membrane.

    water content   The most specific feature of perfluorocarboxylated membranes is the low water uptake in caustic solution of high concentration.

Figure 11 shows water content of the membrane which is expressed as mole number per unit equivalent of carboxylic acid. It is to be noted that the variation of water content with the change in ion exchange capacity is substantially small.

As a result, the perfluorocarboxylated membrane with a high ion exchange capacity attains a high fixed ion concentration, which is defined as milliequivalent of carboxylic acid group per a gram of absorbed water in the membrane.

Water content of the membrane is also dependent upon the concentration of caustic solution in which it is soaked, as is shown in Figure 12.

From these water contents, fixed ion concentration is calculated and plotted in Figure 13 against caustic concentration.

Such a high fixed ion concentration is quite effective to prevent migration of anions into the membrane, which leads to high permselectivity in ionic transport.

    solvent swelling   The sodium type membrane is remarkably swollen by organic polar solvents as shown in Table II while the ester type membrane is quite inactive.

Swelling of the sodium type membrane by solvents is attributed to the presence of ion pairs ($-COO^-Na^+$) in the membrane.

    material transfer rate   Owing to the low water content, the perfluorocarboxylated membrane in aqueous solution exhibits significantly low transfer rate for both water and solutes.

Figure 14 shows osmotic transfer rate of water through the membrane where a NaCl solution is separated from a caustic solution by the membrane.

Low leakage coefficient of NaCl and NaOH are also shown in Figure 15 and in Figure 16, respectively.

Table I   Tensile strength of the membrane

temperature (℃)	tensile strength (Kg/mm²)	
	sodium type	ester type
25	3.2	2.5
50	2.5	0.4
90	2.3	0.07

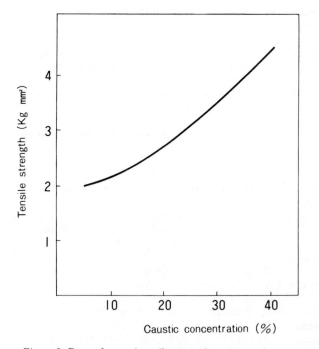

Figure 9. Dependence of tensile strength upon caustic concentration.

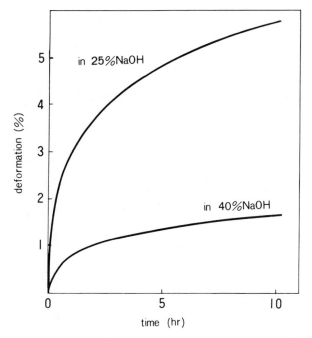

Figure 10. Creep behavior of membrane. Conditions: 0.33 kg/mm² load at 90° C.

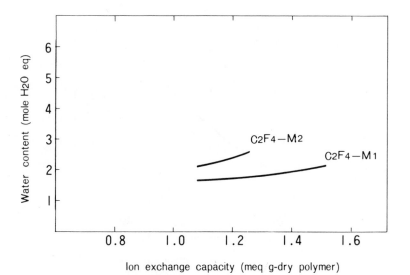

Figure 11. Dependence of H₂O content upon ion exchange capacity of membrane. Conditions : 35 wt% NaOH at 90° C.

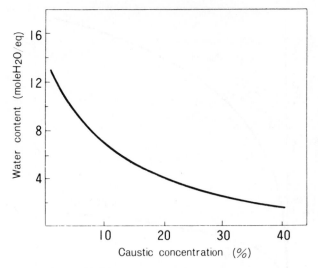

Figure 12. Water content and caustic concentration.

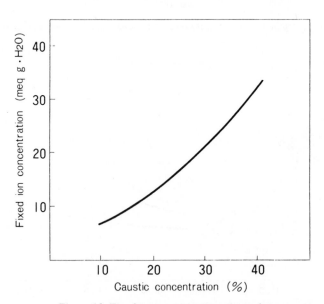

Figure 13. Fixed ion concentration of membrane.

Table II   Solvent swelling
after one day at 25℃

immersing solvent	swelling (%)	
	sodium type	ester type
methanol	800	1
ethanol	15	
propanol	7	
ethylenglycol	80	
propyleneglycol	130	
aceton	10	11

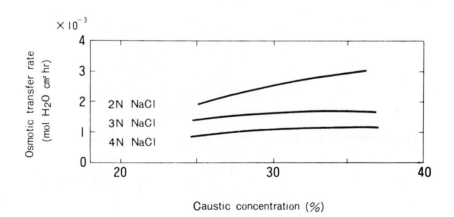

Caustic concentration (%)

Temperature : 85 ℃

Figure 14. Dependence of osmotic transfer rate upon caustic concentration.

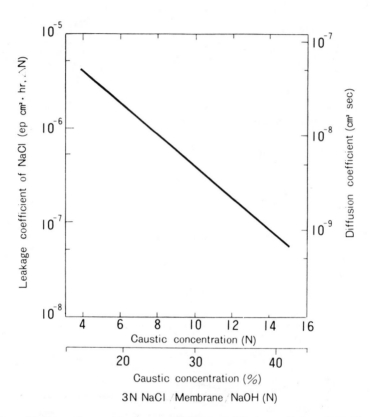

Figure 15. Dependence of leakage coefficient and diffusion cofficient of NaCl upon caustic concentration.

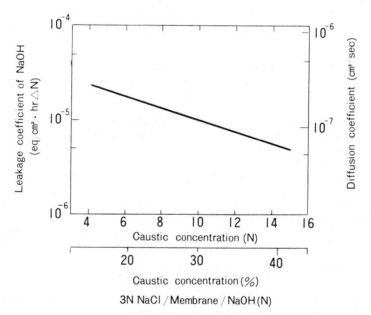

Figure 16. Dependence of leakage coefficient and diffusion coefficient of NaOH upon caustic concentration.

Electrochemical properties

The characteristics of the membrane renders much influence on the state of the mobile ions within it, which determines electrochemical property of the membrane. Electrochemical properties of the membrane relating to chlor-alkali process are as follows.

conductivity   Conductivity of the sodium type membrane is shown against ion exchange capacity in Figure 17. Irrespective of the low water content of the membrane, high conductivity required for electrochemical application is attained by choosing the high ion exchange capacity.

Figure 18 shows the influence of caustic concentration upon conductivity of the membrane. The decrease of conductivity with the increase in caustic concentration is ascribed to the decrease in mobility of sodium ions caused by the dehydration of the membrane. The increase of apparent activation energy for ionic conductance along with caustic concentration as is given in Table III reflects the existance of increasing interaction between sodium ion and the fixed ion in the membrane.

transport number   Transport number obtained through the measurement of membrane potential is directly related to the permselectivity in ionic transport.

Figure 19 shows the dependence of the transport number of the sodium ion upon the caustic concentration. High value of more than 0.9 is achieved at caustic concentrations beyond 25 wt%. This characteristic behavior is explained by the high fixed ion concentration within the membrane.

Thus, high permselectivity combined with high conductivity is the outstanding feature of the perfluorocarboxylate membrane.

Application of Flemion in chlor-alkali process

The Flemion membrane was applied for the use in the electrolysis of sodium chloride solution. In Fig. 20, electrolytic performance of the membranes having different ion exchange capacity (AR, meq/g) of 1.44 and 1.23 are shown against the concentration of caustic soda produced in the cathod chamber.

Current efficiency is dependent upon the caustic concentration and exhibits a maximum value at a certain concentration, which shifts to the higher concentration region with increasing the ion exchange capacity of the membrane. Current efficiency as high as 95% was obtained in high caustic concentrations. Although cell voltage increases with the caustic concentration, the membrane of higher ion exchange capacity give sufficiently low value even in strong caustic soda. Accordingly, by using the membrane of 1.44 meq/g ion exchange capacity, caustic soda of as high as 35 ~ 40% concentration is advantageously produced with low electric power consumption. In addition, the leakage of sodium chloride was proved to lie in significantly low level of less than 50 ppm.

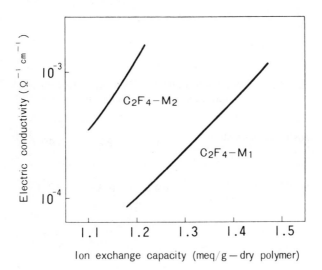

Figure 17. Conductivity of membrane against ion exchange capacity. Conditions: 35 wt% NaOH at 90° C.

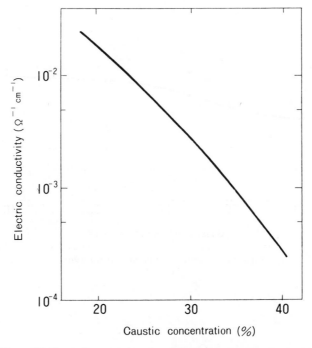

Figure 18. Dependence of electric conductivity upon caustic concentration.

Table III   Apparent activation energy for ionic
            conductance in the membrane

Caustic concentration (%)	15	25	40
Apparent activation energy (kcal/mole)	3.9	9.1	17.1

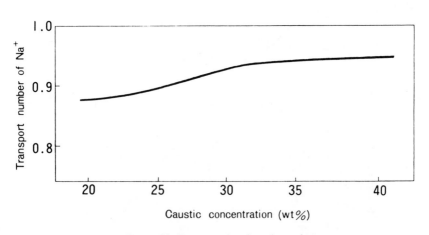

Figure 19. Transport number of membrane.

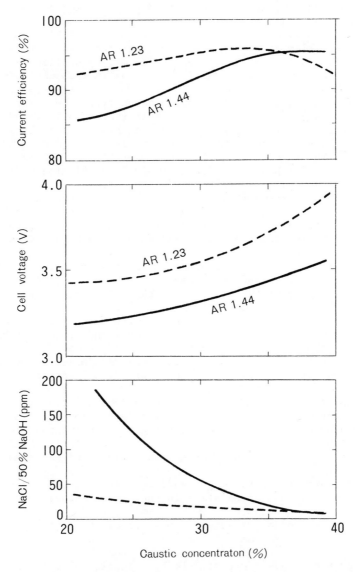

Figure 20. Electrolytic performance of membrane. Conditions: brine concentration, 3.5 N; current density, 20 A/dm²; and 90° C.

The membrane has also demonstrated continuously the current efficiency of 95% and the cell voltage of 3.45 volt for more than a thousand days. Thus, the membrane having the ion exchange capacity of 1.44 meq/g was selected as a standard one to produce strong caustic soda of 35 ~ 40 wt%, being named the Flemion 230.

Another Flemion which is suitable for the case where a caustic solution as low as 20% concentration may be utilized directly from an electrolysis plant on site, such as in the pulp industry, named Flemion 430, was also developed. This has an asymmetric structure, that is, the ion exchange capacity of the membrane surface facing a cathode is made lower than that of bulk membrane. Such a structure of the membrane leads to high current efficiency with a catholyte of 20% caustic soda and low ohmic drop of the membrane. The electrolytic performance of Flemion 430 is shown in Fig. 21, along with that of Standard Flemion. Flemion 430 can consume lesser energy than Standard Flemion 230.

In the electrolysis of potassium chloride, the performance of the membrane is somewhat different from that in the electrolysis of sodium chloride, due to the difference in nature between sodium ion and potassium ion. A noticable aspect is a high diffusibility of potasium chloride through the membrane. However, as shown in Fig. 22, with Flemion, the leakage of potasium chloride is suppressed efficiently, and decreases with the lowering of the ion exchange capacity of the membrane. Taking into consideration of the balance between cell voltage and the quality of caustic, we have selected the membrane with an ion exchange capacity of 1.34 meq/g for the production of caustic potash, and named it Flemion 330. When 35% of caustic potash is produced, a caustic potash containing less than 50 ppm of potassium chloride based on 50% caustic is obtained with a current efficiency of 97% and at a cell voltage of 3.4 V.

We have also accomplished the design and construction of an electrolyzer which is to work most efficiently for the membrane. Our membrane chlor alkali process using Flemion and the electrolyzer is named as the Flemion process. Two commercial plants are in operation in Japan, and another one in Thailand has also started up.

Responding to increasing demand for Flemion, we have expanded the membrane manufacturing plant in the Chiba factory in the summer of 1981.

Moreover, we contrived an innovative electrolytic system named AZEC, which will be commercialized in near future.

The Flemion process is an economic process with no environmental problem, and is expected to be a leading process in the chlor-alkali industry.

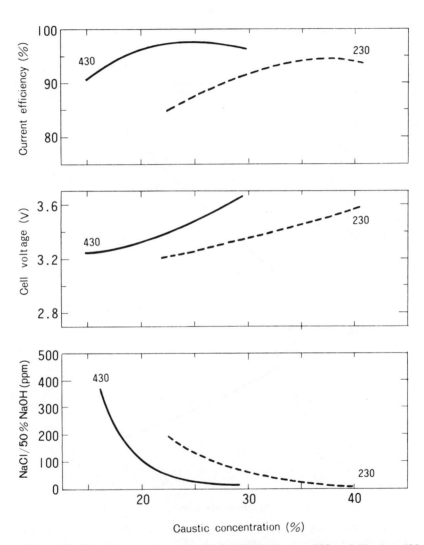

Figure 21. Electrolytic performance of Standard Flemion 230 and Flemion 430. Conditions: brine concentration, 3.5 N; current density, 20 A/dm²; and 90° C.

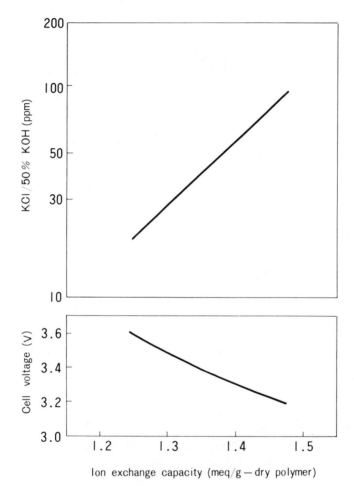

Figure 22. Relationship between ion exchange capacity of Flemion and electrolytic performance. Conditions: concentration of KOH, 35 wt%; current density, 20 A/dm²; and 90° C.

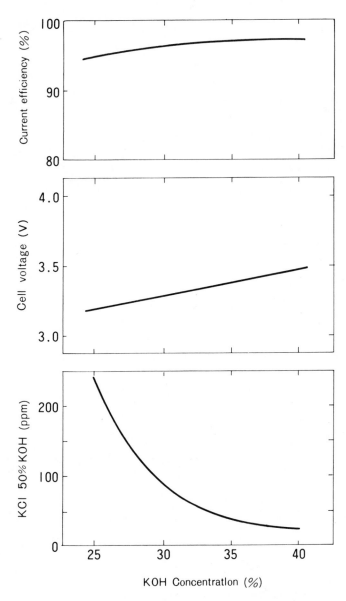

Figure 23. Electrolytic performance of Flemion 330 for caustic potash. Conditions: current density, 20 A/dm²; and 90° C.

RECEIVED September 8, 1981.

Applications of Perfluorosulfonated Polymer Membranes in Fuel Cells, Electrolyzers, and Load Leveling Devices

RICHARD S. YEO

The Continental Group, Incorporated, Energy Systems Laboratory, 10432 North Tantau Avenue, Cupertino, CA 95014

The basic components of an electrochemical cell are a pair of electrodes and the electrolyte. A cell separator is often used to inhibit direct physical mixing of reactant/product for better cell performance. The separator could be either a microporous diaphragm or an ion exchange membrane. Scientists at General Electric had the great idea that ion exchange membranes could serve as both the separator and the sole electrolyte for electrochemical cells, especially fuel cells (1). Extensive work by Grubb (2) and by others (3,4) has proved the idea to be a sound one. The subject has been reviewed by Niedrach and Grubb (5) and by Maget (6).

During the last two decades, tremendous work has been conducted on properties of ion exchange membranes, their behavior under various environmental conditions, and their interaction with electrodes as well as in determining the limiting factors influencing operation and cell life.

Nature of Ion Exchange Membrane

Ion exchange membranes include in their polymeric structure many ionizable groups. One ionic component of these groups is fixed into or retained by the polymeric matrix while the other ionic component is a mobile, replaceable ion which is electrostatically associated with the fixed component. The ability of the mobile ion to be replaced under appropriate conditions by the other ions imparts ion exchange characteristics to these materials.

An ion exchange membrane can be either the cation exchange or the anion exchange type. The use of the cation exchange type for electrochemical cells is far more important than that of the anion exchange type because of its better thermal stability (7).

Ion exchange membranes are outstanding separators for use in electrochemical cells since the membranes are permeable to one kind of ion while resisting the passage of direct flow of liquids and ions of opposite charge; the membranes are

0097-6156/82/0180-0453$05.25/0

self-supporting and can be reinforced to produce membranes having
high mechanical strength, and the membranes can be prepared as
thin sheets of large area which are necessary for favorable cell
geometry. In practice, the thickness of the membranes is preferably as small as possible, for example, from about 0.05 to 1 mm.

Membrane Properties with Regard to the Electrochemical Applications

The essential properties of the membrane for the
electrochemical applications are the following:
   a) adequate chemical and electrochemical stability in
       the cell operating environment;
   b) good mechanical integrity and structural strength to
       ensure dimensional stability under tension;
   c) surface properties compatible with the bonding of
       catalytic electrodes to the membrane;
   d) high water transport to maintain nearly uniform water
       content or to prevent localized drying;
   e) low (preferably zero) permeability for reactant/pro-
       ducts to achieve high current efficiency;
   f) high ionic conductivity for large current densities
       and low internal resistances.
In many cases, a compromise between these properties will
be required because specific properties may be emphasized
according to the application requirements.

Perfluorinated Membranes: Advantages and Disadvantages

The important feature of perfluorinated materials for electrochemical application is their excellent or, perhaps, improved, chemical inertness and mechanical integrity in a
corrosive and oxidative environment (8,9). In contrast, the
hydrocarbon type material is unstable in this environment, due
to the cleavage of the carbon-hydrogen bonds, particularly the
α-hydrogen atom where the functional group is attached (9,10).
The thermal stability of perfluorinated material is excellent as evidenced by the higher glass transition temperature
over their respective non-fluorinated analogues (11). Accordingly, these perfluorinated materials can be used in electrochemical cells at an elevated temperature for better cell
efficiency because of high conductivity and fast kinetics (12).
The relatively high cost of the perfluorinated membrane
limits its application in many electrochemical cells when cost-
effectiveness is a major concern.
Nafion (perfluorosulfonic acid) membranes are currently
used in cells with a corrosive environment and high temperature.
Many of these cells are designed with the solid polymer
electrolyte (SPE) configuration. The merits of the solid polymer
electrolyte technology will be discussed in the next section.

Solid Polymer Electrolyte Technology

The development of solid polymer electrolyte cells is being
actively conducted at General Electric Co. (13) and at Brown
Boveri Research Center, Baden, Switzerland (14). As the name
implies, the solid polymer electrolyte technology uses a solid
polymer sheet as the sole electrolyte in the cells. It also
acts as the cell separator. The majority of the present applications use Nafion with a thickness of 10-12 mils (13). Selected physical and chemical properties of Nafion 120 membranes are
given in Table I. The membrane is equilibrated in water to
approximately 30% water content prior to fabrication into a cell
assembly. The hydrated membrane is highly conductive to
hydrogen ions. It has excellent mechanical strength, and it is
very stable in many corrosive cell environments.

Table I

Physical and Chemical Properties of Nafion-120 Membranes at 25°C

Equivalent weight (EW)	1200
Ion exchange capacity	0.83 meq/g dry polymer
Ionic (H_3O^+) resistance	0.46 Ω cm^2
Tensile at break	2500 psi
Elongation at break	150%
Mullen burst strength	150 psi
H_2 permeability	5.6 x 10^{-4} cm^3 cm cm^{-2} hr atm
O_2 permeability	3.0 x 10^{-4} cm^3 cm cm^{-2} hr atm
Hydrodynamic H_2O permeability	2.7 x 10^{-5} cm^3 $cm\, cm^{-2}$ hr atm
Electro-osmotic permeability	7.5 x 10^{-4} cm^3 C^{-1}

Cells usually have a bipolar configuration. The electrocatalysts are bonded to each side of the membrane (15), and the
resulting SPE is a structurally stable membrane-electrode
assembly as shown in Figure 1. A multi-layer package of
expanded metal screens which presses up against the electrode on
one side serves as the current collector and fluid distributor.
    Although water is the only liquid used, the environment is
essentially highly acidic (14,16,17). This is because the electrodes are in contact with the strongly acidic groups at the
membrane surface (14,17).
    Method of Fabrication. The electrocatalyst, in the form of
fine powder (15), can be produced by the Adams method (18). It
is first mixed with PTFE emulsion solution such that the
catalyst/PTFE ratio is 80:20 (19). The mixture is placed on a
metal foil and sintered at 345°C for more than one hour (20).
After cooling, the Teflon-bonded catalyst is transferred from
the foil to the membrane. The sintering process can be

porous sublayer   impervious interior

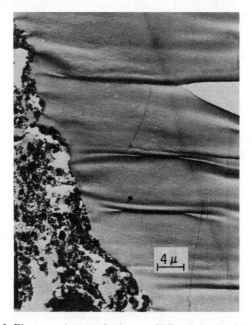

Figure 1. Electron micrograph of a new SPE cell, RuOˣ/SPE Interface (14).

eliminated when the electrodes do not require wet-proofing
treatment. (20).

The catalyst can be bonded to the membrane surfaces by many
different methods such as vacuum deposition (21). However, the
performance of the membrane-electrode assembly produced by the
vacuum deposition method is poor because a smooth metallic layer
is formed while electrode material for an electrochemical cell
should be rough (high surface area).

The catalysts can also be bonded onto each face of the
membrane under pressure and at a temperature (22) usually between
the glass transition temperature and the thermal degradation
temperature of the membrane (17,23,24). At such temperatures the
membrane softens and can flow under pressure, such that the
adhesion force of the membrane is at a maximum, and an intimate
contact between the catalyst and the membrane can be achieved
(17). The heating process is rather short, so that the membrane
is not over-dehydrated. A dehydrated membrane gives poor bonding
(17).

Advantages
a)   High cell efficiency: The SPE cell is unique in that
     the electrocatalysts pressed onto the membrane are in
     the form of fine particles exhibiting extremely high
     surface area (e.g., 200 m^2/g). The interelectrode spac-
     ing is very narrow and is approximately equal to the
     membrane thickness, which is generally not more than
     0.3 mm. The maximization of active surface area of the
     the electrodes in couple with the minimization of in-
     terelectrode spacing allows operation of high current
     density with low ohmic losses. This is particularly
     true when Nafion is used as the SPE because of its good
     conductivity.
b)   Simplicity of electrolyte: In the SPE cell, the reac-
     tant fluid need not be conductive to perform electro-
     chemical synthesis; thus, water and other non-ionic
     species and sparingly soluble gases can be employed as
     reactants without the requirement for a supporting
     electrolyte. Furthermore, since pure water is the only
     liquid in the system and the electrolyte is a solid
     polymer, there is no need for any electrolyte
     conditioning or normality controls. Also, it is not
     necessary to include electrolyte scrubbers in the
     product gas streams.
c)   Flexibility of operation: SPE cells can be operated
     over a wide range of cell environments. The cell can be
     operated below the normal freezing point of aqueous
     electrolytes and up to 150°C. The cells are very amen-
     able to high pressure operation (up to 3000 psig), pro-
     viding an additional process variable.

Types of Applications

Applications in Hydrogen-Oxygen Fuel Cells. A fuel cell is a device which converts the latent chemical energy of fuel directly into electricity. This involves a constant temperature electrochemical energy conversion process, and its efficiency is not limited by Carnot's theorem (25). Fuel cells find many applications in space missions and military power sources. Most recently, it is considered as an ideal contender for the uses in transportation and utility sectors. For further details see reference 25.

The perfluorosulfonic acid (Nafion) membrane found its application in fuel cells long before its introduction to the chlor-alkali industry (26-28). The Nafion membrane is used as the solid polymer electrolyte (separator/electrolyte) in fuel cells. Figure 2 shows the schematic of such an SPE fuel cell.

During the fuel cell reaction, the electrochemical reaction taking place at the cathode of an SPE cell is

$$O_2 + 4 \ H^+ + 4e^- \rightarrow 2 \ H_2O \qquad [1]$$

and at the anode,

$$H_2 \rightarrow 2 \ H^+ + 2e^-$$
$$[2]$$

the overall reaction being

$$H_2 + 1/2 \ O_2 \rightarrow H_2O + \text{Electrical Power} + \text{Heat} \qquad [3]$$

Charge carriers in the hydrated Nafion membrane are hydroxonium ions ($H^+ \cdot x \ H_2O$) which migrate through the membrane by passing from one sulfonic acid group to the adjacent one. The sulfonic acid groups are chemically bound to the perfluorocarbon backbone and do not move; thus, the concentration of hydrated ions remains constant within the membrane.

A small amount of water, coupled with the hydrogen ions, transports across the membrane because of the electro-osmotic effect. At 100°C, for example, there are about 3.5 to 4.0 water molecules transported with each hydrogen ion (13).

With near ambient pressure air as oxidant, the optimum temperature of operation is about 75°C. Because the air (and fuel) streams must be presaturated to ensure membrane stability, higher temperatures cause excessive dilution of the oxygen in the air stream by water vapor. When oxygen is employed, operational temperatures up to 150°C are possible.

For certain applications it is desirable to add unbound supporting electrolyte to the membrane. This can be accomplished by equilibrating the membrane in a suitable aqueous solution of acid or base. Prior to assembly into a cell, the

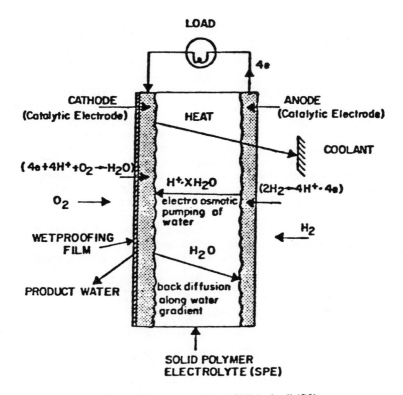

Figure 2. Representation of SPE fuel cell (26).

membrane can be blotted dry and no free flowing fluid need be present in the cell. In other words, the membrane can be considered as a gel-like matrix for the unbound electrolyte. The performance of the membrane so treated can improve because of better conductivity. Also, the performance of the electrodes may also improve. However, during continuous operation, provision must be made to remove the product water or to continuously replace the unbound electrolyte in the membrane because such supporting electrolyte, being unbound, will be leached away by rejected water.

Applications in Water Electrolyzers. With the technology achieved in the Gemini hydrogen-oxygen fuel cell, General Electric sought to carry out the reverse reaction, i.e., water electrolysis, in a cell of similar design (29-32). A Nafion membrane with a thickness of 10-12 mils and EW of 1100 or 1200, is the only membrane currently used in these electrolyzers. At present, small-scale SPE water electrolyzers are commercially available for providing oxygen for space life-support systems in spacecrafts and submarines or as hydrogen generators for laboratory users (e.g., for gas chromatographs). More recently, kilowatt-range cells are being developed at General Electric for large-scale hydrogen production with off-peak electricity either for energy storage/transmission use or for chemical/metallurgical processes. For more details, see reference 30.

Figure 3 represents the schematic of two SPE single cells connected in series (M=membrane, E=electrode). The electrochemical reaction taking place at the anode of an SPE cell is

$$6 \; H_2O \; (1) \rightarrow 4 \; H_3O^+ + 4e^- + O_2(g)$$

and at the cathode, the hydroxonium ions are discharged to produce hydrogen gas as

$$4 \; H_3O^+ + 4e^- \rightarrow 4 \; H_2O \; (1) + 2 \; H_2(g)$$

the overall reaction being

$$H_2O \rightarrow H_2 + 1/2 \; O_2$$

Again, the charge carriers in the membrane, similar to the case of the fuel cell, are hydroxonium ions. The membrane serves as the electrolyte as well as the separator. While operating an SPE water electrolyzer, pure water is circulated at a sufficiently high flow rate (to remove the waste heat) over the anode where it is decomposed electrochemically, producing oxygen gas, hydroxonium ions, and electrons. The hydroxonium ions move through the membrane and then recombine with electrons, which pass via the external circuit, to form the hydrogen

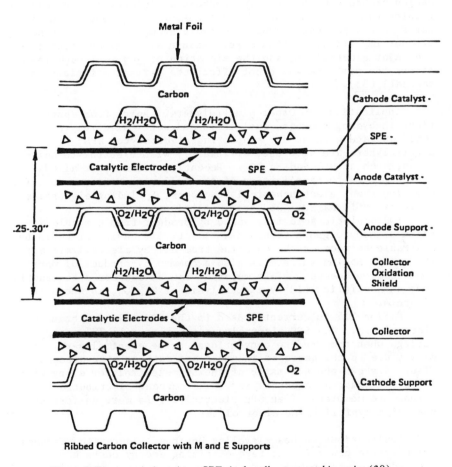

Figure 3. Representation of two SPE single cells connected in series (30).

gas at the cathode. H_2 and O_2 gases are generated at a
stoichiometric ratio at any desired pressure.

A recent study (12) has shown that Nafion is also suitable
for use in water electrolyzers with alkaline solution as the
supporting electrolyte. The major charge carrier is the alkali
metal ion because of the negligible H^+ ion concentration in
alkaline solution and the OH^- ion rejection capability of the
cation exchange membrane. The current efficiency of the cell is
related to the inhibition of the transport of gaseous products
across the separator. Thus, the ionic groups of the membrane
are not important, in this case, because alkaline solution is
the major electrolyte, and the migration of any ionic species
across the membrane would not affect the current efficiency of
the cell (33).

Applications in Other Electrolyzers. Apart from applications in hydrogen-oxygen fuel cells and water electrolyzers
which operate without any supporting electrolyte, SPE
electrolyzers are also used efficiently with electrolyte solutions, such as HCl and Na_2SO_4. Recently, LaConti, et al (34),
have reported the application of the cell with Nafion as SPE for
some important electrochemical processes, including electrolysis
of water, HCl, Na_2SO_4, and brine solution.

Hydrochloric acid is a waste by-product from the chlorin-
ation of many organic compounds. In certain cases, it would be
desirable to recover the chlorine from HCl by electrolysis of
the acid. Sodium sulfate is also a common by-product of many
commercial processes. This material can be converted to caustic
soda and sulfuric acid by electrolysis of Na_2SO_4, as shown in
Figure 4.

Nafion-315 is currently used in the SPE cell for brine
electrolysis. The SPE electrolyzer exhibits a 15-20% energy
savings when compared to conventional brine electrolyzers, primarily due to the decrease in ohmic and cathode overvoltages.
Figure 5 shows the schematic of the SPE electrolyzer along with
a typical membrane electrolyzer. The current distribution
across the membrane of an SPE electrolyzer is more uniform than
that of a typical brine electrolyzer.

Applications in Load-leveling Devices. Increasing demands
for electric power in the face of rising energy costs have
created an acute problem for the electric utility industry in
meeting the intensive peak power demands of industry. Leveling
or peak-shaving these loads and using off-peak stored energy
allow more efficient utilization of base-load energy, with a
consequent significant savings in fossil fuels as required for
base-load power generation. Electrochemical cells are one of
the ideal energy storers because of their high efficiency. The
secondary batteries and regenerative fuel cells that have been
considered for the load leveling/peak-shaving applications are

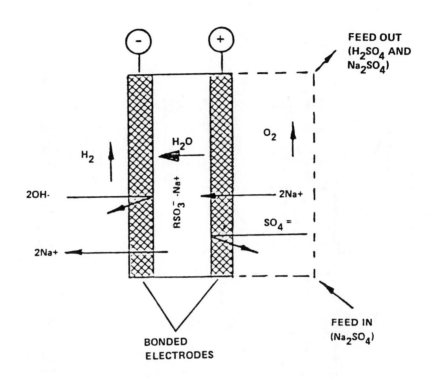

Figure 4. Representation of SPE Na_2SO_4 electrolysis cell (34).

$\underline{Na_2SO_4}$ **(SPE ELECTROLYSIS CELL)**

ANODE: $Na_2SO_4 + H_2O = 2\,Na^+ + 2H^+ + SO_4^= + 1/2\,O_2 + 2e$

CATHODE: $2\,H_2O + 2e = 2\,OH{\text -} + H_2$

OVERALL: $Na_2SO_4 + 3\,H_2O = 2\,NaOH + 1/2\,O_2 + H_2 + H_2SO_4$

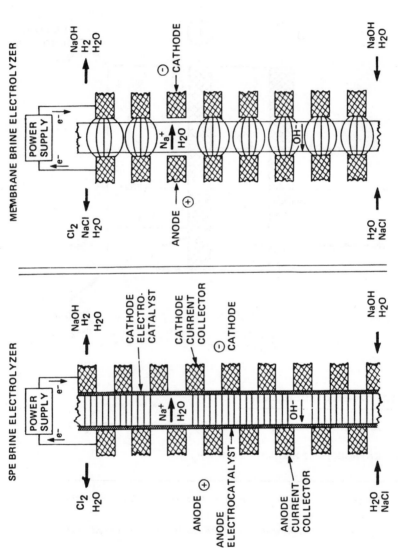

Figure 5. Representation of SPE and typical membrane electrolyzer (34).

the sodium-sulfur battery (35), the lithium-metal sulfide
battery (36), the zinc-chlorine battery (37), the zinc-bromine
battery (38-41), the hydrogen-chlorine cell (42-49), the
hydrogen-bromine cell (50), the iron-chromine redox cell
(51,52), the iron-ferric redox cell (53,54), and the
zinc-ferrocyanide redox cell (55).

The Nafion membrane has been used as a separator in both
the hydrogen-halogen cell (45,50) and in the zinc-bromine
cell (38) because Nafion is a highly stable perfluorinated
material which is not affected by strong acids and halogen.

The function of the separator in these batteries, similar
to that of the water electrolyzer, is the separation of products
(like hydrogen, chlorine, bromine, and zinc) which cause self-
discharge and efficiency loss when they diffuse across the separator. A non-ionic membrane can be used since the transport of
the ionic species, like proton, chloride, bromide, and zinc
ions, do not affect the cell efficiency.

Bromine can complex with bromide ions (56)

$$Br_2 + Br^- \rightleftharpoons Br_3^- \qquad K \simeq 16$$

$$2Br_2 + Br^- \rightleftharpoons Br_5^- \qquad K \simeq 40$$

whereas for chlorine in hydrochloric acid (57), it is

$$Cl_2 + Cl^- \rightleftharpoons Cl_3^- \qquad K = 0.2$$

The use of Nafion is advantageous since the negatively
charged membrane inhibits negative ion migration. Hence,
halogen permeation in Nafion is lower than in the case of the
micro-porous separator (45).

Important Membrane Properties Determining Cell Performance

Cell efficiency is very often related to the transport
properties of the membrane. This is particularly true for those
load-leveling devices in which the kinetics of the cell reactions are fast. The ohmic overvoltage and current efficiency
are largely determined by the membrane conductivity and reactant
permeation, respectively.

Water Content. The transport properties of the membrane
are strongly influenced by the water (or electrolyte) content of
the membrane (12,38,45,50). As discussed in Chapter 1 of this
volume, the water content of the membrane increases with increasing temperature and with decreasing EW. Figure 6 shows the
performance of the hydrogen-chlorine regenerative cell at various temperatures. Besides, the water content can be influenced

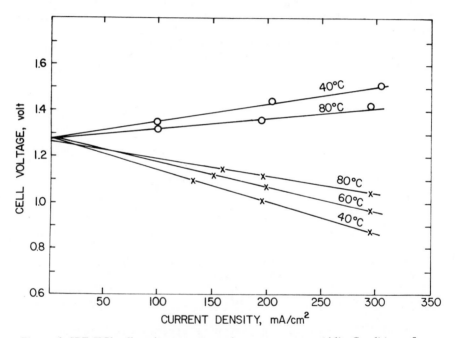

Figure 6. SPE HCl cell performance at various temperatures (46). Conditions: 5 atm Cl pressure; flow rate, 8 cm³/s; and 10% HCl.

by the type and concentration of the electrolyte. The history
of pre-treatment of the membrane is also important in increasing
the water content. The water content is set prior to cell
fabrication by equilibrating the membrane at high temperatures
and pressures. Once equilibrated at a temperature above the
operating temperature of the cell, the water content of the
membrane is fixed and will not change as a function of any
normal operating condition. The only condition that can
change the water content of the membrane in a cell is the drying
force which could remove water by evaporation. To eliminate
such a condition the heat is removed at the electrode, aiding
the recycle of product water formed at the cathode into the
polymer and by presaturation of the inlet reactants to prevent
localized drying of the cell. The incorporation of prehumidi-
fied reactants has eliminated the cell failures experienced in
some of the early ion exchange membrane fuel cells.

Thickness. The present Nafion membrane used in SPE water
electrolyzers is 12 mils thick and has an electrolytic resistance of 2.4×10^{-4} ohm-ft^2 at 80°C, accounting for considerable
voltage losses at high current densities. Considerable voltage
reductions can be achieved through the use of the 5 mil Nafion
membranes. For instance, at 2000 A/ft^2, the improvement in performance compared with the standard 12 mil membrane is more than
0.25 V. With thinner membranes, parasitic losses due to gas
permeability will generally be higher; thus, a careful trade-off
study must be made for a given application.

Membrane Hysteresis Effect. The diffusion of water in the
Nafion membrane is fast (only one order of magnitude less than
the self-diffusion of water). The diffusion of electrolyte in
the membrane decreases as the acid concentration increases. So
the electrolyte content of the membrane will not reach equilibrium when the cell is operated in an unsteady-state fashion,
as shown in Figure 7. The electrolyte content in the membrane
is less than the equilibrium value when the cell is initially
started with 45% HBr and then charged to 7% HBr in 10 hours. On
the other hand, the membrane would absorb more electrolyte than
the equilibrium value if the cell were started with 7% HBr and
discharged to 45% HBr in 10 hours. There is a strong hysteresis
effect (50). This effect increases with decreasing cell cycle
time and an increasing acid concentration range.
     The overvoltage of the hydrogen-halogen cell and the zinc-
bromine cell is mainly contributed by the membrane resistivity,
and this resistivity is a strong function of the electrolyte
content in the membrane. It is clear that the membrane
hysteresis effect will influence the cell performance.

Chemical Stability. Active species, such as peroxide
radicals, form during cell reaction, particularly oxygen

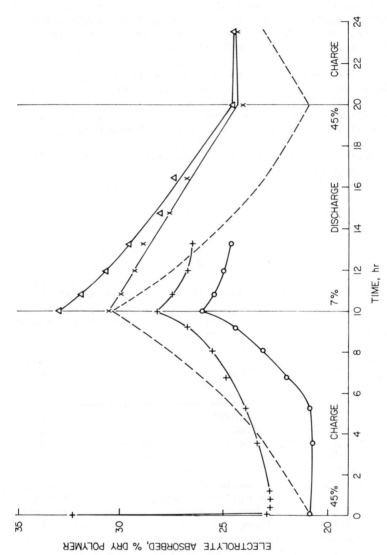

Figure 7. The electrolyte sorption of Nafion during the charge and discharge cycles of HBr cell under various initial conditions (50). Key: $+$, *presoaked in* H_2O, *operation started with charge;* $\bigcirc$, *pre-soaked in 45% HBr, started with charge;* $\triangle$, *presoaked in* H_2O, *started with discharge;* $\times$, *pre-soaked in 7% HBr, started with discharge;* $--$, *steady-state operation.*

reduction.   These peroxyl species appear to have a fairly long
lifetime and readily attack C–H bonds of the membrane and cause
membrane degradation.   The degradation rate has been found to be
aggravated by temperature and presence of certain ions such as
ferrous ions.   The Nafion membrane exhibits excellent resistance
against degradation in environments with 20 ppm Fe^{++} and
30% H_2O_2.   However, Nafion does undergo some degradation
resulting in the formation of HF, CO_2, and low molecular weight
perfluorocarbon species (13).   Figure 8 shows the relative degradation rate of early Nafion material (before 1969) and of an
improved membrane.   The chemical stability of the membrane can
be improved by platinizing the membrane on the surface layer
such that it forms a discontinuous layer within the membrane.
The platinization of the membrane can increase membrane life by
an order of magnitude.

Future Trends

   Development of Alternative Membranes.   Nafion membranes
show considerable promise with respect to their performance
characteristics, low resistivity and long-term stability.   However, the present cost (about $30/ft^2) of Nafion membranes is
rather expensive for the SPE cell to be cost-effective for
industrial and utility applications.   The ultimate goal is to
reduce the membrane cost to about $3/ft^2.   Alternative membranes
have been evaluated for use in the SPE cell.   Only fluorocarbon
membranes have been considered in this application (10), since
the cell environment is corrosive, and cell tempperature is high
(from 80°C to 150°C).   α, β, β - trifluorostyrene-sulfonic acid
membrane has shown excellent performance in SPE water
electrolyzer comparative with that used in the Nafion membrane.
However, the candidate membranes lack long-term stability under
water electrolysis operating conditions, presumably due to the
non-perfluorinated nature of the aromatic ring.
   A recent study has shown that (58) membranes with carboxylic acid groups cannot be used without alkaline supporting electrolyte because of low swelling and poor conductivity in pure
water.   It is clear that the alternate membrane for the SPE cell
should be a perfluorinated sulfonic acid membrane.   To date
(1981), Nafion still retains its uniqueness for this electrochemical application.
   The hydrated Nafion membrane currently used in SPE cells
provides a highly acidic environment, equivalent to a 10 wt%
H_2SO_4 solution (13).   Thus, noble metals or noble metal oxides
have to be used as electrocatalysts.   It is of particular
interest to develop an anion exchange membrane which will transport hydroxyl ions under water electrolysis or fuel cell conditions.   This implies that the cell environment would be alkaline, which would enable the substitution of these expensive

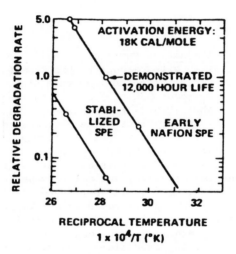

Figure 8. Degradation rates of perfluorinated sulfonic acid membrane (13).

materials with relatively cheap ones. Development of the anion exchange membranes are rather limited by their poor thermal stability.

Application of SPE Technology in Chlor-alkali Cells. The ohmic loss of the chlor-alkali cell can be substantially reduced by the use of the SPE configuration. However, the SPE brine electrolyzer currently employs Nafion-315 as the separator. The capability of this sulfonic acid membrane to reject hydroxyl ions at high caustic concentrations is poorer than that of those membranes consisting of weak acid groups such as perfluorocarboxylic acid or modified sulfonamide Nafion, which are used in many advanced chlor-alkali cells. Naturally, a new generation chlor-alkali cell with high efficiency awaits the merging of the merits of these two fascinating technologies.

Literature Cited

1. Grubb, W.T. U.S. Patent, 2,913,511, 1959.
2. Grubb, W.T., J. Electrochem. Soc., 1959, 106, 275.
3. Grubb, W.T.; Niedrach, L.W., J. Electrochem. Soc., 1960, 107, 131.
4. Cairns, E.J.; Douglas, D.L.; Niedrach, L.W.; A.I.Ch.E. Journal, 1961, 7, 551.
5. Niedrach, L.W.; Grubb, W.T.; in "Fuel Cells;" Mitchell, W., Jr., Ed.; Academic Press: New York, 1963, p. 253.
6. Maget, H.J.R.; in "Handbook of Fuel Cell Technology;" Berger, C., Ed.; Prentice-Hall: Englewood Cliffs, N.J., 1967.
7. Yeo, R.S.; Eisenberg, A.; J. Macromol. Sci-Phys., 1977, B13, 441.
8. D'Agostino, V.; Lee, J.; Cook, E.; U.S. Patent, 4,012,303, 1977.
9. D'Agostino, V.; Lee, J.; Cook, E.; U.S. Patent, 4,107,005, 1978.
10. Hodgdon, R.B.; Boyack, J.R.; LaConti, A.B., "The Degradation of Polystyrene Sulfonic Acid," TIS Report No. 65DE5, Jan. 1966.
11. Shen, M.C.; Eisenberg, A.; in "Progress in Solid State Chemistry," Vol. 3; Pergamon Press, New York, 1966, pg. 407.
12. Yeo, R.S.; McBreen, J.; Kissel, G.; Kulesa, F.; Srinivasan, S.; J. Applied Electrochem., 1980, 10, 741.
13. LaConti, A.B.; Fragala, A.R.; Boyack, J.R.; in "Proceedings of the Symposium on Electrode Materials and Processes for Energy Conversion and Storage," McIntyre, J.D.E.; Srinivasan, S.; Will, F.G., Eds.; The Electrochemical Society, Inc.: Princeton, N.J., 1977, p. 354.
14. Stucki, S.; Menth, A.; in "Proceedings of Symposium on Industrial Water Electrolysis," Srinivasan, S.; Salzano,

F.J.; Landgrebe, A., Eds.; The Electrochem. Soc.: Princeton, N.J., 1978, p. 180.

15. Weininger, J.L.; Russell, R.R.; J. Electrochem. Soc., 1978, 125, 1482.

16. Yeo, R.S.; Orehotsky, J.; Visscher, W.; Srinivasan, S.; J. Electrochem. Soc., 1981, 128.

17. Yeo, R.S.; unpublished result.

18. Adams, R.; Schriner, R.L.; J. Am. Chem. Soc., 1923, 45, 2171.

19. Demczyk, B.G.; Liu, C.T.; J. Power Sources, 1981, 6, 185.

20. Bevan, H.C.; Tseung, A.C.C.; Electrochim. Acta, 1974, 9, 201.

21. Sedgwick, T.O.; Lydtin, H., Eds., "Proceedings of the Seventh Intern. Conf. on Chemical Vapor Deposition," The Electrochemical Soc., Princeton, N.J. 1979.

22. Miekka, R.G.; Lyons, E.H., Jr.; Dempsey, R.M.; U.S. Patent, 3,356,538, 1967.

23. Yeo, R.S.; Srinivasan, S.; J. Electrochem. Soc., (Abstract) 1979, 126, 378C.

24. Yeo, R.S.; Ph.D. Thesis, McGill University, 1976.

25. Tilak, B.V.; Yeo, R.S.; Srinivasan, S.; in "Comprehensive Treatise on Electrochemistry," Volume 3; Bockris, J. O'M.; Conway, B.E.; Yeager, E.B.; White, R.E., Eds.; Plenum Press: New York, N.Y.; 1981, pp. 39-122.

26. Chapman, L.E., in "Proceedings of Seventh Intersociety Energy Conversion Engineering Conference," American Nuclear Society; La Grange Park, Ill., 1972, p. 466.

27. McElroy, J.F.; in "National Fuel Cell Seminar Abstracts," Courtesy Associates, Inc.: Washington, D.C., 1978;   pp. 176-179

28. Fickett, A.P.; "Advances in Ion Exchange Membranes for Rechargeable Fuel Cells," paper presented at the Columbus Meeting of the Electrochemical Society, Feb. 1970.

29. Lu, P.W.T.; Srinivasan, S.; J. Applied Electrochem., 1979, 9, 269.

30. Tilak, B.V.; Lu, P.W.T; Colman, J.E.; Srinivasan, S.; in "Comprehensive Treatise on Electrochemistry," Volume 2; Bockris, J. O'M.; Conway, B.E.; Yeager, E.B.; White, R.E., Eds; Plenum Press: New York, N.Y., 1981, p 1.

31. Nuttall, L.J.; Intern. J. Hydrogen Energy 1977, 2, 395.

32. Nuttall, L.J.; Russell, J.H.; Intern. J. Hydrogen Energy, 1980, 5, 75.

33. Yeo, R.S.; in "Proceedings of the Symposium on Ion Exchange," Yeo, R.S.; Buck, R.P., Eds; The Electrochemical Soc.: Pennington, N.J., 1981, p. 311.

34. LaConti, A.B.; Balko, E.N.; Coker, T.G.; Fagala, A.G.; in "Proceedings of the Symposium on Ion Exchange," Yeo, R.S.; Buck, R.P., Eds.; The Electrochem. Soc.: Pennington, N.J., 1981, p. 318.

35. Dunn, B.; Breiter, M.W.; Park, D.S.; J. Appl. Electrochem., 1981, 11, 103.
36. Pollard, R.; Newman, J.; J. Electrochem. Soc., 1981, 128, 491.
37. Kim, J.T.; Jorne, J.; J. Electrochem. Soc., 1980, 127, 8.
38. Will, F.G.; J. Electrochem. Soc., 1979, 126, 36.
39. Will, F.G.; Spacil, H.S.; J. Power Sources, 1980, 5, 173.
40. Lim, H.S.; Lackner, A.M.; Knechtli, R.C.; J. Electrochem. Soc., 1977, 124, 1154.
41. Eustace, D.J. ; J. Electrochem. Soc., 1980, 127, 528.
42. Beaufrere, A.; Yeo, R.S.; Srinivasan, S.; McElroy, J.: Hart, G.; in "Proceedings of the Twelfth Intersociety Energy Conversion Engineering Conference," American Nuclear Society: La Grange Park, Ill.; 1977, p. 959 .
43. Yeo, R.S.; Srinivasan, S.; in "Proceedings of the Symposium on Load Leveling," Yao, N.P.; Silman, J.R, Eds. The Electrochemical Soc.: Princeton, N.J., 1977, p. 396.
44. Chin, D-T.; Yeo, R.S.; McBreen, J.: Srinivasan, S.; J. Electrochem. Soc., 1979, 126, 713.
45. Yeo, R.S., McBreen, J.; J. Electrochem. Soc., 1979, 126, 1682.
46. Yeo, R.S.; McBreen, J.; Tseung, A.C.C.; Srinivasan, S.; McElroy, J.; J. Applied Electrochem., 1980, 10, 393.
47. Yeo, R.S.; Zeldin, A.; Kukacka, L.E.; J. Applied Polym. Sci., 1981, 26, 1159.
48. Hsueh, K.L.; Chin, D-T.; McBreen, J.; Srinivasan, S.; J. Applied Electrochem., 1981, 11.
49. Balko, E.N.; J. Applied Electrochem., 1981, 11, 91.
50. Yeo, R.S.; Chin, D-T.: J. Electrochem. Soc., 1980, 127, 549.
51. Ling. J.S.; Charleston, J.; in "Proceedings of the Symposium on Ion Exchange," Yeo, R.S.; Buck, R.P., Eds.; The Electrochem. Soc., Pennington, N.J., 1981, p. 334.
52. Warshay, M.; Wright, L.O.; J. Electrochem. Soc., 1977, 124,173.
53. Lee, J.; D'Agostino, V.; Zito, R.; in "Proceedings of the Symposium on Ion Exchange," Yeo, R.S.; Buck, R.P., Eds.; The Electrochem. Soc., Pennington, N.J., 1981, p. 355.
54. Hruska, L.W.; Savinell, R.F.; J. Electrochem. Soc., 1981, 128, 18.
55. Adams, G.B.; Hollandsworth, R.P., in "Proceedings of Second Annual Battery and Electrochemical Technology Conference," Courtesy Associates, Inc.: Washington, D.C., 1978.
56. Mussini, T.; Faita, G.; Ric. Sci., 1966, 36, 175.
57. Cerquetti, A.; Longhi, P.; Mussini, T.; Natta, G.; J. Electroanal. Chem. Interfacial Electrochem., 1969, 20,411.
58. Yeo, R.S.; Chan, S.F.; Lee, J.; J. Membrane Sci., 1981, 9.

RECEIVED August 3, 1981.

General Applications of Perfluorinated Ionomer Membranes

BRIAN KIPLING

University of Calgary, Department of Chemistry, Calgary, Alberta T2N 1N4 Canada

Described here are applications other than industrial chloralkali, water electrolysis and fuel cells. The applications are divided into six general classes.

Separators

The use of Nafion as a separator in cells for chloralkali electrolysis is described elsewhere. The exceptional chemical inertness and thermal stability coupled with favourable electrical conductivity being of particular advantage in this application. These properties have been exploited in a number of other electrochemical applications.

The RAI Research Corporation also offers a range of battery separators under the name of Permion (1). These membranes are made by radiation grafting of a suitably active group onto an inert base film. The active groups include weak acids such as acrylic and substituted acrylic acid and stronger acidic groups such as sulfonated styrene. The base film can be Teflon R, polyethylene or polypropylene. They are thus not strictly perfluorinated membranes as is Nafion, but in chemical inertness and in many physical properties such as electrical conductivity and ion flux are useful as separators in batteries.

Grot (2, 3, 4) has summarized the properties of some perfluorinated membranes used as separators, mainly for chloralkali electrolysis, and also described regeneration of chromic acid solutions. Solutions of chromium(VI) in sulfuric acid are used in a number of industrial processes. For example in etching plastics prior to metallizing. The used solution contains chromium(III) which can be electrolytically reoxidized to chromium(VI) using lead anodes. Nafion in tubular form (80 mm diameter) is used as the separator in cylindrical cells. Sulfuric acid is used as the catholyte, the permselectivity of the Nafion ensuring that little sulfate ion migrates to the chromium(VI) solution. In the regeneration of chrome plating solutions even small amounts of sulfate migrating across the separator would upset the critical ratio

0097-6156/82/0180-0475$05.00/0

H_2SO_4:CrO_3 (usually around 1:100) and an organic acid is used as catholyte, the small amounts of organic anion which do migrate across the membrane are oxidized by the chromic acid. An additional problem in the regeneration of chrome plating solutions is removal of low concentrations of cations such as Fe^{3+}, Cu^{2+}, Ni^{2+}. The Nafion separator allows dialysis of the non-oxidizable cations to the cathode, and even though efficiency of removal is low due to the relatively high concentration of hydrogen ions, (the hydrogen ion carries the bulk of the current), the process is still economically attractive because the amounts of the unwanted cations are usually small.

In chromic acid solutions containing large amounts of unwanted cations, e.g. solutions after use for etching printed circuits contain large amounts of copper(II) ions and regeneration of the spent solutions with no pretreatment results in very low efficiencies in terms of copper(II) ion removal. About 8-13% of the copper has been removed when all the chromium(III) has been reoxidized to chromium(VI) (4). Regeneration cells for chromic acid solutions based on the cylindrical geometry using tubular Nafion are commercially available from AMJ Chemicals. (54). A similar process, though the nature of the membrane was not specified was described by Belobaba (6) for removing cations such as chromium(III) and nickel(II) from spent electroplating solutions.

Nafion in cylindrical form is also supplied by C.G. Processing Inc. (5), for use in electrowinning of gold. The membrane serves several purposes in electrolysis cells where gold in concentrated chloride (or cyanide) solution is deposited on steel wool cathodes. The membrane prevents migration of the anionic gold complex to the anode, so that current is carried almost entirely by sodium ions from the sodium hydroxide anolyte. Chloride ions are also prevented from reaching the anode, thus avoiding corrosion problem with the stainless steel electrodes, and finally oxygen gas, evolved at the anode is kept out of the catholyte where it would interfere with the deposition of the gold (53). The relatively high temperatures and high current densities limit the range of available membranes to those of the Nafion type which have the appropriate combination of properties.

On a somewhat different scale is the construction of separator tubes described by Harrar and Sherry (7). These tubes are made from Nafion (tubular form) with either Kel-F or glass plus Tefzel and are intended for use in electrolysis cells for controlled potential electrolysis or for coulometry. Nafion tubing 4.3 mm diameter (internal) is attached to a length of glass using Tefzel. The Nafion end of the tube is sealed with a plug of borosilicate glass rod producing a composite of a glass tube with a section of Nafion tube (Figure 1) which serves to isolate the counter electrode from the working electrode compartment. Electrochemical characteristics of the tubes were measured and the use of these tubes in coulometry and voltammetry indicated (see also refs 35, 37).

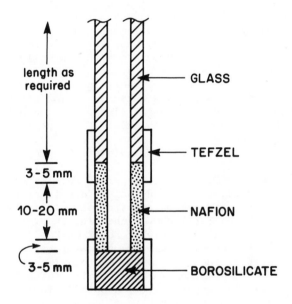

Figure 1. Nafion/glass/Tefzel separator.

Sakagami, Kato, Hirai and Murayana (8) have used Nafion 110
as a separator and collector in the electrodialysis of protein.
A solution, or homogenized dispersion of the protein, is contained
in a compartment separated from the electrodes by sheets of the
Nafion membrane. (A cylindrical geometry with tubular Nafion
would seem a more convenient arrangement here). On passage of
current the protein migrates, (direction depends upon the protein
and the pH of the dispersion), and collects as a homogenous film
on the membrane surface - from which it could be easily removed.

Energy Sources

Although the membranes used in some of the applications listed here are not of the perfluorinated variety, the chemical inertness, high electrical conductance and relatively high solute flux
confer technical advantages over most non-fluorinated membranes.
The use of membranes in energy sources can be divided into two
categories. As electrode separators in conventional batteries,
the Nafion membranes would seem especially useful for nonaqueous
electrolyte systems because of their inertness to most organic solvents. The Permion membranes based on teflon with sulfonate exchange sites seem to have potential application here too. Dampier
(9) evaluated a number of membranes with respect to electrical resistance, transference numbers and interdiffusion rates for lith-
ium(I), copper(II), bromide, perchlorate in propylene carbonate,
but did not at that time consider Nafion or Permion membranes.
Lopez, Kipling and Yeager (10) reported results of transport
studies for sodium, cesium and iodide ions in propylene carbonate
which showed that nonaqueous aprotic solvents may remove the advantage of high ion flux, though the flux in protic solvents such
as methanol is still quite high. Later work (11) suggested that
the presence of even small amounts of water may have a pronounced
effect on rates of ion transport. Kratochvil and Betty (12) used
an anion exchange membrane in acetonitrile for a battery based on
copper(II)-(I) and copper (0)-(I) couples. The membrane was
effective in preventing migration of copper cations. The cell was
intended as a power source at low temperatures.

In addition to uses in conventional type batteries, the use
of ion exchange membranes in dialytic batteries has been proposed
by several groups. Based on early experiments of Manecke (13) the
dialytic battery is a device for extracting the free energy of
dilution of saline water. The principles were discussed by Clampitt and Kiviat (14) who proposed a scheme for direct generation
of electricity using an electrochemical concentration cell. Fresh
water and saline solutions are separated by an ion exchange membrane. A reversible chloride electrode is placed in each compartment and the E.M.F. of such a cell is a function of the difference
in concentration between the two solutions. Ion transport across
the membrane gives rise to a current flow. This proposal was
examined in more depth by Weinstein and Leitz (15), who amongst

other factors looked at power output as a function of salt concentration on the freshwater side. A compromise must be struck, given a constant saline concentration for sea water, between maximum concentration difference (giving maximum potential difference) and increasing internal resistance as the salt concentration of the freshwater is reduced to a small value. An optimum of 0.026 M sodium chloride for the freshwater side is somewhat higher than average river water (0.01 M). Other factors such as membrane resistance, water transport etc. were examined in lesser detail but Weinstein and Leitz attempted an economic prognosis based on current costs and concluded that major advances in membrane technology were needed if the dialytic battery were to become a commercial possibility. The availability of adequate quantities of river water may also be a limiting factor. Forgacs and O'Brian (16) have proposed a similar model for a dialytic battery and made more optimistic predictions, though without specifying membrane types or costs. A device based upon the difference in salt concentration between fresh water and sea water but utilizing osmotic pressure rather than a potential generation was proposed by Norman (17). His osmotic pump using a reverse osmosis membrane is theoretically equivalent to a waterfall of height 225 m, and a device for extracting useful work from this osmotic pressure was described.

Electrodes

Ion selective electrodes using ion exchange membranes have been investigated over a long period of time (18). Two major problems of such electrodes are lack of selectivity and short life times. Polymer membrane electrodes have usually incorporated a plasticizer to increase ionic mobility in the membrane and an electroactive species. The rate of leaching of these components controls the lifetime of the electrode (19, 20). The unique morphology of Nafion polymers (11, 21, 22) coupled with, or perhaps resulting in, high selectivities (23) prompted Martin and Freiser to follow their earlier work on a dinonylnaphthalenensulfonate membrane (24) with an investigation of Nafion 120 as a membrane in ion selective electrodes (25). Ionic mobilities in Nafion are sufficiently high to obviate the need for plasticizers and the ion exchange sites are covalently bound to the polymer backbone so that in principle at least lifetimes should be infinite. Martin and Freiser prepared electrodes by sealing a disc of Nafion 120 to the end of a glass tube. The tube was then filled with appropriate internal reference solution. They report Nernstian response for electrodes for cesium ion and for tetrabutylammonium ion and present selectivity data for these electrodes in presence of various other ions. Similarities in selectivity sequences to the DNNS electrode indicate that solvent extraction considerations may be significant in determining selectivity, perhaps from partition between hydrophilic and hydropholic regions within

the Nafion membrane. The use of a strong cation exchange membrane
in an ion selective electrode was reported by Inokuma, Ochiar,
Endo and Hiiro (26) for control of pickling bath solutions containing nitric and hydrofluoric acids. A perfluorinated membrane is
particularly suitable here because of its chemical inertness.

Analytical Preconcentration Techniques (Donnan Dialysis)

The permselectivity of ion exchange membranes can be exploited as a preconcentration technique. Thus suppose two solutions (1
and 2) are separated by a cation exchange membrane which is totally impermeable to anions. Initially solution 1 contains A^+ ions
and solution 2 contains B^+ ions. It can be shown (27, 28) that at
equilibrium the following relationship holds

$$\frac{a_{A,1}}{a_{A,2}} = \frac{a_{B,1}}{a_{B,2}}$$

where a denotes activities, A and B denotes species A^+ and B^+ in
solutions 1 and 2 on different sides of the membrane. If the
initial concentration of A^+ is large and initial concentration of
B^+ very small, then at equilibrium a relatively small fraction of
A^+ will have moved from solution 1 to solution 2 but a relatively
large fraction of B^+ will have moved from solution 2 to solution 1.
By suitable choice of volumes for the two solutions a high degree
of concentration of B^+ can be achieved. Similar considerations
apply to anions using an anion exchange membrane. This technique
is often referred to as Donnan dialysis. In practice total impermeability of coion (anion in the case of a cation exchange
membrane) is not achieved and so although the cation equilibrium
situation described is observed it is eventually superseded by
complete ionic equilibrium as anions slowly permeate the membrane
until both cation and anion concentrations are equal on each side
of the membrane. Permselectivity is a result of relative diffusion rates of cations and anions through the membrane and holds
only under Donnan exclusion conditions, i.e. is most effective in
low ionic strength solutions.

Blaedel and Haupert (28) demonstrated the feasibility of
using this phenomenon as a preconcentration technique using isotope
tracer studies on the ions Na^+, Cs^+, Zn^{2+}, and later Blaedel and
Christensen (29) extended the work to include the anions I^- and
HPO_4^{2-}. They found anion transport to be much slower than the previously reported cation transport. Coion transport in the anion
exchange membranes was much higher and apparently dependent on the
anionic charge of the bulk electrolyte. Further studies with more
recently available membranes (1) seem to be needed.

Blaedel and Kissel (30) used electrodes responsive to the
desired ion wrapped with an ion exchange membrane. For example a
glass electrode wrapped with a cation membrane, Permion P1010,

responds to hydrogen ions when immersed in a solution. The small
volume of solution contained between the membrane and the glass
bulb provides enrichment factors of about one hundred and thus
extends the sensitivity range for the electrode.

The use of ion exchange membranes as a preconcentration device assumes no selective binding of the ions of interest. Blaedel
and Niemann (31) examined three commercially available membranes
by a variety of techniques and found evidence of impurity groups
in the membranes "capable of binding cations (specifically copper
(II)) very strongly by mechanisms other than the ion exchange
mechanism". Such effects would vitiate the use of these membranes
in preconcentration techniques. No definitive identification of
the groups was presented but carboxylate and olefinic groups were
detected spectroscopically.   Of the three membranes examined
Nafion XR170 appeared to be the most homogeneous with respect to
exchange site and binding of copper(II) ions.

Preconcentration of copper at concentrations of around 10^{-5}M
prior to determination by atomic absorption was reported by Cox
and Dinunzio (32). Using Nafion 125, Permion P1010 or Permion
4010 small volume receiver cells were separated from larger volume
samples. After a fixed time the solution in the receiver was
analyzed for copper, and by suitable adjustment of conditions the
copper concentration of the receiver solution was shown to be a
linear function of copper content in the original sample. The
authors found curious effects on adding certain cations to the
sample solution. Mg^{2+} or Al^{3+} caused marked acceleration in the
rate of Cu^{2+} transport, an effect which it is tempting to link as
does Twardowski (37) with the Blaedel and Niemann impurity sites
(31), but which Cox and DiNunzio claim is not feasible because
potassium transport is similarly accelerated by addition of sodium
or magnesium ions to the sample solution. Optimum conditions with
respect to composition of receiver solution, sample solution,
temperature and stirring rate were established in order to provide
a precise method for ion preconcentration.

The voltammetric determination of nitrate following a preconcentration step using Permion 1025 was developed by Cox,
Lundquist and Washinger (33). Nitrate is exchanged from the sample solution through the membrane into a small volume cell contain
ing 0.1 M potassium chloride and 0.01 M lanthanum chloride. The
method eliminates cationic interferences with the voltammetry but
does not successfully deal with anionic interferences such as a
sulfate or nitrite.

Donnan dialysis of conjugate bases of weak acids presents an
additional factor for control, namely pH. Cox and Cheng (34) investigated preconcentration of a number of anions in this cate-
gory. Sample pH was adjusted to ensure that a major fraction of
the weak acid system was in anionic form. Maximum enrichment
factor, defined as ratio of receiver concentration to sample concentration after some given, fixed time (usually 30 minutes), was
obtained when pH of the receiver solution was much less than the

pK of the corresponding acid. For the anions chloride, phosphate, arsenate, chloracetate and pyruvate the enrichment factor of about seven was obtained, suggesting that transfer rate across the membrane-receiver solution boundary may be the rate determining step. The much lower enrichment factor of about 3 for sulfate is attributed to strong interaction with the methyl pyridinium exchange sites of the Permion 1025 membrane. The presence of sulfate ion in the sample solution also has the effect of decreasing enrichment factors for other anions. Donnan dialysis not only serves as a preconcentration process but in addition provides matrix normalization for analytical methods where the matrix is important. Cox and Twardowski (35) have described a voltammetric method of analysis for the metal ions Cd(II), Pb(II), Zn(II) and Cu(II) in which a preliminary dialysis using Permion 1010 removes interferences from various surfactants such as Trilon X-100 and ligands such as humic acid. The dialysis technique is similar to that described by Cox and DiNunzio (32) the receiver solution being analyzed by anodic stripping voltammetry after a fixed contact time, via the membrane, with the sample solution. In presence of complexing agents (humic acid) pH control of the sample solution to control proportion of uncomplexed metal ion is also important, possibly allowing for speciation in addition to matrix normalization and concentration enrichment. The availability of Nafion in tubular form, Nafion 811, was utilized by Cox and Twardowski (36) in a dynamic Donnan dialysis system where solution is pumped through Nafion tubing, a coil of the tubing being immersed in the sample solution. The use of tubing allows an increase of surface area of membrane to receiver volume ratio and minimizes concentration polarization on the receiver membrane surface thus giving much enhanced enrichment ratios in shorter contact times. In a typical experiment receiver solution, 0.2 M MgSO$_4$ plus 5×10^{-4} M Al$_2$(SO$_4$)$_3$ was pumped through the tubing, diameter 0.63 mm, length 10 m at a rate of about 6.0 mL min^{-1} giving enrichment factors of up to 27 in dialysis times of about 15 minutes. Considerably more efficient than factors of 5-10 in 1 hour of static dialysis.

Cox and DiNunzio (32) established an optimum composition for the receiver solution, 0.2 M Mg(II), 5×10^{-4} M Al(III). Other receiver solutions show a less favourable cation transport across membranes such as permion 1010, a grafted teflon base onto styrene with sulfonic acid exchange sites. For example 0.1 M Na$^+$ shows enrichment factors only 50% that of the above solution. This lower cation transport rate is attributed to interaction between the mobile cations and the fixed exchange sites. The function of the multivalent ions is to provide a shield between exchange sites and mobile ions and thus lower residence times in the membrane. Cox and Twardowski (38) in an elegant experiment provide evidence supporting this view by applying on AC field across the membrane during dialysis with a sodium nitrate receiver. A 5V cm^{-1} sine wave at frequencies ranging from 10^{-1} KHz to 10^3 KHz was used during transport of the ions Cu(II), Pb(II), Cd(II), Zn(II). It

was argued that increasing frequency would increase the rate of
dissociation of cation-sulfonate bonds and ultimately residence
times of the mobile cations would reach a diffusion controlled
limit. The transport of ions through the membrane was measured in
terms of enrichment factors (see earlier) and reached a limiting
value at about 10^2 KHz of around 4.0. The value observed using
the multivalent receiver solution. For the latter on applied
field has relatively little effect on the enrichment factor.

The use of tubular Donnan dialysis systems has stimulated
attempts to provide predictive models, principally for industrially oriented applications. Ng and Snyder (39) have recently published one such attempt applied to dialysis of nickel(II) into a
sulfuric acid receiver solution. Correlations between mass transport coefficient and Reynolds number are reported, and the factors
controlling transport over a range of nickel concentrations are
discussed.

Preconcentration in a slightly different way is described by
Eisner and Mark (40) who equilibrated small areas of cation exchange membranes with sample solutions and then used the membrane
as a source of ions for deposition in an anodic stripping voltammetry system. The concentration of the ion in the membrane is
linearly related to its concentration in the bulk sample solution.
The pre-equilibrated membrane was also analyzed by neutron activation thus extending the range of ions for which the technique is
useful. Data are quoted for Ag^+, Cu^{2+}, Zn^{2+}, Co^{2+} and In^{3+}, all
of which show favourable distribution for the membrane phase.
Equilibration times are inversely proportional to concentration
ranging from several minutes at 10^{-4} M to one day or more at 10^{-6}
M. The method affords a convenient separation from nonionic and
anionic species which interfere with the measurement technique.
A similar preconcentration process was developed by Lochmuller,
Galbraith and Walter (41) for the analysis of water for trace
metals. The membrane after equilibration with the water sample is
in this case analyzed by proton induced X-ray emission. Claimed
advantages of the latter technique are a wider range of applicability than neutron activation, easier applicability to rapid
routine analysis than anodic stripping and greater sensitivity
than conventional X-ray fluorescence spectroscopy.

Catalysis

Superacids such as Magic acid (42), a system containing
antimony pentafluoride and fluorosulfonic acid and designated a
superacid because it is a more ready electron pair acceptor than
anhydrous aluminium chloride (43), have useful catalytic properties in synthetic organic chemistry where the reaction involves a
carbocation intermediate. The extremely low nucleophilicity of
the counter ion of such acids make it possible to prepare cations
such as the tertiary butyl $(CH_3)_3C^+$ with an appreciably long life
time in superacid solutions whereas such cations are too reactive

to exist in less acidic media. The highly acidic character of
superacids also promotes protonation of very weak bases such as
alkanes as a preliminary stage of isomerization, alkylation and
polymerization reactions (43).

The wide applicability of superacids encouraged attempts to
prepare them in the solid phase by absorbing antimony fluoride
and fluorosulfonic acid on supports such as fluorinated graphite
or a fluorinated polyolefine resin (45). Nafion is an obvious
candidate for such systems and has been used as a catalyst for a
variety of reactions. Olah (43) reviewed some of the earlier
applications to isomerization, polymerization, transbromination,
nitration, acetalization and hydration, and used Nafion (in the
hydrogen ion form) for gas phase alkylation of benzene and alkyl-
benzenes. Nafion showed higher catalytic activity then other
solid phase superacid catalysts thus enabling lower temperatures
and pressures to be used and hence "cleaner" products. Catalyst
lifetimes were also longer at moderate temperatures (below 200°C)
but at 220°C Nafion decomposed with loss of sulfonate groups and
catalytic activity. The same catalyst has also been reported to
be effective in solution phase esterification reactions (46).
Beltrame, Carniti and Nespoli (47) used Nafion as a catalyst for
the isomerization of m-xylene and Olah (48) has used Nafion as a
catalyst for Friedel-Craft reactions of toluene and phenol with
alkylchloroformates and oxalates. The chemical inertness (at
least below 220°C) of this heterogeneous catalyst confers obvious
advantages over more conventional homogeneous superacid catalysts
(43, 49).

Permeation Distillation

Nafion exhibits a relatively high diffusion rate for water
vapour and this property has been used for removal of water from
gas streams (50). The technique of Permeation distillation is a
form of counter current extraction with a Nafion membrane acting
as the boundary between a moist sample gas stream flowing in one
direction and a dryer purge gas stream flowing in the counter
direction. The principle is shown in Figure 2. Moist sample gas
flows through the Nafion tube which is enclosed in a shell. The
shell may be of steel, polyethylene or any suitable material. A
dry purge gas is pumped into the shell at C. Water vapor from the
moist gas permeates the walls of the Nafion tube and is removed by
the flow of dry gas on the shell side of the tube. The gas stream
exiting from the tube at B is of much lower water content than the
initial gas at A. The wet purge gas at D may be dried for reuse
or discarded. Suitable control of temperature, pressure and gas
flow rates plus selection of tube length and diameter enables a
high degree of drying to be attained by this simple technique. A
commercially available unit based on this concept is marketed by
Perma Pure Products Inc. (51). In this unit a bundle of Nafion
tubes is contained within each shell in order to provide a large

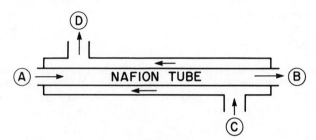

Figure 2. Perma pure gas dryer.

capacity and tubes of various lengths are available (tube length, for a given flow rate determines the extent of drying). The unit is designed to operate at ambient temperatures and is adaptable to various pressure ranges for both sample and purge gases.

An application of the unit has been described by Baker (52). Infra-red spectroscopy of air samples at pressures up to ten atmospheres requires removal, selectively, of water vapour. Using dry nitrogen gas as a purge Baker found that absorbance due to water in air samples could be reduced about ten fold, thus allowing use of regions of the spectra for detection of trace constituents without reference beam compensation. Water bands in the region 1900 to 1300 cm^{-1} which were extremely strong in untreated samples were reduced to such low values that traces of sulfur dioxide could be detected by its 1370 cm^{-1} band. The permeation technique is not specific for water, the lower alcohols (up to hexanol), esters, ethers, amines and some ketones also diffuse through the Nafion tube. Many small inorganic molecules however such as CO, CO_2, SO_2, CS_2 do not diffuse as well as most hydrocarbons and larger organic molecules. The dryer does not remove particulates or condensable oil vapors or mists.

Acknowledgements

The author acknowledges with gratitude the considerable assistance provided by Dr. W.G. Grot of E.I. Du Pont De Nemours and Co. Inc. and by Dr. H.L. Yeager of the University of Calgary in the preparation of this review.

Literature Cited

1.   RAI Research Co., Permion and Raipore Brochures.
2.   Grot, W.G. Chemie Ing. Tech. 1975, 47, 617.
3.   Grot, W.G. Chemie Ing. Tech. 1978, 50, 299.
4.   Grot, W.G. Case Western Symposium, October 1980.
5.   C.G. Processing Inc., Box 133, Rockland, D.E. 19732.
6.   Belobaba, A.G.; Pevnitskaya, M.V.; Kozina, A.A.; Nefedova, G.Z.; Freidlin, Y.G. Izv. Sib. Otd. Akad. Nauk. SSSR. Ser Khim Nauk 1980, (4), 161.
7.   Harrar, J.E.; Sherry, R.J. Anal. Chem. 1975, 47, 601.
8.   Sakagami, T.; Kato, T.; Hirai, T.; Murayama, N. Eur. Pat. Appl. 11504, 28 May, 1980.
9.   Dampier, F.W. J. Appl. Electrochem. 1973, 3, 169.
10. Lopez, M.; Kipling, B.; Yeager, H.L. Anal. Chem. 1977, 49, 629.
11. Yeager, H.L.; Kipling, B. J. Phys. Chem. 1979, 83, 1836.
12. Kratochvil, B.; Betty, K.R. J. Elect. Soc. 1974, 121, 851.
13. Manecke, G. Z. Phys. Chem. 1952, 201, 1.
14. Clampitt, B.H.; Kiviat, F.E. Sci. 1976, 191, 719.
15. Weinstein, J.N.; Leitz, F.B. Sci. 1976, 191, 557.

16. Forgacs, C.; O'Brien, R.N. Chem. in Can. 1979, 19.
17. Norman, R.S. Sci. 1974, 186, 350.
18. Parsons, J.S. Anal. Chem. 1958, 30 (7), 1262.
19. Cutler, S.G.; Meares, P.; Hall, D.G. J. Electroanal. Chem. 1977, 85, 145.
20. Oesch, W.; Simon, W. Anal. Chem. 1980, 52, 692.
21. Yeo, R.S.; Eisenberg, A. J. Appl. Polym. Sci. 1977, 21, 875.
22. Gierke, T. 152nd Meeting of Electrochem. Soc., Atlanta, 1977.
23. Steck, A.; Yeager, H.L. Anal. Chem. 1980, 52, 1215.
24. Martin, C.R.; Freiser, H. Anal. Chem. 1980, 52, 562.
25. Martin, C.R.; Freiser, H. Anal. Chem. 1981, 53, 902.
26. Inokuma, Y.; Ochiar, T.; Endo, J.; Hiiro, K. Nippon Kagaku Kaishi 1980 (10), 1469.
27. Helfferich, F. "Ion Exchange", McGraw Hill, New York, 1962.
28. Blaedel, W.J.; Haupert, T.J. Anal. Chem. 1966, 38, 1305.
29. Blaedel, W.J.; Christensen, E.L. Anal. Chem. 1967, 39, 1262.
30. Blaedel, W.J.; Kissel, T.R. Anal. Chem. 1972, 44, 2109.
31. Blaedel, W.J.; Niemann, R.A. Anal. Chem. 1975, 47, 1455.
32. Cox, J.A.; DiNunzio, J.E. Anal. Chem. 1977, 49, 1272.
33. Cox, J.A.; Lundquist, G.L.; Washinger, G. Anal. Chem. 1975, 47, 320.
34. Cox, J.A.; Cheng, K.H. Anal. Chem. 1978, 50, 601.
35. Cox, J.A.; Twardowski, Z. Anal. Chim. Acta. 1980, 119, 39.
36. Cox, J.A.; Twardowski, Z. Anal. Chem. 1980, 52, 1503.
37. Twardowski, Z. Ph.D. Thesis, S.I.U. 1980.
38. Cox, J.A.; Twardowski, Z. Anal. Letters 1980, 13, 1283.
39. Ng, P.K.; Snyder, D.D. Proc. of Symposium on Ion Exchange, Yeo, R.S.; Buck, R.P. Eds, Electrochem. Soc. 1981.
40. Eisner, U.; Mark, H.B. Talanta, 1969, 16, 27.
41. Lochmuller, C.H.; Galbraith, J.W.; Walter, R.L. Anal. Chem. 1974, 46, 440.
42. Registered Trade Mark of Cationics Inc., Columbia, S.C.
43. Olah, G.A.; Prakash, G.K.S.; Sommer J. Sci. 1979, 206, 13.
44. Olah, G.A.; Baker, E.B.; Evans, J.C.; Toglyesi, W.S.; McIntyre, J.S.; Bastieau, J.J. J.A.C.S. 1964, 86, 1360.
45. Olah, G.A.; Kaspi, J.; Bukala, J. J. Org. Chem. 1977, 42, 4187.
46. Olah, G.A.; Keumi, T.; Meidar, D. Synth. 1978 (2), 929.
47. Beltrame, P.; Beltrame, P.L.; Carniti, P.; Nespoli, G. Ind. Eng. Chem. Prod. Res. Dev. 1980, 19, 205.
48. Olah, G.A.; Meidar, D., Mulhotray, R.; Olah, J.A.; Narang, S.C. J. Catal. 1980, 61, 96.
49. Olah, G.A. Aldrichimica 1979, 12, (3), 43.
50. Kertzman, J. ISA Aid 73425 (1973), 121.
51. Perma Pure Products Inc., Box 70, Oceanport, New Jersey 07757.
52. Baker, B.B. J. Am. Ind. Hyg. Ass. 1974, Nov.
53. Personal communication from Anglo American Corporation of South Africa Ltd.
54. A.M.J. Chemical Inc.

RECEIVED August 7, 1981.

INDEX

INDEX

Jacket design by Kathleen Schaner.
Production by Robin Giroux and Karen Gray.

Elements typeset by Trade Typographers, Inc., Washington, D.C.
Printed and bound by The Maple Press Company, York, PA.